DATE DUE		

OXYGEN

OXYGEN

The molecule that made the world

Nick Lane

Great Clarendon Street, Oxford OX2 6DP

Oxford University Press is a department of the University of Oxford.
It furthers the University's objective of excellence in research, scholarship,
and education by publishing worldwide in

Oxford New York

Auckland Bangkok Buenos Aires Cape Town Chennai
Dar es Salaam Delhi Hong Kong Istanbul Karachi Kolkata
Kuala Lumpur Madrid Melbourne Mexico City Mumbai Nairobi
São Paulo Shanghai Taipei Tokyo Toronto

Oxford is a registered trade mark of Oxford University Press
in the UK and in certain other countries

Published in the United States
by Oxford University Press Inc., New York

First published 2002

British Library Cataloguing in Publication Data

Data available

Library of Congress Cataloguing in Publication Data

ISBN 0-19-850803-4

10 9 8 7 6 5 4 3 2 1

Typeset in Stone Serif
by Footnote Graphics, Warminster, Wiltshire
Printed in Great Britain
on acid-free paper by
TJ International Ltd, Padstow, Cornwall

Contents

For Ana

Acknowledgements

I am especially grateful to three people, without whom this book may never have passed from concept to reality. First, to Dr John Emsley, whose beguiling science writing and generous, enquiring mind have been an inspiration to many aspiring chemists and writers. I am grateful to John for introducing me to Oxford University Press, and for many invigorating discussions about science, society and language. Second, to Dr Michael Rodgers of OUP, whose sharp eye and masterly editorial skills have cultivated a generation of science writers. I am grateful to him for seeing potential in my first attempts to put together a story for this book, for his timely and gentle nudges towards a clear and unembellished writing style, and for his encouragement and support throughout. Finally, I thank my wife, Dr Ana Hidalgo, who has lived and breathed this book with me. With her wide-ranging knowledge and sure perspective, she has pointed out my foolish errors with a smile, and reinforced my conviction in the central ideas. I am chastened by her intolerance for opacity, which at times was hard to bear, but without which I fear this book would have made no sense to anyone but the author.

I am indebted to many people who have taken the time to read and criticize parts of the manuscript with an expert eye. I am especially grateful to the academics who received an e-mail from me out of the blue, and who responded with alacrity and detailed comments. In this regard, I thank Robert Berner, Professor of Geology and Geophysics at Yale University; Donald Canfield, Professor of Ecology at the University of Southern Denmark in Odense; Dr José Castresana, in the Biocomputing Unit at the European Molecular Biology Laboratory in Heidelberg; Dr David Bremner, Reader in Applied Chemistry at the University of Abertay, Dundee; Tom Kirkwood, Professor of Biological Gerontology at the University of Newcastle; John Allen, Professor of Plant Cell Biology at the University of Lund, Sweden; and Gustavo Barja, Professor of Physiology at the Complutense University in Madrid. I am also very grateful to several colleagues for many stimulating discussions and insights: in particular Dr Barry Fuller, Reader in Surgical Sciences at the Royal Free Hospital, London;

Dr Erica Benson, Reader in Plant Biochemistry and Biotechnology at the University of Abertay, Dundee; Dr Roberto Motterlini and Dr Roberta Foresti, pioneers of haem oxygenase research at the Northwick Park Institute for Medical Research, London; and Professor Colin Green, inspiring head of the Northwick Park Institute and research dynamo.

I also thank several friends who have read and commented on large parts of the manuscript, for giving me a better sense of what the reader might enjoy, or at least tolerate. I thank especially Vince Desmond, Ian Ambrose, Allyson Jones, Paul Asbury, Dr Malcolm Jenkins, and Mike Carter. I thank my parents and brother Max for their impassioned debates about the use of language, their willingness to engage with science from the other side of the cultural divide, and their unstinting support. Without them I would never have started.

With all this help, one might think that few errors or infelicities could have survived. Some did, and I am lucky to have benefited from the learning and literary skills of the copy editor, Eleanor Lawrence, who has tightened and clarified the manuscript with many judicious changes. Finally, I thank Abbie Headon, of OUP, who has responded quickly and helpfully to my many queries about the editorial process. Any remaining faults are all my own.

CHAPTER ONE

Introduction

Elixir of Life — and Death

OXYGEN DEFIES EASY CLASSIFICATION. Ever since it was discovered in the 1770s, its properties and chemistry have been squabbled over by scholars and charlatans alike. The controversy persists today. Oxygen is hailed as the Elixir of Life — a wonder tonic, a cure for ageing, a beauty treatment and a potent medical therapy. It is also purported to be a fire hazard and a dangerous poison that will kill us in the end. The popular health press is contradictory. Inhaling pure oxygen in cosmopolitan 'oxygen bars' and health clinics is said to work wonders, yet the opposite — 'high-altitude therapy' — is claimed to eliminate superfluous oxygen, conferring the health benefits of austerity. So-called 'active' oxygen treatments, meaning ozone and hydrogen peroxide, are touted as miraculous scourges of bacterial infection, or as cures for cancer; yet at the same time we are told that the secret of a long life is to eat plenty of antioxidants, to protect us against the very same 'active' forms of oxygen. Oxygen seems to attract nonsense and misinformation like a magnet.

However muddled these accounts, they agree about one thing: oxygen is important. After all, if we stop breathing it, we will be dead in minutes. Our bodies are beautifully designed to deliver oxygen to each of our 15 million million cells. All the symbolism of red blood ultimately rests in the simple chemical bonding between oxygen and haemoglobin in our red blood cells. Suffocation and drowning — the physical deprivation of oxygen — are among the darkest of human fears. If we think of a planet without oxygen, we think of a sterile place pockmarked with craters, a place like the Moon or Mars. The presence of oxygen in a planetary atmosphere is the litmus test of life: water signals the potential for life, but oxygen is the sign of its fulfilment — only life can produce free oxygen in the air in any abundance. If pressed for an unemotional reason for not cutting down the rain forests or polluting the oceans, we may argue that these great resources are the 'lungs' of the world, ventilating the Earth with life-giving oxygen. This is not true, as we shall see, but illustrates the reverence in which we hold oxygen. Perhaps it is not entirely surprising that we seek mystical or healing properties in a colourless, odourless gas.

This book is about life, death and oxygen: about how and why life produced and adapted to oxygen; about the evolutionary past and future of life on Earth; about energy and health, disease and death, sex and regeneration; and about ourselves. Oxygen is important in ways that most of us hardly even begin to imagine, ways that are far more fascinating than the loud claims of health features. But before we begin our journey, we need to mark out the playing field. Is oxygen an elixir or a poison, or both? And how can we tell the difference? The easiest way to find out is to go back in time, to the beginnings of our own understanding.

Even the discovery of oxygen was controversial. Credit is usually split between the English clergyman and chemist Joseph Priestley, the Swedish apothecary Carl Scheele and the French tax-collector and father of modern chemistry Antoine Lavoisier. Scheele was first of the three, but he put too much trust in the shifting sands of aristocratic patronage and floundered without publication for six years. Priestley had no such problems. In 1774, he made oxygen by focusing sunlight onto an oxide of mercury, and was quick to write three volumes on the subject. Scheele and Priestley would have taken all the credit, except that neither grasped the full significance of the new gas. Both appreciated that burning was more vigorous in pure oxygen — Scheele even called it 'fire-air' — yet both thought of combustion in terms of an erroneous theory, the idea that an invisible substance known as phlogiston was released into the air during burning (rather than oxygen being taken up). They thought of oxygen as pure 'dephlogisticated' air, from which all the contaminating phlogiston had been removed.

Lavoisier, the chemical revolutionary and political conservative, cast off this twisted theory in the years before the French Revolution. It was

Lavoisier who bequeathed the name oxygen, and he who finally proved that oxygen was the reactive constituent of air.[1] Combustion, said Lavoisier, was the reaction of oxygen with carbon or other substances. In a famous demonstration, he showed that the Holy Roman Emperor's diamonds (made of carbon) could be vaporized if heated in the presence of oxygen (to form carbon dioxide), but that diamonds are impervious to heat if oxygen is excluded. Diamonds are forever only in the absence of oxygen. Lavoisier went further. By collecting gases and taking precise measurements with his ultra-sensitive balances, he showed that combustion and human respiration are fundamentally the same process — both consume oxygen and materials containing carbon and hydrogen, to give off carbon dioxide and water.

Lavoisier was still engaged in weighing the gases emitted by breathing and perspiration when he was interrupted by soldiers of the Revolutionary Tribunal, accompanied by an uncontrollable mob. He had made some powerful enemies, among them the revolutionary leader Jean Paul Marat. Convicted on the ludicrous charge of watering the soldiers' tobacco, and of appropriating revenues belonging to the state, Lavoisier was beheaded in May 1794. As the great mathematician Lagrange said, "it took but a moment to cut off that head, though a hundred years perhaps will be required to produce another like it."

Curiously, this famous story of the discovery of oxygen is probably wrong. Not only did the alchemists get there first, but they also had a clear appreciation of the significance of oxygen. In 1604, 170 years before Scheele, Priestley and Lavoisier, the Polish alchemist Michael Sendivogius wrote that "Man was created of the Earth, and lives by virtue of the air; for there is in the air a secret food of life... whose invisible congealed spirit is better than the whole Earth." Sendivogius proposed that this 'aerial food of life' circulated between the air and earth by way of an unusual salt — nitre, or saltpetre.[2] When heated above 336°C, nitre decomposes to release oxygen, which the alchemists knew as *aerial nitre*. Sendivogius

[1] The name 'oxygen' was derived from the Greek for 'acid-former', in the mistaken belief that oxygen is necessary for the formation of all acids. It is necessary for some, such as sulphuric and nitric acids, but not for others, like hydrochloric acid.

[2] Nitre (potassium nitrate, KNO_3), gives every appearance of condensing from the air as a white crust on well-manured soils (containing nitrogen). It has some remarkable properties. Besides being a fine fertilizer, it was used for preserving meat and as a folk remedy. When added to drinks, nitre cools them down as effectively as ice, yet when taken as a medicine produces a strong warming effect. The acid of nitre (*aqua regia*, or 'queen of waters') could dissolve gold, an attribute that appealed to alchemists. Nitre is also the chief component of gunpowder, which was invented by Chinese alchemists in the ninth century.

evidently believed he had discovered the Elixir of Life, "without which no mortal can live, and without which nothing grows or is generated in the world." His ideas went beyond theory. Sendivogius almost certainly produced oxygen by heating nitre, and may well have explained his methods to the Dutch inventor and fellow alchemist Cornelius Drebbel, a lost hero of Renaissance science.

Drebbel gave a brilliant demonstration of the practical utility of oxygen in 1621. After constructing a solar-powered perpetual-motion machine and a variety of refrigerators and automata for King James I of England, Drebbel built the world's first submarine. King James, accompanied by thousands of his subjects, stood on the banks of the River Thames to watch its maiden voyage from Westminster to Greenwich, a distance of ten miles. Manned by twelve oarsmen, the wooden submarine apparently stayed under water for three hours. Much of the interest centred on how Drebbel had managed to refresh the air for the rowers during this time. According to eye-witness accounts, discussed some years later (in 1660) by the great chemist Robert Boyle, Drebbel had used a bottle of 'liquor' (others referred to it as a gas) to refresh the 'vital' parts of the air:

> Drebbel conceived that it is not the whole body of the air, but a certain quintessence, or spirituous part of it, that makes it fit for respiration, which being spent, the remaining grosser body, or carcass (if I may so call it) of the air, is unable to cherish the vital flame residing in the heart.... For when, from time to time, he [Drebbel] perceived the finer and purer part of the air was consumed... he would, by unstopping a vessel full of this liquor, speedily restore to the troubled air such a proportion of vital parts, as would make it again, for a good while, fit for respiration.

Presumably, Drebbel had succeeded in bottling oxygen gas, following the instructions of his mentor Sendivogius, by heating nitre. Sendivogius, Drebbel and Boyle clearly thought of the air as a mixture of gases, one of which was the vital gas oxygen. They recognized that in confined spaces the air was depleted of oxygen by breathing or burning. Boyle certainly saw respiration and combustion in similar terms — the vital flame residing in the heart — even if he did not appreciate how exactly analogous the two reactions really were. Boyle's contemporary and Fellow of the Royal Society, John Mayow, went even further: he showed that aerial nitre (oxygen), when breathed into the lungs, gave arterial blood its red colour. Aerial nitre, he said, was a normal constituent of the air, from where it "becomes food for fires and also passes into the blood of animals by respiration.... It is not to be supposed that the air itself, but only its more active and subtle part, is the igneo-aerial food." In other words, despite his archaic vocabulary, Mayow had a strikingly modern concept of oxygen as early as 1674.

Against this background, Priestley's attachment to the phlogiston theory (the idea that burning releases an invisible substance) a century later looks almost comical, although he was by no means alone. Guided by false phlogiston, the study of airs had been groping at shadows for the best part of a century. To account for experimental results, phlogiston was assigned a positive weight, no weight at all, or a negative weight, depending on the requirements. Even those who credit Priestley with the discovery of oxygen accept that his adherence to this theory blinded him to the true meaning of his discovery.[3] In another sense, though, Priestley was uncannily prescient. He foresaw not only the medical applications of oxygen (which he persisted in calling dephlogisticated air) but also its potential danger. In his *Experiments and Observations on Different Kinds of Air*, published in 1775, Priestley mused on his own experiences of breathing pure oxygen:

> The feeling of it to my lungs was not sensibly different from that of common air; but I fancied that my breast felt peculiarly light and easy for some time afterwards. Who can tell but that, in time, this pure air may become a fashionable article in luxury.... From the greater strength and vivacity of the flame of a candle, in this pure air, it may be conjectured, that it might be peculiarly salutary to the lungs in certain morbid cases, when the common air would not be sufficient to carry off the putrid effluvium fast enough. But, perhaps, we may also infer from these experiments, that though pure dephlogisticated air [oxygen] might be very useful as a medicine, it might not be so proper for us in the usual healthy state of the body; for, as a candle burns out much faster in dephlogisticated than in common air, so we might, as may be said, *live out too fast* [Priestley's italics] and the animal powers be too soon exhausted in this pure kind of air. A moralist, at least, may say, that the air which nature has provided for us is as good as we deserve.

[3] To be fair to Priestley, he was perfectly aware of the problems with the phlogiston theory. He compared phlogiston with light and heat, which could not be weighed either (then as now).

Anyone inhaling pure oxygen in an 'oxygen bar' today might smile at Priestley's quaint analogy and moral sentiments, but few researchers would disagree with the substance of these remarks. Strikingly, Priestley's words contain the first suggestion (to my knowledge) that oxygen might accelerate ageing. This caution was lost on his contemporaries, who went on to embrace the medical potential of oxygen before the end of that century. Despite suspicions, its toxicity was not documented for a further hundred years.

The first person to use pure oxygen therapeutically, on a large scale at least, was Thomas Beddoes. Beddoes founded the Pneumatic Institute for inhalational gas therapy in Bristol in 1798, in which he employed the brilliant young chemist Humphry Davy. The pair aimed to treat diseases hitherto found incurable. Unfortunately, they were overly ambitious in their choice of patients, and their treatments offered little clinical benefit. Worse, impurities in the gas frequently caused inflammation of the lungs. (Ironically, the inflammation may not have been caused only by impurities; pure oxygen also inflames the lungs.) Faced with these problems, as well as unreliable supplies of oxygen, the institute closed its doors in 1802. Davy later described his work there as "the dreams of misemployed genius, which the light of experiment and observation has never conducted to truth".

This pattern of hopes and failures persisted throughout much of the nineteenth century. Problems with impurities, and diverse methods of delivery to the patient, meant that a clinical consensus never emerged. Sometimes oxygen was inhaled directly from a mask or bag. In other cases, the gas was bubbled through a bucket of water placed near the bed, and the air of the room fanned towards the patient. Failure must have been assured. With such disparate procedures, and little in the way of systematic comparisons, it is hardly surprising that outcomes were discrepant. Advocates of oxygen therapy claimed miraculous cures (which may have been true in conditions such as pneumonia) but the voices of mainstream medicine were for the most part unimpressed, arguing that any perceived benefits were transitory, palliative or imaginary. The gap was forced even wider by the usual quacks and charlatans, who peddled a secret preparation known as 'compound oxygen' to a gullible public. Some claims made in the 1880s were remarkably similar to those made today by proponents of 'active oxygen' therapies. They were dismissed then, as today, by ethical practitioners of oxygen therapies.

Medical interest in oxygen therapies picked up after a number of anecdotal reports suggested that higher oxygen pressures really did affect health. For example, patients with pneumonia living at high altitude in cities like Mexico City, where the oxygen pressure is low, were found to have a better chance of recovery if rushed down to the plains, where oxygen pressure is higher. Similarly, patients with cardiovascular disease generally fared better at sea level than at high altitude. Impressed by these reports, the American physician Orval Cunningham reasoned that still higher barometric pressures might amplify the effect. Following a number of apparent successes, a grateful client helped finance construction of the largest-ever hyperbaric chamber in Cleveland, Ohio, in 1928 — a million-dollar hollow steel ball, 20 metres [65 feet] in diameter and five stories tall, pressurized to about twice the atmospheric pressure at sea level.

Cunningham fitted out his giant steel ball as a hotel, with a smoking lounge, restaurant, rich carpeting and private rooms. Unfortunately, he used compressed air, not oxygen, so the total oxygen pressure was no higher than could have been achieved with a mask, at a fraction of the cost. Worse, rather than treating people with conditions such as pneumonia and cardiovascular disease, who might have benefited, he overstepped the mark by treating patients with diabetes, pernicious anaemia and cancer, on the fallacious grounds that all these conditions were caused by anaerobic (oxygen-hating) bacteria. Both his objectives and his results failed to impress the American Medical Association, who condemned the scheme as "tinctured much more strongly with economics than with scientific medicine". The steel ball lasted but a few years before being dismantled and sold for scrap in 1942, contributing to the American war effort.

Cunningham should have known better. Despite the equivocal history of oxygen therapy, the field had finally been put on a scientific footing by the distinguished Scottish physiologist John Scott Haldane (father of the biologist J. B. S. Haldane) in the early years of the twentieth century. Haldane was an expert in diving medicine and had spent the First World War using oxygen to treat injuries caused by chlorine gas. He summed up his experiences in the groundbreaking book, *Respiration*, published in 1922, in which he argued that some patients with respiratory, circulatory and infectious conditions could be *cured* by continuous oxygen inhalation. Given properly, he said, oxygen therapy was not just palliative, but could break the vicious cycle of degeneration, giving the body an opportunity to recover its own healthy equilibrium.

Haldane's tenets underpin modern oxygen therapies, yet even today we do not have a clear appreciation of how beneficial these therapies can be. A large clinical trial, reported in the prestigious *New England Journal of Medicine* in January 2000, showed that inhalation of 80 per cent oxygen for two hours halved the risk of wound infection after colorectal surgery, compared with routine practice (30 per cent oxygen for two hours). The finding that a simple treatment can make a big difference is encouraging; but the fact that a treatment available in essentially the same form for 200 years can still make medical headlines at the start of the twenty-first century is salutary. If nothing else, it illustrates just how far the progress of science can be impeded by a professional knee-jerk response to the inflated claims of quacks and charlatans.

Another reason for caution was spelled out by Haldane early in the twentieth century — the possibility of oxygen toxicity. Haldane himself had written that:

> The probable risks of prolonged administration of pure oxygen must be borne in mind and if necessary balanced against the risks of allowing the oxygen-want to continue. No fixed rule can be given. The proper course to pursue must be determined by the physician after careful observation of the patient, and in the light of experience and knowledge.

It is understandable that physicians prefer to err on the side of caution; but what are these risks? Haldane's careful wording makes them sound a little theoretical, but oxygen, especially under pressure, can cause shockingly physical reactions, as Haldane knew well from his own research in diving medicine.

The toxicity of oxygen is slow-acting, or hidden from view, in normal circumstances. Many people receive oxygen therapy in hospital, or spend days, sometimes weeks, in oxygen tents, or inhale oxygen in bars with no ill effects. Astronauts often breathe pure oxygen for weeks on end, though in space the capsule is pressurized to only one third of atmospheric pressure, which makes it equivalent to breathing 33 per cent oxygen. The difference that pressure makes to the oxygen concentration in the atmosphere explains why three astronauts died when Apollo 1 caught fire in 1967, as they were completing tests on the ground. In space, the inside of the capsule is always pressurized to a higher pressure than the surrounding vacuum, which means that spacecraft are built to withstand a greater pressure inside than outside. To maintain this pressure differential, Apollo 1 was pressurized to above atmospheric pressure while on the ground. Unfortunately, the spacecraft was still being ventilated with pure oxygen. This meant that instead of an atmosphere equivalent to 33 per cent oxygen, the astronauts were actually breathing the equivalent of 130 per cent oxygen. In this oxygen-rich atmosphere, a spark from the electrical wiring led to an uncontrollable fire, which reached a temperature of 2500°C within minutes.

But oxygen is more than just a fire risk: it is toxic to breathe. This toxicity depends on the concentration and duration of exposure. Most people can breathe pure oxygen for a day or two, but we cannot breathe it for longer without risk. If the concentration of oxygen is increased even more by compressing the gas, then the toxic effects become dramatic.

The realization that oxygen is toxic came from the experiences of the earliest scuba divers, towards the end of the nineteenth century. (The word scuba was a later coinage, and stands for self-contained underwater breathing apparatus.) Scuba divers were vulnerable because they carried their breathing apparatus with them, and usually breathed pure oxygen. The oxygen in the apparatus could be compressed by water pressure. Breathing pure oxygen at depths below about 8 metres [26 feet] causes seizures similar to an epileptic grand-mal — a disaster if the diver loses consciousness underwater.

Oxygen convulsions were first described systematically by the French physiologist Paul Bert, professor of physiology at the Sorbonne in Paris. In his celebrated 1878 monograph on barometric pressure, Bert discussed the effect of oxygen on animals subjected to different pressures in a hyperbaric chamber. Very high oxygen concentrations caused convulsions and death in a matter of minutes. The following decade, in 1899, the Scottish pathologist James Lorrain Smith showed that lower levels of oxygen could have an equally deadly, but delayed, effect. Animals exposed to 75 per cent oxygen or more (at normal air pressure) had such serious inflammation of the lungs after a few days that they died. For this reason, oxygen dosages in hospitals are always strictly controlled. Convulsions and lung injury became familiar worries to scuba divers, however. Both Paul Bert and James Lorrain Smith are still commemorated in diving terminology. Unfortunately for Smith, his unusual name, along with his habit of styling himself J. Lorrain Smith, frequently turns the tribute into the 'Lorraine Smith' effect.

While many divers were careful not to dive too deep while breathing pure oxygen, the Navy could not always afford to be so cautious. In the British Royal Navy Submarine Escape Handbook, published in 1942, seamen were instructed to watch out for the symptoms of oxygen poisoning: "tingling of the fingers and toes, and twitching of the muscles (especially around the mouth); convulsions followed by unconsciousness and death if a remedy is not taken." Naval divers during the war invented a mythical monster, Oxygen Pete, who lurked at the bottom of the sea waiting to molest unwary divers. Oxygen toxicity 'hits' during this time were referred to as "getting a Pete".

A more rigorous understanding of oxygen toxicity, human limits and gas mixtures was clearly needed, and J. B. S. Haldane was commissioned by the Royal Navy to follow in the footsteps of his father. Always an advocate of being one's own rabbit, Haldane subjected himself and his colleagues to various oxygen concentrations under different pressures, noting how long it took before convulsions set in.[4] Exposure to pure oxygen at seven atmospheres pressure led to convulsions within five minutes. He later wrote that:

> The convulsions are very violent, and in my own case the injury caused to my back is still painful after a year. They last for about two minutes and are followed by flaccidity. I wake in a state of extreme terror, in which I may make futile attempts to escape from the steel chamber.

Nonetheless, his efforts were successful. The Royal Navy secretly developed various nitrogen/oxygen (nitrox) mixtures, which lowered the risk of both oxygen toxicity and nitrogen narcosis (the 'bends'). These nitrox mixtures were used by British commandos defending Gibraltar in the Second World War, and were kept such a close secret that even the US Navy did not find out until the 1950s. Using nitrox mixtures the British divers could operate at greater depths. A major element of British strategy was to lure the combatants into deep waters until they were overwhelmed by convulsions. Mugged by oxygen: perfidious Albion indeed!

---

[4] In a celebrated essay on self-experimentation, "On being one's own rabbit", published in 1928, Haldane wrote that "to do the sort of things to a dog that one does to the average medical student requires a license signed in triplicate by two archbishops." He also thought it peculiar that so few chemists wondered what it actually *felt* like to become more acid or alkaline, or dilute.

Breathing oxygen at high concentration is obviously toxic. Above about two atmospheres of pressure, pure oxygen causes convulsions and sometimes death. Oxygen accounts for about a fifth of atmospheric pressure, so pure oxygen at two atmospheres pressure is ten times our normal exposure. At lower concentrations, oxygen is unlikely to cause convulsions, but breathing pure oxygen at normal atmospheric pressure (five times our normal exposure) for a few days can still cause life-threatening lung damage. Such serious inflammation of the lungs prevents us from breathing properly. Ironically, we then cannot pass oxygen into the blood stream, so we actually die from oxygen starvation to the rest of the body. At lower levels of oxygen (40 or 50 per cent oxygen, or about twice our normal exposure), the lungs can normally withstand injury and continue to function, though they may become damaged in the end. In these circumstances, the rest of the body adapts by slowing down the heart beat and producing fewer red blood cells. These adaptations are the opposite of the changes that take place to oxygen deprivation at high altitudes. The result, in both cases, is that the tissues receive the same amount of oxygen as before, no more nor less. Such adaptations illustrate the importance of unchanging oxygen levels in the body. They also mean that we cannot gain any long-term benefit from either high or low levels of oxygen, except when we are sick and pathologically oxygen-deprived.[5]

I imagine that most people are comfortable with the idea that too much oxygen can be bad — in effect, that it is possible to have too much of a good thing. Similarly, there is nothing challenging about the idea that we respond to moderate perturbations by re-establishing the physiological *status quo*. It is a very different proposition to say that 21 per cent oxygen is toxic and will kill us in the end. This is as much as to say that, despite millions of years of evolution, we still cannot adapt to the concentration of oxygen that nature has provided for us. This statement is counter-intuitive to say the least, yet it is the basis of the so-called *'free radical'* theory of ageing. In essence, this theory argues that ageing, and so death, is caused by breathing oxygen over a lifetime. Oxygen is thus not only necessary for life, but is also the primary cause of ageing and death.

[5] Athletes training at altitude must come down to sea level and race within days or weeks or else the benefits are lost. When we train at altitude, we generate more red blood cells to absorb extra oxygen from the thin air. When we return to sea level, we adapt back to the higher levels of oxygen by producing fewer red blood cells. The benefits never outlast the adaptation.

Many people have heard of free radicals, even if they have only a hazy idea of what they actually are. Most free radicals of biological importance are simply reactive forms of molecular oxygen, which can damage biological molecules (we will consider them in detail in Chapter 6). Regardless of whether oxygen causes convulsions and sudden death, or slow lung damage, or ultra-slow ageing, it always acts in exactly the same way: all forms of oxygen toxicity are caused by the formation of free radicals from oxygen. As the great sixteenth-century alchemist Paracelsus said, the poison is in the dose. Convulsions are caused by a massive excess of free radicals acting on the brain, lung damage by a smaller excess acting on the lungs. But free radicals are not just toxic. Fire is impossible without free radicals. So too is photosynthesis or respiration. When we use oxygen to extract energy from food we have to produce free-radical intermediates. The secret to all the chemistry of oxygen, whether we think of it as 'good' or 'bad', is the formation of free radicals.

As conventionally stated, the idea that breathing oxygen causes ageing is disarmingly simple. We produce free radicals continuously inside every cell of our body as the cells respire. Most of these are 'mopped up' by antioxidant defences, which neutralize their effects. The trouble is that our defences are not perfect. A proportion of free radicals slip through the net and these can damage vital components of cells and tissues, such as DNA and proteins. Over a lifetime, the damage gradually accumulates until it finally overwhelms the ability of the body to maintain its integrity. This gradual deterioration is known as ageing.

According to this conventional, if simplistic, explanation, the more antioxidants we eat, the more we can protect ourselves against damage from free radicals. This is why fruit and vegetables are good for us: they contain lots of antioxidants. Nowadays, many people supplement their diet with potent antioxidants in the belief that their diet cannot provide an adequate supply. The implication is that if we eat enough of the right kind of antioxidants, we can postpone ageing and the diseases of old age indefinitely. This has been touted as 'the antioxidant miracle'.

The truth is rather more complicated, but far more interesting. I shall argue that oxygen free radicals *do* cause ageing, but that the implications are almost exactly the opposite of what we might expect. We will never extend our lives significantly, to 150 or 200, by loading ourselves with even the most potent antioxidant supplements. On the contrary, antioxidant supplements might actually make us more vulnerable to some diseases. Antioxidants are bit players in the large cast of adaptations that life has made to the presence of oxygen in the air. We can only understand their role if we consider them in the context of the play as a whole. The response of life to the threats and the possibilities of oxygen include adaptations that have had the most profound consequences.

Let's just consider a few examples. Take photosynthesis — the formation of organic matter by plants, algae and some bacteria using the energy of sunlight — which today supports almost all life on Earth. It is probable that photosynthesis (which generates oxygen as a waste product) could only have evolved because life had already adapted to provide itself with defences against the oxygen free radicals produced by ultraviolet radiation in the environment. This may explain why life took off on Earth but never did on Mars. Take the abundance of large animals and plants characteristic of our world. The first multicellular organisms probably evolved from clumps of cells which clustered together to deal collectively with the rising tide of atmospheric oxygen produced by photosynthesis. Without the threat of oxygen toxicity, life would never have advanced beyond a green slime. Even gigantism relates to oxygen. Giant size offers an escape from the threat of oxygen, as metabolic rate is slower in very large animals, and explains the evolution of monster dragonflies, with a wingspan as broad as a seagull, and possibly the rise and fall of the dinosaurs. Think about the sexes. Why should there be only two sexes? Why not one, or three, or many? The evolution of two sexes may have been a way of coping with oxygen. We shall see that babies can only be born young if they are born of two sexes, otherwise oxygen causes the birth of degenerate offspring, destined to age prematurely. This may explain why cloned animals tend to die young. Dolly the sheep, for example, already has arthritis at the age of five, betraying a 'real' age of eleven. Finally, think of powered flight. Birds and bats have exceptionally long lives for their sizes. Why? Flying demands metabolic adaptations to oxygen that also confer a long lifespan. If we want to extend our own lifespan, we must look to the birds.

These are grand statements, which I shall explain and defend later in the book. They are all part of our journey to find out how oxygen affects our own lives and deaths.

This is unashamedly a book about science. It is not a catalogue of dry facts about how the world works; rather, like science itself, it is full of quirks, experiments, oddities, speculations, hypotheses and predictions. Science is often presented as 'the facts', frequently in short sound-bites. The scientific method is described as a methodical unravelling of 'the truth', which, if this were true, would bore most people, including most scientists, to tears. The impression that science gives access to an objective reality (as opposed to the subjective world of ethics) sets it up in opposition to religion as an ethical system and gives scientists an air of preaching. In fact, science gives vivid insights into the workings of nature, but falls short of objective reality. Too often, scientific 'facts' turn out to be wrong or misleading — we are told that there is 'no risk' of a Frankensteinian disaster, only to see it come true before our eyes. At other times, scientists squabble about the meaning of obscure research findings, discrediting their colleagues in public. It is hardly surprising that the general public views science and scientists with growing scepticism. Apart from the unfortunate schism this opens up in society, it means that fewer young people dream of becoming scientists. This is a tragedy. I wonder if the tragedy might be averted to some extent if people had a better idea of the workings of science — of the fun, creativity and adventure.

The real interest of science lies in the unknown, the excitement of charting new terrain. Poking around in the unknown rarely generates a perfect picture of the world — we are more likely to construct a kind of medieval map, a distorted but recognizable picture of reality. Scientists try to link together the contours of a story through experiments that fill in a detail here or there. Much of the joy of science lies in devising and interpreting experiments that test these hypothetical landscapes. I have therefore been careful to explain the experiments and observations that underpin the story of this book. I have tried to show how it is that science can be interpreted in different ways, and I have presented the evidence itself, along with its flaws, so that you may judge for yourself whether my own interpretation is convincing. I hope this approach will help you to share the spirit of adventure along the border of the known and the unknown.

Science, then, generates hypotheses based on evidence that is specific but limited in scope – islands of knowledge in a sea of unknowing. Very often, individual results only make sense when seen in the context of a bigger picture. All scientific papers have a discussion section, whose purpose is to place the new results in perspective. But science is nowadays highly specialized. It is rare for a medical researcher to refer to the studies of geologists and palaeontologists in the discussion, or for a chemist to be much concerned with evolutionary theory. For most of the time this matters little, but in the case of oxygen, perspective is obliterated by too confined a view. In this case, geology and chemistry have a great deal to say on evolutionary theory, and palaeontology and animal behaviour have much to contribute to medical science. All these fields offer insight into our own lives and deaths.

If an understanding of oxygen's role in life and death requires a multidisciplinary approach, it also offers fresh perspectives on each of these fields. Looking at evolution and health through the prism of oxygen solves some long-standing conundrums. I have already mentioned one example: the evolution of two sexes. If we start with the dilemma itself — why did two sexes evolve — it is difficult to discriminate between one hypothesis and another. We can't even eliminate the possibility that things 'just happened' that way. Thinking about the role of oxygen in ageing may seem to be irrelevant to this problem, but it actually forces us to conclude that two sexes are necessary for reproduction if a species produces motile sex cells that must search for a mate; and it generates a number of predictions. Thinking about life in this way also explains why we cannot extend our lives just by taking antioxidant supplements, and points us to more realistic ways of postponing ageing and the ailments of old age. Oxygen thus acts as a magnifying glass, enabling us to scrutinize life from some unusual angles. That means that this book is about life, death and oxygen, and not just about oxygen.

I have tried to write for a wide audience who may have little knowledge of science, and hope to be accessible to anyone prepared to make a little effort. The argument works out over the book as a whole, and you'll have to read to the end to get the full story! Each chapter, however, tells a story of its own, and I have not assumed much prior knowledge from previous chapters. We shall see that life's adaptations to oxygen, which began nearly 4 billion (4000 million) years ago, are still written in our innermost constitution. We shall see that radiation poisoning, nuclear reactors, Noah's flood, photosynthesis, snowball Earths, giant insects, predatory monsters, food, sex, stress and infectious diseases are all linked by oxygen. We shall see that an oxygen-centric view gives striking insights into the nature of ageing, disease and death. We shall see that oxygen, a simple colourless, odourless gas, made the world in which we live, framing our own passage across the stage. We shall see all this by thinking about how and why oxygen has influenced the evolution of life from the very beginning.

CHAPTER TWO

In the Beginning

The Origins and Importance of Oxygen

IN THE BEGINNING THERE WAS NO OXYGEN. Four billion years ago, the air probably contained about one part in a million of oxygen. Today, the atmosphere is just less than 21 per cent oxygen, or 208 500 parts per million. However this change might have come about, it is pollution without parallel in the history of life on Earth. We do not think of it as pollution, because for us, oxygen is necessary and life-giving. For the tiny single-celled organisms that lived on the early Earth, however, oxygen was anything but life-giving. It was a poison that could kill, even at trace levels. A lot of oxygen-hating organisms still exist, living in stagnant swamps or beneath the seabed, even in our own guts. Many of these die if exposed to an oxygen level above 0.1 per cent of present atmospheric levels. For their ancestors, who ruled the ancient world, pollution with oxygen must have been calamitous. From dominating the world they shrank back to a reclusive existence at the margins.

Oxygen-hating organisms are said to be *anaerobic* — they cannot use oxygen and, in many cases, can only live in its absence. Their problem is that they have nothing to protect them against oxygen poisoning: they possess few, if any, antioxidants. In contrast, most living things today tolerate so much oxygen in the air because they are stuffed full of antioxidants. There is a paradox hidden in this progression. How did modern organisms evolve their antioxidant protection? According to the standard textbook view, antioxidants could not have been present in the first cells that began emitting oxygen as a toxic waste product: how could they have adapted to a gas that had not existed before? Yet if this assumption is true — that antioxidants evolved after the rise in atmospheric oxygen — then the huge rise in atmospheric oxygen must have posed a very serious challenge to early life. If oxygen had anything like the effect on the first anaerobic cells that it does on their descendants today, then there ought to have been a mass extinction of anaerobic organisms that would put the fall of the dinosaurs in the shade.

Why should we care? According to the free-radical theory of ageing discussed in Chapter 1, oxygen toxicity sets limits on our lives. If this is true, the ways in which life has adapted to oxygen over evolutionary time should be revealing. Did the rise in atmospheric oxygen really cause a mass extinction? How did life adapt? If ageing and death are caused by an ultimate failure to adapt, can we learn anything from how the survivors of this putative holocaust coped? Can we somehow 'do more' of whatever they did? In the next few chapters, we will attempt to answer some of these questions by charting the response of organisms to changing oxygen levels over the aeons.

The origins and early history of life have attracted renewed research interest in the past few decades. Some of our most basic ideas about the genesis of life have been turned on their heads. Yet so persuasive and ingrained was the old view that even recent biology textbooks cling to its tenets. Many scientists working in other fields seem oblivious to the rewriting of their gospels. The old story is worth recounting here because the role ascribed to oxygen emphasizes its toxicity.

In the 1920s, J. B. S. Haldane in England and Alexander Oparin in Russia independently began to think about the possible composition of the Earth's original atmosphere, on the basis of gases known to be present in the atmosphere of Jupiter (which could be detected by their optical spectra). Haldane and Oparin argued that if the Earth condensed from a cloud of gas and dust, along with Jupiter and all the other planets, then the original atmosphere of the Earth ought to have contained a similarly noxious mixture of hydrogen, methane and ammonia. Their ideas stood the test of time, and formed the basis of a famous series of experiments by Stanley Miller and Harold Urey in the United States during the 1950s.

Miller and Urey passed electric sparks (simulating lightning) through a gaseous atmosphere comprising the three Jupiter gases, and collected the end-products. They found a complex mixture of organic compounds, including a high proportion of amino acids (from which all living things make proteins, the building blocks of life). Such reactions, they said, could have turned the early oceans into a thin organic soup containing all the precursors of life. The only other ingredients needed for life to congeal out of this soup were chance and time, both of which seemed to be virtually limitless: the planet is 4.5 billion years old, and the first fossils of large animals date from half a billion years ago. Four billion years should have been long enough.

The choice of the gases to work with made good practical as well as theoretical sense. Hydrogen, methane and ammonia do not last long in the presence of oxygen and light. The mixture becomes oxidized, and when this happens the yield of organic compounds quickly falls off. Chemically speaking, *oxidation* refers to the removal of electrons from an atom or molecule. The reverse process is called *reduction*, which involves the addition of electrons.

Oxidation is named after oxygen, which is good at stripping electrons from molecules; to help you remember, think of oxygen as being caustic or destructive, like a paint-stripper. Oxidation strips off the electron paint, whereas reduction has the blanketing effect of a fresh coat of paint.[1] The point is that oxygen can strip organic molecules of electrons, often shredding the molecules, which give up their own electrons as a sacrificial offering, in the process. Today, cells counter this kind of damage with antioxidants, but in the beginning there were no antioxidants. Free oxygen would have been an insurmountable problem, because any organic molecules, or incipient forms of life, would have been shredded if much oxygen was present. The fact that life did start can only mean that oxygen was not present in any abundance.

The first cells, then, presumably evolved in an oxygen-free atmosphere, and in turn must have generated energy without the aid of oxygen. This seemed a reasonable proposal, as at the close of the nineteenth century Louis Pasteur had described fermentation as 'life without oxygen', and subsequent research had proved him right. Because yeasts and many other single-celled organisms depend on fermentation for their

[1] There is also an old mnemonic, 'LEO the lion says GER': Loss of Electrons is Oxidation, Gain of Electrons is Reduction).

energy, and are simple in structure, it was an easy extrapolation to assume that they were relics of ancient life. This single-celled life must have lived by fermenting organic compounds dissolved in the oceans, the theory went, until they were superseded by the evolution of the first oxygen-evolving photosynthetic bacteria — the cyanobacteria (once, inaccurately but poetically, called the blue-green algae).

The cyanobacteria learnt to harness the energy of the Sun. Microscopic they may have been, but as the aeons ticked away, inconceivably large numbers of cyanobacteria (several billion fit in a droplet of water) silently polluted their environment with toxic oxygen waste. To begin with, this oxygen would have reacted with minerals dissolved in the oceans or eroded from the rocks, oxidizing them and locking up the oxygen in mineral compounds. These enormous natural resources acted as a buffer against free oxygen for hundreds of millions of years. In the end, however, the buffer became completely oxidized. With nothing left to take up the slack, the atmosphere and oceans became abruptly (in geological terms) contaminated with excess oxygen. The cost was terrible — an oxygen holocaust. Here is Lynn Margulis, distinguished professor of biology at the University of Massachusetts, Amherst, writing in 1986:

> This was by far the greatest crisis the earth has ever endured. Many kinds of microbes were immediately wiped out. Microbial life had no defence against this cataclysm except the standard way of DNA replication and duplication, gene transfer and mutation. From multiple deaths and an enhanced bacterial sexuality that is characteristic of bacteria exposed to toxins came a reorganisation of the superorganism we call the microcosm. The newly resistant bacteria multiplied, and quickly replaced those sensitive to oxygen on the Earth's surface as other bacteria survived beneath them in the anaerobic layers of mud and soil. From a holocaust that rivals the nuclear one we fear today came one of the most spectacular and important revolutions in the history of life.

According to this view, the success of the new world order stemmed not just from the ability of microorganisms to withstand oxygen toxicity, but from a stunning evolutionary *tour de force* in which cells became dependent on the very substance that had been a deadly poison. The inhabitants of this brave new world were energized by oxygen.

The old theory continues: our dependency on oxygen obscures the fact that it is a toxic gas, intimately linked with ageing and death, to say nothing of being a serious fire risk. Over evolutionary time, the reactivity of oxygen has served to modulate its own accumulation in the atmosphere. We are told that ever since the explosion of multicellular life, some 550 million years ago, atmospheric oxygen has hovered around 21 per cent, the outcome of a sustainable natural balance. If its concentration strays too high, then oxygen toxicity suppresses plant growth. As a result, the amount of oxygen produced by photosynthesis falls, and this lowers atmospheric levels again. In an atmosphere with more than about 25 per cent oxygen, we are told, even wet rain forests would flare up in vast conflagrations. Conversely, if oxygen levels were to fall below about 15 per cent, animals would suffocate and even dry twigs would fail to light. The continuous record of fossil charcoal in sedimentary rocks over the past 350 million years suggests that fires have continuously swept the Earth. If so, then oxygen levels could never have fallen below 15 per cent. Thus, the biosphere has regulated atmospheric oxygen at a level congenial to itself throughout the modern age of plants and animals.

This is the story I grew up with, and much of it is still widely accepted, or at least unquestioned. Although based on somewhat limited evidence, most of the claims sound biologically plausible. To summarize: life evolved through chemical evolution in a primordial soup, formed from a planetary atmosphere containing methane, ammonia and hydrogen. The first cells fermented this soup until they were displaced by cyanobacteria, which used solar energy to power photosynthesis, giving off oxygen as a toxic waste product. This poisonous gas oxidized the rocks and oceans, and finally accumulated in the atmosphere, causing an apocalyptic extinction, an oxygen holocaust. From the ashes a new world order emerged, which depended on the very gas that had wiped out most of its ancestors. The new order was energized by oxygen. Even so, the toxicity and reactivity of oxygen constrained the biosphere to regulate its atmospheric content at 21 per cent.

So firmly entrenched was this story in my own mind that I was exasperated to hear a claim on television that oxygen levels once reached 35 per cent — during the Carboniferous period, around 300 million years ago. Nonsense, I thought! Everything would burn! Plants wouldn't grow! I was not alone. The idea, although advanced seriously by geochemists of international standing, had at first been derided by the wider geological and biological communities. It was not until I began to research the subject that I became convinced that the revisionists were right. Many of the ideas are still controversial, and most of the individual pieces of evidence are flawed, but they have one redeeming feature: in the last two decades we have stepped from the realms of 'geopoetry' into a new era of molecular evidence, which underpins the new models of global change. Taken together, I find the weight of evidence convincing, even if the new story flies in the face of oxygen toxicity and indeed, sometimes, common sense.

Before examining the evidence, and asking how it affects the lives we lead today, we should reorientate ourselves to the emerging picture. Almost every step of my previous summary has been reversed. Far from coalescing from a primordial soup, the new story goes, life might have begun in hot sulphurous vents known as black smokers, deep in the mid-ocean trenches. Paradoxically, the last common ancestor of all known life, tenderly known as LUCA (the Last Universal Common Ancestor), is thought to have used trace amounts of oxygen to respire, even before her descendants learnt to photosynthesize (at least to generate oxygen). Instead of muddling along by fermentation, the first cells are thought to have extracted energy from a range of inorganic elements and compounds including nitrate, nitrite, sulphate and sulphite — and oxygen. If so, LUCA was already resistant to oxygen toxicity before there was any free oxygen in the air. Presumably, her descendants, such as the cyanobacteria, were similarly protected against their own waste product, and so did not succumb to an oxygen holocaust.

In fact, there is no solid evidence that oxygen ever caused a mass extinction. Instead of rising swiftly to reach an equilibrium controlled by the biosphere, the oxygenation of the Earth seems to have proceeded in a series of sharp jerks or pulses, each one precipitated by non-biological factors such as plate tectonics and glaciation. Each rise in atmospheric oxygen has been linked with prolific biological 'radiations', in which life expanded to fill vacant ecological niches, in much the same way that the empty prairies propelled the colonization of the American West. An injection of oxygen into the air immediately preceded the rise of single-celled *eukaryotes* — cells containing a nucleus — which are the cellular ancestors of all multicellular organisms, including ourselves. Similar oxygen injections preceded the explosion of multicellular plants and animals at the beginning of the Cambrian period (which began 543 million years ago), and the evolution of giant insects and plants during the Carboniferous and early Permian (320–270 million years ago); perhaps even the rise of the dinosaurs. Conversely, several mass extinctions are associated with periods of falling oxygen levels, including the 'mother of all extinctions' at the end of the Permian (around 250 million years ago). The inescapable conclusion, that oxygen is a Good Thing, may give few people a sleepless night, but will certainly help constrain our ideas of oxygen toxicity in ageing and disease.

The first sacred cow to be sacrificed was the Jupiter-like composition of Earth's primordial atmosphere. In fact, life must have evolved under an atmosphere that contained very little methane, hydrogen or ammonia. The evidence for this is direct and comes from geology.

The Earth and the Moon were formed just over 4.5 billion years ago. The age of the craters on the Moon, dated from rock samples brought back by the Apollo astronauts, suggests that our planetary system was bombarded by meteorites for at least 500 million years. The bombardment ended around 3.8 to 4 billion years ago. The oldest sedimentary rocks on Earth, which were laid down along what is now the west coast of Greenland, have been reliably dated to an age of 3.85 billion years — a mere 700 million years after the formation of the Earth and certainly not long after the end of the bombardment.

Despite their antiquity, these ancient rocks speak of an atmosphere and a hydrological cycle surprisingly similar to our own. The fact that these rocks were once sediments implies that they were laid down under a large body of water. The sediments were presumably eroded by rainwater from a land mass. This constrains the possible temperatures to within a range compatible with evaporation, cloud formation and precipitation. The mineral content of the rocks allows an informed guess about the composition of the air at the time. There are carbonates present, which probably formed from carbon dioxide reacting with silicate rocks, as happens today; we may presume then that carbon dioxide was present. There are also various iron oxides in the rocks, which from a chemical point of view could not have formed under Jupiter-like conditions, but equally could not exist if more than a trace of oxygen was present. We may take it that oxygen was no more than a trace gas at the time. Finally, we must assume that nitrogen was the main component of the air then, as today, because nitrogen is almost inert as a gas and cannot be generated in large amounts by life. No known chemical or biological process could have produced an atmosphere so rich in nitrogen, so it must have been there all along. Thus, the Earth's atmosphere, nearly 4 billion years ago, probably consisted mostly of nitrogen, as today, with some carbon dioxide and water vapour, and trace amounts of other gases including oxygen. There was essentially no methane, no ammonia, and no hydrogen.

These predictions, based on the composition of early rocks, are supported by a second line of evidence, which provides a clue to the origin of this early atmosphere. This is the rarity of inert unreactive gases, particularly neon, in the Earth's atmosphere today. Neon is the seventh most abundant element in the Universe. It was abundant in the clouds of dust and gas from which the Earth and the other planets of the Solar System condensed. As an inert gas, neon is even more unreactive than nitrogen. If any of the Earth's original atmosphere had survived the meteorite bombardment, it should have contained about the same amount of neon as nitrogen. In fact, the ratio of neon to nitrogen is 1 to 60 000. If there ever had been a Jupiter-like atmosphere on Earth, then it must have been swept away during that first ferocious period of meteorite bombardment.

Where, then, did our modern atmosphere come from? The answer seems to be volcanoes. As well as emitting sulphurous fumes (which would have precipitated in the rain), volcanic gases include nitrogen and carbon dioxide (in about the right balance), tiny amounts of neon, and almost no methane, ammonia or oxygen.

Where did the oxygen come from? There are only two possible sources of the oxygen in the air. By far the most important is *photosynthesis*, the process in which plants, algae and cyanobacteria use the energy of sunlight captured by the green pigment chlorophyll to 'split' water. The splitting of water releases oxygen, which is discharged into the atmosphere as a waste product, while the chemical energy derived from the split is used to bind carbon dioxide from the air and package it into the sugars, fats, proteins and nucleic acids that make up organic matter. Photosynthesis therefore uses sunlight, water and carbon dioxide to produce organic matter. It gives off oxygen as a waste product.

If photosynthesis were the only living process on the planet, oxygen would continue to build up in the air until the plants used up all the available carbon dioxide. Then everything would grind to a halt. Clearly this has not been the case. In fact, there are a number of processes that can consume oxygen, including reactions with minerals in the rocks and oceans and with volcanic gases. Today, however, almost all the oxygen produced by plants is used up by the *respiration* of animals, fungi and bacteria, which use oxygen to 'burn up' or oxidize the organic material they take in as food, extracting energy for the organism's use and releasing carbon dioxide back into the air.[2] Because animals, bacteria and fungi all consume organic matter that comes from another organism, they can be classed together as *consumers*. By definition, consumers gain their energy through the respiration (the controlled burning) of the sugars, fats and proteins made by the primary photosynthetic *producers*. The overall reaction of respiration, in which oxygen and sugars are consumed and the waste products carbon dioxide and water are produced, is almost exactly the opposite of photosynthesis, and consumes essentially the same amount of oxygen that is being produced by photosynthesis. As well as using up oxygen, burning the food we eat regenerates the carbon dioxide needed for photosynthesis to continue; however much we may feel like parasites, the plants need us as much as we need them.

If the consumers were to devour all the organic matter made by primary producers, then all the oxygen released into the air would be consumed by respiration. Perhaps surprisingly, this is very close to what actually happens. The oxygen released by the photosynthesizers is almost completely (99.99 per cent) used up by the animals, fungi and bacteria which feed on the remains of the producers, or on each other. The apparently trivial 0.01 per cent discrepancy, however, is in fact responsible for all life as we know it. It represents the organic matter that is not burnt, but is instead buried under sediments. Over several billion years, this adds up to a vast amount of buried organic matter.

If organic remains are buried rather than eaten, then the complete re-uptake of oxygen by consumers is prevented.[3] The left-over oxygen accumulates in the atmosphere. Almost all our precious oxygen is derived from a 3-billion-year mismatch between the amount of oxygen generated by the primary producers and the amount used up by consumers. The vast amount of dead organic matter buried in the rocks dwarfs the total carbon content of the living world. The Yale University geochemist Robert Berner estimates that there is 26 000 times more carbon buried in

[2] Plants, algae and cyanobacteria also respire, using some of the oxygen released during photosynthesis to burn the carbohydrates produced by photosynthesis and extract the stored energy from them.

[3] For the chemically minded, the overall equation for photosynthesis can be given as $CO_2 + H_2O \rightarrow CH_2O$ (organic carbon in the form of carbohydrate) $+ O_2$; respiration can be given as the reverse reaction. This means that for every molecule of CH_2O (or its equivalent in other organic matter) that is buried and not burned up by respiration, one molecule of O_2 is left in the air.

the crust than is present in the entire living biosphere. Put another way, this means that the entire living world accounts for just 0.004 per cent of the organic carbon currently present on or in the Earth. If all this organic matter reacted with oxygen, then there would be no oxygen left at all. If a mere 0.004 per cent of total organic carbon — in other words, just the living biosphere — reacted with oxygen, then 99.996 per cent of the atmospheric oxygen would be left over. Thus, even the most foolhardy destruction of world forests could hardly dint our oxygen supply, though in other respects such short-sighted idiocy is an unspeakable tragedy.

Buried organic matter takes the form of coal, oil and natural gas, as well as less obvious remains mixed with sediments and minerals such as iron pyrites or fool's gold. Ordinary sandstone rocks, which do not appear to have any trapped carbon at all, typically contain a few per cent organic carbon by weight. Because these rocks are so abundant, they actually account for most of the organic carbon buried in the Earth's crust. Only a small proportion of buried matter is accessible in the form of fossil fuels. This means that, even if we succeeded in burning all the coal, oil and gas trapped in the Earth's crust, we would still only deplete a few per cent of atmospheric oxygen.

The original source of oxygen in the atmosphere was not biological photosynthesis, however, but a chemical equivalent. Few processes show more vividly the importance of the *rate* of a reaction, and the difference that life can make. Solar energy, especially the ultraviolet rays, can split water to form hydrogen and oxygen without the aid of a biological catalyst. Hydrogen gas is light enough to escape the Earth's gravity. Oxygen, a much heavier gas, is retained in the atmosphere by gravity. On the early Earth, most of the oxygen formed in this way reacted with iron in the rocks and oceans, locking it permanently into crust. The net result was that water was lost, because after it had been split, the hydrogen seeped into space and the oxygen was consumed by the crust instead of accumulating in the air.

Over billions of years, the loss of water through the effects of ultraviolet radiation is thought to have cost Mars and Venus their oceans.[4]

[4] The Mars Global Surveyor, which has been orbiting the red planet since April 1999, has sent back detailed pictures of sedimentary rocks that NASA scientists say probably formed in lakes and shallow seas. Erosional channels suggesting the presence of flowing water on Mars sometime in the past were described long ago, but the new images provide the first solid evidence that oceans once existed on Mars. Whether these oceans drained away under the surface of the planet or evaporated into space, or both, is as yet unknown.

Today, both are dry and sterile, their crusts oxidized and their atmospheres filled with carbon dioxide. Both planets oxidized slowly, and never accumulated more than a trace of free oxygen in their atmospheres. Why did this happen on Mars and Venus, but not on Earth? The critical difference may have been the rate of oxygen formation. If oxygen is formed slowly, no faster than the rate at which new rocks, minerals and gases are exposed by weathering and volcanic acitivity, then all this oxygen will be consumed by the crust instead of accumulating in the air. The crust will slowly oxidize, but oxygen will never accumulate in the air. Only if oxygen is generated faster than the rate at which new rocks and minerals are exposed can it begin to accumulate in the air.

Life itself saved the Earth from the sterile fate of Mars and Venus. The injection of oxygen from photosynthesis overwhelmed the available exposed reactants in the Earth's crust and oceans, allowing free oxygen to accumulate in the atmosphere. Once present, free oxygen stops the loss of water. The reason is that it reacts with most of the hydrogen split from water to regenerate water, so preserving the oceans on Earth. James Lovelock, father of the Gaia hypothesis and a rare scientific mind, estimates that today, with oxygen in the air, the rate of hydrogen loss to space is about 300 000 tons per year. This equates to an annual loss of nearly 3 million tons of water. Although this may sound alarming, Lovelock calculates that at this rate it would take 4.5 billion years to lose just 1 per cent of the Earth's oceans. We can thank photosynthesis for this protection. If ever life existed on Mars or Venus, we can be sure that it never learnt the trick of photosynthesis. In a very real sense, our existence today is attributable to the early invention of photosynthesis on Earth, and the rapid injection of oxygen into the atmosphere through the action of a biological catalyst.

How life began on earth is beyond the scope of this book. Interested readers should turn to the writings of Paul Davies, Graham Cairns-Smith and Freeman Dyson, listed in Further Reading. Let us accept that life evolved in the oceans of an Earth shrouded in an atmosphere of nitrogen and carbon dioxide, but with as yet only trace amounts of oxygen. Photosynthesis probably evolved early. We will return in Chapter 7 to the theme of how and why this happened. For now, we wish to chart how life responded to the challenge of rising oxygen levels, as photosynthesis

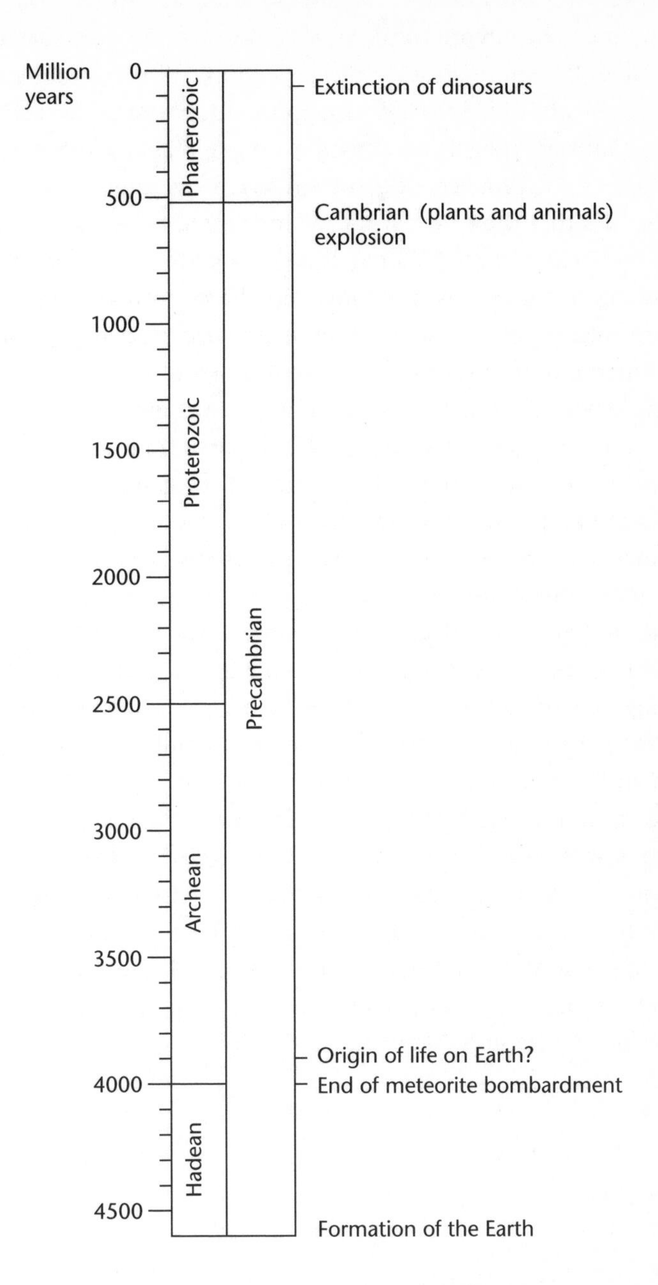

Figure 1: Geological timeline from the formation of the Earth 4.6 billion years ago to the present day. Note the immense duration of the Precambrian era. The first plants and animals appeared around the time of the Cambrian explosion 543 million years ago. The extinction of the dinosaurs was about 65 million years ago.

pumped out oxygen into the air and the oceans. Did oxygen pollution bring about an apocalyptic extinction, as proposed by Lynn Margulis and others, or did it stimulate evolutionary innovations? So long after the event, is there any evidence left to support either interpretation?

The gauntlet was thrown down as long ago as the 1960s by Preston Cloud, one of the pioneers of geochemistry. Even after the large technological strides that the field has taken since then, his work and views still cast a long shadow today. Cloud argued that the major events of early evolution were coupled with changes in the oxygen content of the air. Each time oxygen levels rose, life responded with exuberance. Cloud himself set three criteria to prove this hypothesis: we need to know exactly how and when the oxygen levels changed; we need to show that the adaptations of life happened at exactly the same time; and we need good biological reasons for linking the change in oxygen levels to the evolutionary adaptation. Just how far Cloud's hypothesis is true in the light of new evidence is the focus of the next three chapters.

To make our quest more manageable, we will split Earth's history into three unequal parts (Figure 1). First comes the Precambrian, that long and silent age before there were any visible fossils in the rocks, excepting a faltering experiment in multicellular life in the last few moments. Next comes the so-called Cambrian explosion, when multicellular life exploded into the fossil record like Athena from the head of Zeus, fully formed and wearing armour (shells in their case). Finally the Phanerozoic arrives, our 'modern' age of land plants, animals and fungi, when trilobites, ammonites, dinosaurs and mammals pursued each other in geologically swift succession. The conditions that enabled such an explosion of multicellular life were all set in the Precambrian period. We will deal with this period in Chapter 3, therefore, and with the Cambrian explosion and the Phanerozoic in Chapters 4 and 5, respectively.

CHAPTER THREE

Silence of the Aeons

Three Billion Years of Microbial Evolution

IT IS NEXT TO IMPOSSIBLE FOR US, with our historical perspective honed to decades or centuries, to conceive of the vast tract of time that ebbed away during the Precambrian era. We are dealing with a period that spanned 4 billion years — nine-tenths of the total duration of the Earth. Imagine that we are rocketing backwards through time at a rate of one millennium per second. In two seconds, we will have returned to the time of Christ, in ten seconds to the birth of agriculture; in half a minute we will see the first cave painters, and in less than two minutes we will catch a glimpse of our ape-like ancestors shuffling across the African savanna. Rushing backwards, the catastrophe that wiped out the dinosaurs will unfold before our eyes in 18 hours time; and in 4½ days we will have prime seats for the opening drama of multicellular life in the Cambrian explosion. Then we continue our journey in silence. In 44 days time, we will have returned to the first mysterious stirrings of life, and in 53 days the Earth will condense from a cloud of gas and dust.

For 40 days and 40 nights, in our compressed time scale, the Earth was populated entirely by microscopic single-celled bacteria and simple algae. With no real fossil record to bridle the imagination, it is not surprising that most of the pioneering efforts to understand the early history of life were little better than speculation. How can we have any coherent idea today of biochemical changes taking place in microbes that left little trace in the rocks, or of the oxygen concentration in a fleeting atmosphere long gone? The answer is indeed written in the rocks, sometimes in microscopic fossils, and sometimes in the molecular ghosts of ancient geochemical cycles. More than this, the atavistic genes of modern organisms often betray their evolutionary roots. The script written in the genes is enigmatic, although obviously meaningful. Our only guide, a molecular Rosetta stone, is the way in which the proteins encoded by the genes are used today. If a protein such as haemoglobin, the red pigment of red blood cells, is specifically designed to bind oxygen today, and we know from genetic sequences that some bacteria also have a gene for a similar protein, there is certainly a good possibility that our common ancestor had it too. If so, we can infer that they too used the haemoglobin to bind oxygen. If, instead, they used it for something else, the clue to what that was may still be hidden in the structure of the molecule.

To understand the effect of oxygen on evolution, we need to trace two stories in the rocks and the genes: the evolution of the microbes themselves, and the timing and magnitude of the oxygen build-up in the air. Before we begin, however, we will do well to bury a hatchet in a particularly subversive double-headed ghoul. This is the common misapprehension that evolution necessarily tends towards greater complexity, and that microbes, being microscopic and without brains, are at the bottom of the evolutionary pile. So many evolutionary biologists have attacked the lay concept of evolution as a progression towards a higher plane, and to so little avail, that one begins to wonder whether there is a global conspiracy to thwart them. Two cautionary tales should provide a clearer perspective on Precambrian evolution. The first challenges the assumption that evolution tends towards greater complexity, while the second argues that microbes are far from simple.

In 1967, Sol Spiegelman, a molecular biologist at the University of Illinois, reported a series of experiments designed to establish the smallest unit that could evolve by natural selection. He took a simple virus that replicated itself using only a handful of genes, which consist of a string or 'sequence' of 4500 'letters'. The protein products of these genes subverted the molecular machinery of infected cells to produce new viral particles. Spiegelman wanted to see just how simple the viral life-cycle could become if he provided the virus with all its raw materials in a test tube, instead of a host cell with its complicated molecular machinery. He gave his virus the main enzyme necessary for it to complete its life-cycle, and a free supply of all the basic building blocks needed for it to copy its genes.

The results were spectacular. For a while the virus replicated itself exactly, preserving its original gene sequence. After a period, however, a mutation caused part of one gene to be lost. Because this gene was only necessary for the virus to complete its normal life-cycle in an infected cell, and was not necessary in the test tube, the mutant virus could survive quite happily without it. More than happily, in fact: the new gene sequence was shorter than the old one, so the mutant virus could replicate itself faster than the non-mutant viruses. This faster rate of replication allowed the mutants to prevail over their competitors until they, too, were overtaken by a new mutant, an even slimmer-line virus able to replicate itself still faster. In the end, Spiegelman produced a degenerate population of tiny gene fragments, which became known as 'Spiegelman's monsters'. Each little monster was just 220 letters long. They could replicate at a furious speed in the test tube, but could not hope to survive in the outside world.

The moral of the tale is simple. Evolution selects for beneficial adaptations to a particular environment, and the simplest, fastest or most efficient solution will tend to win out, even if this means that excess baggage is jettisoned and organisms become less complicated. We now realize that many simple single-celled organisms, which we once thought were relics from a primitive age that had never evolved a complex lifestyle, have instead lost their ancient sophistication. We touched on fermentation in the last chapter. Far from being a simple energy-producing system that was later displaced by more efficient mechanisms involving oxygen, it seems that, as in the yeasts, fermentation is often a recent (in evolutionary terms) adaptation to oxygen-free environments, and such fermenters have actually lost their ancestors' ability to use oxygen.

My second cautionary tale illustrates the metabolic sophistication of supposedly simple microbes. Humans and other large animals will quickly suffocate and die without oxygen, because our bulk, a community of some 15 million million cells, precludes the use of any other type of respiration. As a result, we are rather limited in the biochemical reactions we can carry out, albeit very efficient at marshalling our limited resources. Some microbes, however, can live using oxygen to respire, but if deprived of air will simply switch to another way of satisfying their energy requirements and continue without a glitch.

The bacterium *Thiosphaera pantotropha* is one such, and is about as far removed as we can get from our sense of an evolutionary pinnacle: it lives on faeces. Originally isolated in 1983 from an effluent-treatment plant, it applies an extraordinary virtuosity to the extraction of energy from sewage.

When oxygen is present, it extracts energy from a wide range of organic and inorganic substrates by aerobic respiration. When conditions become anaerobic, however, it can extract energy from thiosulphate or sulphide using nitrogen oxides instead of oxygen. The only trick missing from this metabolic cabaret is an ability to ferment. Such biochemical versatility lends the bacterium great flexibility in lifestyle — it can switch from one energy-producing process to another in response to sudden chemical changes, which are brought about by the periodic injections of dissolved oxygen used to speed up the decomposition of effluent in treatment plants.

Curiously, genetic analysis of a wide range of living organisms suggests that LUCA — the hypothetical bacterium proposed as the Last Universal Common Ancestor, the greatest grandmother of all living things in the world today — may have had a similar ability to switch between different types of metabolism, nearly 4 billion years ago. Most of her descendants appear to have lost their illustrious ancestor's flexibility. We will return to this theme in Chapter 8.

The Precambrian, then, was a time of spectacular metabolic innovation. Microbes learnt to harness the power of the Sun, as well as the oxidizing power of oxygen, and to generate energy from an array of sulphur, nitrogen and metal compounds. The chemistry of these life-giving reactions has sometimes left subtle traces — so-called carbon or sulphur signatures — in sedimentary rocks, and occasionally, not at all subtly, in the form of billions of tons of rocks. The metabolism of ancient microbes was directly or indirectly responsible for our most important reserves of iron, manganese, uranium and gold, to say nothing of the gold prospector's false nugget, iron pyrites. These rocks and ores were not deposited continuously or synchronously, but at different times and under different environmental conditions. Their sequence has been carefully reconstructed through precise radioactive dating, and together the findings open a colourful window on oxygen and life in the formative years of our planet.

The first signs of life in rocks are found in the same Greenland rocks that we discussed in Chapter 2, and take the form of an anomaly in the proportions of different carbon isotopes they contain. This important finding was reported in the journal *Nature* in 1996 by a NASA-funded doctoral student, Stephen Mojzsis, and his colleagues at the Scripps

Institution of Oceanography at La Jolla in California. The interpretation of these carbon signatures in rocks is so important to our story that it is worth explaining what they are and why they are there. Not only do carbon isotopes preserve a record of the triumphs and tribulations of life, but their shifting proportions can permit surprisingly quantitative estimates of the changes in the atmospheric composition of the ancient Earth.

There are several different atomic forms of carbon (as opposed to molecular forms such as diamond or graphite). These atomic variants are called *isotopes*. Each carbon isotope has six protons in the nucleus, giving them all an atomic number of six. This means they are all carbon and all have exactly the same chemical properties. But the carbon isotopes differ in the number of neutrons in their nuclei and so vary in their atomic weight. The more neutrons they have, the heavier the atoms. Carbon-12, for example, has six neutrons, giving it an atomic weight of 12 (6 protons + 6 neutrons), whereas carbon-14 has 8 neutrons, giving it an atomic weight of 14 (6 + 8).

Carbon-12 is by far the most abundant carbon isotope on Earth (accounting for 98.89 per cent of the total) and has an honourable place in chemistry as the standard against which the relative weights of all other elements are measured. The carbon-12 nucleus is stable and does not decay. In contrast, carbon-14 is produced continuously in minute amounts (about 1 part in 10^{12}; one part in a million million) in the upper atmosphere, through the bombardment of cosmic rays. The unstable carbon nuclei formed decay through radioactive emissions at a fixed rate. The half-life (the time taken for half the total mass to decompose) is exactly 5570 years. This short time period (in geological terms) makes radiocarbon dating useful for determining the age and authenticity of prehistoric remains or historical documents, such as the Dead Sea scrolls and the Turin shroud.[1]

Fascinating as it is, carbon-14 has no further place in our story. It is the other main isotope of carbon, carbon-13, that concerns us here. Unlike carbon-14, carbon-13 has a stable nucleus and does not decay. In

[1] Carbon-14 is dispersed throughout the atmosphere and absorbed by living plants in photosynthesis, then eaten by animals, in proportion to its abundance in carbon dioxide. This abundance remains roughly constant because in the long term the rate of formation balances the rate of decay. When plants and animals die, however, the cessation of gas exchange or breathing means that their tissues are no longer in equilibrium with the atmosphere, so their carbon-14 content declines in proportion to the rate of radioactive decay; older organic compounds therefore contain less carbon-14.

this respect, it is similar to carbon-12. The total amount of carbon-13 in the Earth and its atmosphere is therefore constant (1.11 per cent of the total). This means that the overall ratio of carbon-12 to carbon-13 on or in the Earth is constant (99.89 to 1.11). In other words, if we add up the total amount of carbon in plants, animals, fungi and bacteria, and buried as coal, oil and gas, and present in the air as carbon dioxide, and dissolved in the oceans and swamps as carbonates, and petrified as carbonate rocks (such as limestone), then we will find that the overall ratio of stable carbon isotopes is 99.89 to 1.11.

Despite this fixed ratio, there are still some small but definite variations in the ratio of carbon-12 to carbon-13 in the carbon buried in the rocks. These variations are brought about by living things, and so far as we know, *only* by living things. The reason is that photosynthetic cells using carbon dioxide from the air or sea to make organic matter prefer to use carbon-12. This is because the lighter carbon-12 atoms form weaker chemical bonds, which are broken down more easily by enzymes than the stronger bonds of the heavier carbon-13 isotope. The faster rate at which carbon-12 bonds are cracked means that organic matter becomes enriched in carbon-12 relative to carbon-13. In fact, the ratio of carbon-12 to carbon-13 is skewed towards carbon-12 by an average of 2 or 3 per cent compared with the unadulterated background ratio.

When the remains of plants, algae or cyanobacteria are buried in sediments, their extra carbon-12 is buried with them. Because the buried organic matter is enriched in carbon-12, it is impoverished in carbon-13. This means that more carbon-13 is left behind as carbonates in the oceans or rocks, or as carbon dioxide in the air. This is called the principle of mass balance – which simply says that what is buried below the ground cannot be found above the ground. The implications of this elementary idea have a surprisingly long reach. Both carbon-12 and carbon-13 are incorporated into carbonate rocks (such as limestone) in a ratio that reflects their relative concentration in the oceans. As more carbon-12 is buried as part of organic matter, more carbon-13 is left behind in the oceans, and so the carbonate rocks have a relatively high content of carbon-13. Thus biological activity is betrayed in two different ways: by an enrichment of carbon-12 in buried organic matter, such as coal, or by an enrichment of carbon-13 in carbonate rocks such as limestone.

Geological periods conducive to carbon burial, such as the Carboniferous (about 300 million years ago) with its huge, low-lying swamps and massive coal seams, leave robust carbon-12 signatures in organic inclusions such as coal in the rocks. The farther back in time we go, the harder it is to read carbon signatures, if only because less and less organic matter survives intact. Eventually, the samples shrink to the size of grains and require sophisticated equipment to read them. With this in mind, Steven Mojzsis and colleagues set about studying the ancient Greenland rocks, determined to think small. Their approach brought swift rewards: they found minute carbon residues trapped inside grains of a calcium phosphate mineral called apatite. Apatite can be secreted by microorganisms, but can also crystallize inorganically from the oceans, so the association of carbon with apatite is, in itself, no more than suggestive of life. When the Scripps team examined the carbon-isotope ratios, however, the results were startling. The carbon inclusions were enriched in carbon-12 by as much as 3 per cent over the normal background ratio. As a leading geochemist, Heinrich Holland, remarked in the journal *Science*: "the most reasonable interpretation of the data is surely that life existed on earth more than 3.85 billion years ago." Not only this, but life may even have discovered the trick of photosynthesis, which is, after all, the main source of carbon signatures today.

Is this credible? Other pieces of evidence fit the same story. Moving forward a mere 300 million years, to the 3.5-billion-year-old rocks in Warrawoona in Western Australia, we find microscopic fossils that resemble modern cyanobacteria. Throughout the Precambrian period, most cyanobacteria lived in communal structures called stromatolites: great domes of living rock, which grew to heights of metres. A few living stromatolites are still found today in the right conditions — in Shark Bay in Western Australia, for example. Nearby, shapes resembling modern stromatolites are imprinted in rocks 3.5 billion years old. There is little evidence of geothermal activity, past or present, in these bays, so it seems likely that the microbes living in these ancient stromatolites gained their energy from photosynthesis, just as they do today. While none of these findings is conclusive on its own, when taken together, the carbon signatures, microfossils and fossil stromatolites do make it look as if photosynthetic bacteria were already colonizing the early Earth at least 3.5 billion years ago.

The earliest definitive evidence for the existence of cyanobacteria must wait another 800 million years. We are now 2.7 billion years before the present, floating in the shallows of an ocean that was soon to precipitate some of the largest iron-ore formations in the world. Today we can visit these iron formations in the Hamersley Range, near Wittenoom in Western Australia. For such old rocks, they have suffered relatively little chemical and physical change, called metamorphosis by geologists. Heat and pressure, the twin forces of metamorphosis, tend to destroy flimsy biological molecules. Because the Hamersley Range had suffered so little metamorphosis, Jochen Brocks and his colleagues at the Australian Geological Survey and University of Sydney, held out hope that a few ancient molecules — characteristic biological fingerprints called biomarkers — might have survived intact in the shales underlying the iron formations. After conducting a painstaking series of extractions and laboratory tests to eliminate the possibility of contamination with more recent molecules, their hopes were rewarded in full when they discovered a rich mixture of recognizable biomarkers. Their work was promptly published in *Science* in August 1999, with a flurry of commentary. Not only had the Australian surveyors found fingerprints diagnostic of cyanobacteria — that is, molecules found only in cyanobacteria — they also found a large number of complex steranes, a family of molecules derived from sterols such as cholesterol, which have only ever been found in the cell membranes of our own direct ancestors, the single-celled eukaryotes.

The finding was a double whammy: proof that oxygen-producing cyanobacteria and the first representatives of our own eukaryotic ancestors coexisted not less than 2.7 billion years ago. The earliest known fossils of eukaryotic cells date to about 2.1 billion years ago, so Brocks and his colleagues had pushed back the evolution of the eukaryotes 600 million years. This is significant in terms of the environment that must have existed to support these cells. Apart from anything else, the biosynthesis of sterols is an oxygen-dependent process, requiring more than a trace of oxygen in the atmosphere. Modern eukaryotes can only synthesize sterols if given at least 0.2 to 1 per cent of the present atmospheric levels of oxygen, and there is no reason to suppose that their ancestors were any different. If the cyanobacteria had indeed evolved between 3.5 and 3.85 billion years ago, as was suggested by the fossil evidence in the Warrawoona rocks and the carbon signatures, it is quite plausible that some free oxygen could have accumulated in the atmosphere by this time. But did this increase in oxygen correspond exactly in time to the evolution of the eukaryotes? And if so, did the rise in oxygen in fact stimulate their evolution?

Trends in carbon-isotope ratios can be used, in principle, to calculate changes in atmospheric oxygen. This is because the burial of organic matter prevents the complete oxidation, by respiration, of the carbon produced by photosynthesis. As photosynthesis and respiration are essentially reverse reactions, the one generating and the other consuming oxygen, any increase in the amount of carbon buried should lead to an equivalent increase in the amount of free oxygen left over in the air. If we know exactly how much carbon was buried at any one time, then, in principle, we can calculate how much oxygen must have been left in the air. In practice, however, unless we can be certain that the rate of oxygen removal by volcanic gases or the erosion of land masses remains constant, all we can say is that there was a qualitative increase in oxygen. During recent geological history, the younger rocks preserve a detailed history of environmental change, and we are sufficiently familiar with most of the important parameters to calculate oxygen levels on the basis of carbon burial, as we shall see in Chapter 5. Unfortunately, this approach is unreliable when dealing with the very ancient Precambrian period — there are so many uncertainties that, at best, we only get a sense of the direction of change. For a more quantitative estimate, we must employ other methods.

One clue to oxygen levels during this period is to be found in the very same iron formations that overlie the shales of the Hamersley Range. Massive sedimentary iron formations were deposited here and around the world in alternating bands of red or black ironstone (haematite and magnetite, respectively), and sediment, typically flint or quartz. The individual bands range in depth from millimetres to metres, while the formations themselves can be up to 600 metres [approximately 2000 feet] thick. Most of these formations were deposited between 2.6 and 1.8 billion years ago, but sporadic outcrops range in age from 3.8 billion to 800 million years.

Today, after the exhaustion of most premium ore deposits, the banded iron formations are by far the world's richest source of low-grade iron ore. According to the US Geological Survey, world iron-ore resources still exceed 800 billion tons of crude ore, containing more than 230 billion tons of iron, much of which comes from Australia, Brazil and China. Of this total, at least 640 billion tons were laid down between 2.6 and 1.8 billion years ago. The Hamersley formation alone contains 20 billion tons of iron ore, with 55 per cent iron content.

Exactly how these iron formations came into being, or why they should be banded, is a mystery. Or rather, there are so many possible explanations, and so little evidence to support one theory over another, that few geologists would be bold enough to attempt a categorical explanation. There have nonetheless been some imaginative attempts. Ancient superstition held that large deposits of haematite (from the Greek 'blood-like') formed from the streams of blood that flowed into the ground after great battles. More scientifically, the banding of ironstones has been attributed to cyclical extinctions of algal populations, overcome by their own toxic oxygen waste. Neither theory has much credence. In fact, there is no reason to suppose that all the formations were produced in the same way, especially those separated by deep gulfs of time. But some general principles do apply to them all, and these reveal something of the conditions under which they must have formed. Most importantly, no banded iron formations have been deposited since atmospheric oxygen approached modern levels. Because iron does not dissolve in the presence of oxygen, the immediate implication is that the oceans were oxygen-free before the deposition of the banded iron formations, and too well aerated to support their formation in later times. To tease the truth out of this implication, we will need to look at the behaviour of iron in a little more detail.

Only the Earth's core, and meteorites, contain pure iron. Tools made from meteoritic iron are an expensive curiosity. All iron in ores from the Earth's crust is oxidized to some extent, although we shall see that iron in the oxidized state does not always imply the presence of oxygen. There are two main forms of iron in nature, ferrous iron (Fe^{2+}), which tends to be soluble, and the more highly oxidized ferric iron (Fe^{3+}), best known in the guise of rust (ferric oxides), which is insoluble.[2] In the presence of oxygen, soluble ferrous iron is oxidized to insoluble rust. Not surprisingly, there is very little iron dissolved in today's well-ventilated oceans, as oxygen snatches electrons from dissolved iron and the ferric oxide compound precipitates out as rust before any iron build-up can occur. One exception is the poorly ventilated floor of the Red Sea, where dissolved iron is enriched to 5000 times normal levels, and only bacteria can survive. The early Precambrian oceans must have been similar in this respect: in the absence of oxygen, dissolved iron from volcanic emissions and erosion could have accumulated to very high levels.

[2] The two forms differ in their degree of oxidation, ferric iron (Fe^{3+}) being more oxidized than ferrous iron (Fe^{2+}).

A second modern example gives an idea of what might have happened next. The Black Sea is the largest body of poorly oxygenated water in the world, and is stratified into two layers. The surface waters are well oxygenated to depths of about 200 metres [656 feet], and if not fished to oblivion support a teeming ecosystem, including the famous caviar sturgeon. In contrast, the deeper waters, which account for 87 per cent by volume of the Black Sea, are stagnant and cannot support animal life (with the sole exception, it seems, of nematode worms, the only known animal that can complete its life cycle in the absence of oxygen). The current state of the Black Sea seems to have developed about 7500 years ago, several thousand years after the end of the last ice age, in an event that has been linked to Noah's Flood by the marine geologists William Ryan and Walter Pitman of Columbia University. As the great land glaciers melted, the sea level around the world rose by several hundred feet. The Black Sea, however, was isolated in its own basin by a land bridge across the Bosphorus, and the glacial meltwater did not affect its depth as much as that of the surrounding seas. The basin was left low and dry, so to speak, well below sea level, as is the Dead Sea today.

Whether as the consequence of an earthquake, or stormy weather, or the pressure of the rising Mediterranean, the land bridge spanning the Bosphorus finally collapsed with a roar that must have sounded like the wrath of God. This, say Ryan and Pitman, was the reality of Noah's flood. Salt water poured into the low Black Sea basin at an estimated rate of 10 cubic miles [42 million cubic metres] per day — a cascade 130 times greater than the Niagara Falls. The villages clinging to the shores were drowned beneath the Mediterranean waters, they say, in a catastrophe whose memory reverberated around the ancient world. An area the size of Florida was added to the existing lake.

Since biblical times, the shallow, tideless straits of the Bosphorus have impeded mixing of the brackish Black Sea water with the saline water of the Mediterranean. The denser saline sinks to the bottom, and the undisturbed bottom waters rarely come into contact with the air. The only living things that thrive in these depths are anaerobic (oxygen-hating) bacteria. Many of these are sulphate-reducing bacteria, which generate the noxious gas hydrogen sulphide as a waste product. Because hydrogen sulphide reacts with any oxygen percolating down, the depths remain anoxic and the stratified system, once established, sustains itself. The build-up of hydrogen sulphide makes the deep waters of the Black Sea stink of rotten eggs, and stains the mud on the bottom black, giving the sea its modern name. Its ancient name, the Euxine, lends itself to the term euxinic, which refers to any foul-smelling sulphidic body of water, lacking oxygen, movement and animal life in the depths.

The Black Sea, although the largest, is not the only euxinic body of water on the planet. Similar conditions occur in some Norwegian fjords that are separated from the open ocean by shallow glacial sills. Even the oceans occasionally develop euxinic conditions. Climatic conditions sometimes conspire to cause an upwelling of nutrient-rich bottom waters to the surface. Here, the combination of plentiful nutrients and bright sunlight stimulates an algal bloom, leading to a massive but transient increase in biomass. As the nutrients are exhausted, the algae die and sink to the bottom. Their decay consumes oxygen faster than it can be replenished by currents or diffusion from the oxygen-rich surface waters. These oxygen-poor conditions stimulate a second bloom, this time of oxygen-hating sulphate-reducing bacteria, which release hydrogen sulphide as they break down the organic matter. Stagnant conditions may set in for periods of months until the supply of decaying organic matter is exhausted. Occasionally, the stagnant waters well up to the surface, releasing hydrogen sulphide gas into the atmosphere. One such upwelling occurred in St Helena Bay near Cape Town in South Africa in 1998, provoking furious and misguided complaints about the smell of rotten sewage in the air.

Such combinations of circumstances may explain the genesis of banded iron formations. Back in Precambrian times, the low levels of atmospheric oxygen must have kept the oceans permanently euxinic. The surface waters, however, were home to photosynthetic bacteria at least 2.7 billion years ago, and perhaps as long as 3.8 billion years ago. As happens today, there must have been frequent upwellings of the bottom waters, bringing dissolved nutrients and iron into contact with the photosynthetic bacteria living in the surface layers. If these bacteria were cyanobacteria, as suggested by the biomarkers in the Hamersley Range, then they would have been producing oxygen as a waste product of photosynthesis. In such oxygen-rich waters, dissolved iron welling up to the surface would have precipitated out as rust, and sunk to the bottom of the ocean to form beds of red haematite and black magnetite.

If this was the case, the banding of ironstones with flint or quartz could have been produced by seasonal influences, such as higher rates of photosynthesis (and therefore oxygen production) in the summer than in the winter, or seasonal upwellings according to climatic variations. The seasonal fluctuations in iron deposition would have been set against a steady precipitation of silica. This could not happen today. There is little dissolved silica in the modern oceans: it is extracted by some algae and lower organisms for use in their 'skeletons'. However, in the days when bacteria ruled the waves, silica was not used in this way, and so must have continuously exceeded its solubility limit of about 14–20 parts per million. It would have precipitated in a steady rain to form thick beds of flint or quartz, alternating with seasonal beds of ironstone.

Although this is, perhaps, the most widely accepted model of banded iron formation, there are still some difficulties with it. The oldest iron formations, 3.8 billion years old, were surely formed before oxygen began to accumulate. Furthermore, most of the ironstones around the world do not consist of simple iron oxides such as haematite, as might be expected if oxygen levels were genuinely high and the reactions were no more than bucket chemistry. There are other biological mechanisms that can oxidize iron without any requirement for free oxygen. One was described in 1993 by Friedrich Widdel and his colleagues at the Max Planck Institute for Marine Microbiology in Bremen. They isolated a strain of purple bacteria from lakeside sediments which could use the energy from sunlight to produce iron ores without requiring free oxygen. The main product of the bacterial reaction is a brownish rust-like deposit, ferric hydroxide, which is commonly found in banded iron formations. Widdel argued that the same seasonal upwellings that brought nutrients and iron to the sunny surface waters could have stimulated great bursts of iron-ore formation by purple bacteria. Thus, while the presence of cyanobacteria and rusting iron in banded iron formations suggests that free oxygen may have played a role in their genesis, Widdel and his colleagues have shown that some iron formations could have been formed by purple bacteria in the absence of oxygen. Despite their promise, then, banded iron formations cannot give us a quantitative estimate of oxygen levels in the air during this period.

One possible solution to the problem of exactly when oxygen levels rose has been put forward by Donald Canfield of the University of Southern Denmark, a leading authority on Precambrian oxygen levels, in a series of papers published in *Science* and *Nature*. Canfield turned, rather elliptically at first sight, to the oxygen-hating sulphate-reducing bacteria that produce hydrogen sulphide under stagnant conditions, to estimate the timing of the increase in atmospheric oxygen. His rationale was founded on two observations.

First, sulphate-reducing bacteria gain their energy from a reaction in which hydrogen reduces sulphate to produce hydrogen sulphide. Although sulphate (SO_4^{2-}) is found at high levels in modern sea water (at about 2.5 grams per litre) it should not have been plentiful in the early Precambrian period, as its formation requires the presence of oxygen. This premise is supported by the absence of sulphate evaporites, such as gypsum, from the early Earth. If sulphate can only form in the presence of oxygen, then the sulphate-reducing bacteria could not have established themselves until there was some oxygen in the atmosphere. We can go further: because low sulphate is a rate-limiting factor for sulphate-reducing bacteria, virtually precluding their growth in freshwater lakes, their activity depends on the concentration of sulphate. This in turn depends on the concentration of oxygen. Put another way, even though sulphate-reducing bacteria are strictly anaerobic — they are actually killed by oxygen — they cannot exist in a world without oxygen, and their activity is ultimately governed by oxygen availability.

The second observation applied by Canfield relates to sulphur isotopes. Just as photosynthesis leaves a carbon signature in the rocks, the sulphate-reducing bacteria similarly discriminate between the two stable isotopes of sulphur, sulphur-32 and sulphur-34. As with carbon isotopes, the lighter sulphur-32 atoms form weaker bonds that are more easily broken by the action of enzymes. Sulphate-reducing bacteria therefore produce hydrogen sulphide gas enriched in sulphur-32, leaving more sulphur-34 behind in the oceans. In some conditions, both the hydrogen sulphide and sulphate can precipitate from the oceans to form rocks. Sulphur signatures can be read in these rocks. In particular, and perhaps surprisingly for those who still do not associate minerals with life, hydrogen sulphide reacts with dissolved iron to form iron pyrites, which then sinks to the bottom sediments. Iron pyrites can be formed by either volcanoes or bacteria. Against the consistent, unadulterated ratio of sulphur isotopes from volcanoes, the hand of biology signs off with a clear signature — in other words, a distortion in the natural balance of isotopes.

Canfield examined the sedimentary iron pyrites deposited during the Precambrian period for sulphur signatures, and found them. The first signs of a skewing in the sulphur-isotope ratios date to about 2.7 billion years ago, implying there was a build-up of oxygen at this time. Interestingly, this is very close to the date given to the first eukaryotic cells in the Hamersley shales by Jochen Brocks and his colleagues. After this, little changed for half a billion years. Then, around 2.2 billion years ago, there was an abrupt rise in the sulphur-32 content of iron pyrites, suggesting that the amount of sulphate in the oceans must have risen to the point where they could support a much larger population of sulphate-reducing bacteria. This, in turn, indicates that much more oxygen must have been available to produce the sulphate. Thus Canfield's work implies that there was a small rise in oxygen levels 2.7 billion years ago, followed by a much larger rise about 2.2 billion years ago.

Unequivocal evidence of free oxygen in the air and oceans requires proof of oxidation on land, as changes wrought by the thin air cannot be obscured or confounded by the rich biology and chemistry of the oceans. More than a billion years before the invasion of the land by plants and animals, the terrestrial populations of microbes could not have compared in abundance or diversity with their marine cousins. The widespread rusting of iron minerals on land is therefore the most tangible evidence we have for oxygen in the atmosphere. These rusting iron minerals are found in fossil soils (palaeosols), and in the so-called continental red-beds.

In a classic series of measurements, the geochemists Rob Rye and Heinrich Holland from Harvard University examined the iron content of ancient fossil soils, and used these measurements to estimate the period when oxygen built up in the air. Their reasoning was as follows. Because iron dissolves in the absence of oxygen, but is insoluble in the presence of oxygen, iron could leach out of very ancient soils (when there was no oxygen in the air) but became trapped in more recent soils (when oxygen was present in the air). By measuring the iron content of fossil soils, Holland and Rye estimated that a large rise in atmospheric oxygen took place between 2.2 and 2 billion years ago. From the amount of iron left in the fossil soils, as well as its rustiness — its oxidation state — they estimated that the concentration of atmospheric oxygen at this time probably reached 5 to 18 per cent of present atmospheric levels.

In terms of timing, these findings are corroborated by the appearance of continental red-beds between 2.2 and 1.8 billion years ago. These sandstone rock formations were probably formed by free oxygen reacting with iron in the rocks during the erosion of mountain ranges. Rivers must have run red as they flowed over the barren surface of the Earth, a scene that conjures up images of nuclear winter. Rather than being washed out to sea, some eroded minerals deposited in valleys and alluvial plains, ultimately forming the beds of red sandstone. Because the red-beds were formed from eroded minerals, however, we cannot use them to estimate the concentration of oxygen in the air, only the timing.[3] The timeline from the first carbon signatures in Greenland rocks to the formation of the red-beds is shown in Figure 2.

A bizarre microbial relic also attests to a rise in free oxygen around 2 billion years ago: the natural nuclear reactors at Oklo, in Gabon, West Africa. The solubility of uranium, like iron, depends on oxygen. But unlike iron, uranium becomes more soluble, rather than less, in the presence of oxygen. The chief uranium mineral found in rocks older than about 2 billion years is uraninite, but this ore is very rarely found in younger rocks. The sudden transition is associated with the rise in oxygen. What seems to have happened is that, as the oxygen levels increased, oxidized uranium salts leached out of uraninite ores in the rocks and washed away in streams. Their concentration cannot have been higher than a few parts per million.

In Gabon, 2 billion years ago, several streams converged on shallow lakes encrusted with bacterial mats, similar to the mats that still exist today in the geyser pools at Yellowstone National Park in the United States and elsewhere. Some of the bacteria that lived in these mats had a penchant for soluble uranium salts as an energy source. They converted the soluble uranium back into insoluble salts, which precipitated out in the shallow water beneath them. Over the next 200 million years or so, the bacterial mats deposited thousands of tons of black uranium ore in their lakes.

There are two main isotopes of uranium, both radioactive, as most of the Cold War generation knows. Uranium-238 has a long half-life of 4.51 billion years. Half the uranium-238 that was present when the Earth condensed from its cloud of radioactive dust is still out there somewhere. Its sister isotope, uranium-235, decays much faster, with a half-life of about 750 million years. Most uranium-235 has therefore already decayed into its daughter elements, by emitting neutrons. If one of these neutrons hits a nearby uranium-235 nucleus, however, the effect is to split the nucleus into one or more additional neutrons, plus large fragments of roughly equal mass, with a liberation of energy equal to the total loss of

[3] The red colour of the continental red-beds shows that the iron was completely oxidized, as would be expected for deposits of eroded debris that had been exposed to the air for an indeterminate, but probably lengthy, period. Because there is no spectrum of oxidation, we cannot estimate the atmospheric oxygen levels from the red-beds.

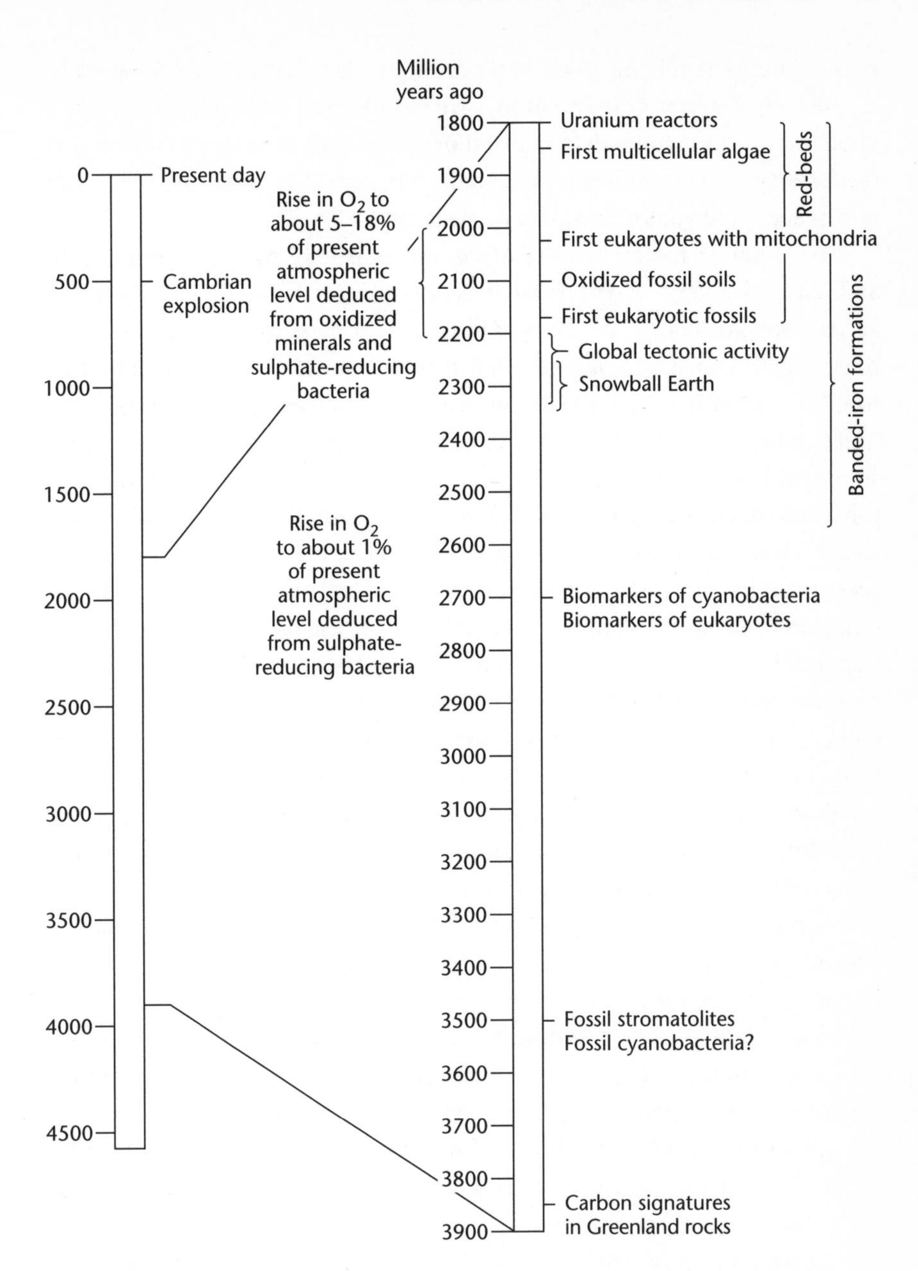

Figure 2: Geological timeline expanding the mid-Precambrian period (Archaean and early Proterozoic). Note the burst of evolutionary activity in the period 2.3 to 2 billion years ago, as oxygen levels rose to about 5–18% of present atmospheric levels.

mass. (Energy is related to mass according to Einstein's famous equation $E = mc^2$.) If the uranium-235 atoms are closely packed together, there is a good chance that the newly emitted neutrons will hit more uranium-235 nuclei. In these circumstances, a chain reaction — nuclear fission — can take place, potentially causing a nuclear explosion.

For nuclear fission to take place, uranium-235 must be enriched to at least 3 per cent of the total mass of uranium. Today, uranium-235 accounts for only 0.72 per cent of uranium by mass, so we must enrich it ourselves if we wish to build a nuclear power station or an old-fashioned uranium atom bomb. Two billion years ago, however, less uranium-235 had already decayed. Its content in uranium ores would have been higher — in fact about 3 per cent. The uranium-loving bacteria in Gabon therefore stockpiled enough ore enriched in uranium-235 to start a nuclear fission chain reaction. This, at any rate, was the conclusion of the French secret service in 1972. There had been something of a panic when uranium ores mined along the Oklo River, near the border with the Congo Republic, turned out to be depleted in uranium-235. Some consignments had less than half the expected 0.72 per cent uranium-235. In an Africa still emerging from colonial rule and beset by civil unrest, the implication that some tribal group had stolen enough uranium to make a nuclear bomb did not bear thinking about. The French threw everything at the problem, and it was not long before a large team of scientists from the French Atomic Energy Commission solved the case.

Samples of the Oklo ores showed clear relics of spent radioactive fission, even when they were extracted from undisturbed seams. Instead of decaying naturally, tons of uranium-235 had fissioned away in half a dozen separate locations, producing a million times the power of natural decay. The natural reactors in Gabon had apparently been sustained for millions of years by a steady flow of water from the streams that fed into the ancient uranium lakes. Water slows the speed of neutrons, reflecting them back into the core of the reactor, so instead of quelling the inner fires, water actually promotes nuclear fission. The streams did more than this, however — they also acted as safety valves against nuclear explosion. Whenever the chain reactions approached danger levels, water boiled off, allowing neutrons to escape. This scuttled the chain reactions and shut down the reactors until flow was re-established. There is no evidence of a nuclear explosion. The entire system was finally buried beneath sediments where it remained undisturbed until the arrival of the French, a testament to the ingenuity of bacteria 1.8 billion years before Enrico

Fermi and his Chicago team applied their genius to making the first man-made atomic bomb; and indeed a testament to the potential long-term safety of burying nuclear waste.

What of the catastrophic mass extinction, the oxygen holocaust described by Lynn Margulis (see Chapter 2, page 19)? There is no trace of a holocaust in the rocks. Far from being a profound and debilitating challenge, the appearance of oxygen seems to have driven the evolution of new forms of metabolism, and new branches in the tree of life, as argued by Preston Cloud in the 1960s (see Chapter 2). But why did it take so long for oxygen to accumulate, despite more than a billion years of continuous production by cyanobacteria? To put it into context, this interlude is twice as long as the entire modern era of plants and animals (the Phanerozoic), or for that matter 15 times as long as the period since the demise of the dinosaurs. Is this long gestation perhaps hidden evidence of a difficult adaptation, concealing the throes of life as it struggled to cope with a poisonous gas? It seems unlikely. A number of speculations can explain the delay; for example, iron-loving bacteria may have dominated the ecosystem until the iron ran out, or the cyanobacteria may have been restricted to shallow-water stromatolite communities that absorbed as much oxygen as they produced, because of the presence of non-photosynthetic oxygen-respiring bacteria. The most likely explanation is simply that there was no change for a billion years because a stable equilibrium persisted for that time.

The long stasis was finally shattered by an apocalyptic climate change about 2.2 to 2.3 billion years ago. The Earth plummeted into the first ever ice age. This was no trivial ice age, to be compared with the recent Pleistocene cold snap, but a global freeze that may have covered the tropics in glaciers a kilometre [3280 feet] thick — in Joseph Kirschvink's memorable phrase, a 'snowball Earth'. What made the pleasant Precambrian climate collapse so violently is not known. One theory, argued by the sometime NASA geochemist James Kasting, is that the appearance of free oxygen itself brought about the freeze. As it built up in the air, oxygen would have reacted with methane (produced in large amounts by bacteria), and so removed this important greenhouse gas from the early atmosphere. As the greenhouse effect was undermined, temperatures plummeted and the Earth succumbed to the grip of an ice age. Kasting's theory has been advocated by James Lovelock among others, who claims an important role for methane-producing bacteria in his books on Gaia, but at present the theory suffers from lack of strong supporting evidence.

Whatever the reason, there is no doubt that the Earth plunged into a serious ice age about 2.3 billion years ago. It was to last for 35 million years. Hard on the heels of this ice age the planet was racked by a period of heightened tectonic activity, leading to major continental rifting and the uplift of mountain belts on a scale comparable with the Andes.

Joseph Kirschvink, a specialist in palaeomagnetism at Caltech (the California Institute of Technology) is a leading advocate of the snowball Earth theory, and one of its most thoughtful commentators. He argues that, after the glaciers finally melted, the stones and mineral dust scoured out by glacial erosion would have filled the oceans with minerals and nutrients, stimulating a cyanobacterial bloom and a rise in oxygen. As evidence for this claim, Kirschvink and his co-workers cite a huge deposit of manganese ore in the Kalahari desert in southern Africa, dated to right after the end of the snowball Earth. The Kalahari manganese field contains some 13.5 billion tons of manganese ores, or about 4 billion tons of manganese, making it by far the world's largest economic reserve of this element.

In comparison with iron, manganese is not easily oxidized, so manganese oxide ores are unlikely to have deposited from the oceans until the dissolved iron had already been exhausted; and indeed, the Kalahari manganese field overlies a rich bed of haematite, the most highly oxidized iron ore, in the Hotazel iron formation. Such a complete deposition of iron and manganese seems to demand a surplus of oxygen. In modern waters, manganese deposition is almost invariably brought about by algal or cyanobacterial blooms, which can generate very high levels of oxygen in a short period. Considered together then, Kirschvink argues, the nutrients from the melting snowball Earth stimulated a cyanobacterial bloom, followed by a precipitate oxidation of the surface oceans, ultimately aiding the accumulation of free oxygen in the atmosphere.

The drama is in the speed. If the underlying rate of change is less than the buffering capacity of the environment to absorb that change, the system as a whole can maintain a pernicious chemical equilibrium. The tendency to approach a stable equilibrium is antithetical to life, which might almost be defined as a state of dynamic disequilibrium. In Chapter 2, we saw that the Earth was saved from the sterile fate of Mars by an injection of oxygen from photosynthesis into the atmosphere, preventing the oceans from ebbing away into space with the loss of hydrogen gas. After this, however, the world sank into a second period of stasis, in which the oxygen produced by cyanobacteria was balanced by the uptake from bacterial respiration, and reaction with rocks, dissolved minerals and gases. This new equilibrium lasted from about 3.5 billion years ago until 2.3 billion years ago, nearly a quarter of the Earth's history. Life on Earth was saved from an interminable ecological balance between iron-loving bacteria, stromatolites and cyanobacteria by the sudden punctuation of the snowball Earth, a shock that rocked it from slumbering complacency with a second big injection of oxygen.

The history of the next billion years lends support to this view of life: not a lot happened, at least to the naked eye. After the deposition of the vast banded iron formations, the dramatic climate swings, the tectonic movements, the oxidation of the surface oceans and the rusting of the continents, Earth seems to have settled down once more to a period of equilibrium, in which a new balance was established. If isotope ratios and fossil soils are to be believed, oxygen levels remained more or less constant at 5 to 18 per cent of present atmospheric levels throughout this period — more than enough for oxygen metabolism to become widespread among our ancestral eukaryotic cells. Better oxygenation would also have increased the concentration of sulphate, nitrate and phosphate in the oceans, lifting these particular brakes on growth. We begin to see simple multicellular algae in the fossil record, and a better preservation of a wider range of eukaryotic cells, suggesting that there may have been a blossoming of genetic variety.

The evolutionary success of our eukaryotic ancestors may well have been linked directly with the higher oxygen levels. We shall see in Chapter 8 that eukaryotes are a hotchpotch of different components. Each individual cell is crammed with hundreds or even thousands of tiny organs, known as organelles, which carry out specialized tasks such as respiration or photosynthesis. Modern life would be unthinkable without these organelles, yet they are aliens within. Some of them show signs of independent origins. One type, called *mitochondria,* evolved from a strain of purple bacteria. They are the sites at which the oxygen-requiring steps of respiration are carried out in all eukaryotic cells, including those of plants and algae. Photosynthesis in plant and algal cells takes place in another organelle — the *chloroplast* — which is derived from cyanobacteria.

Eukaryotic cells are thought to have developed from their primitive precursors into a kind of internal marketplace during the long period of environmental stability beginning around 2 billion years ago. Small bacteria were engulfed by the primitive eukaryotic cells, but somehow survived inside the larger cells like Jonah in the whale. As a result, the eukaryotes eventually became a community of cells within cells.[4] The stalemate must have encouraged the trading of metabolic wares in exchange for shelter. This intimate symbiotic relationship was ultimately so successful that the internalized bacteria are now barely recognizable as once-independent entities. The long-term success of the relationship, however, conceals an interesting paradox. Let us take the mitochondria as an example.

Imagine: 2 billion years ago a small purple bacterium was engulfed by a larger cell, which then had a case of indigestion. Whether the larger cell was predatory, or the invading cell infective, is immaterial. The fact that the insider deal persisted at all means that it was never seriously detrimental. The fact that it finally dominated, to the extent that virtually all eukaryotes have mitochondria, means that it must ultimately have been beneficial. The advantage is obvious in today's world: mitochondria use oxygen to generate energy, by far the most efficient means of biological energy generation known. In those days, though, it ought to have been a different matter. The problem is as follows. The energy currency of all cells is a compound called *ATP* (adenosine triphosphate). Cells use ATP either directly or indirectly to power most of the metabolic reactions that maintain life and make new material for the cell to grow. Both the symbiotic bacteria and their hosts would have produced ATP independently, by fermentation in the case of the eukaryotes, and by burning carbohydrate 'fuel' using oxygen in the case of the bacteria. The bacterial method was much more efficient, so they could produce much more ATP. Like all currencies, ATP is exchangeable. Any ATP produced by the bacteria could in principle be consumed by the host. For the host to benefit in this way, though, the bacteria would have had to export ATP to the

[4] There is some dispute about whether the eukaryotes were produced in a single fusion event between different types of bacteria, or whether a series of engulfments took place. With the exception of chloroplasts, evidence is beginning to favour a single event, or concentrated series of events, which occurred before the deepest evolutionary branches of the eukaryotes.

host cell. Modern mitochondria have pores in their bounding membranes that enable this to happen; but free-living bacteria do not have an ATP export mechanism. On the contrary, free-living bacteria are protected by membranes and cell walls specifically designed to keep the outside world out and the inside world in. Genetic studies indicate that the ATP-export mechanism in mitochondria evolved later, albeit before the major evolutionary branches of the eukaryotes. But if the hosts could gain no extra energy from their guests, how *did* they benefit? Why did this symbiosis flourish?

Evidence from similar symbiotic relationships today suggests that, while the host cell may have gained no energetic benefit, it might instead have been protected from within by its oxygen-guzzling guests. By converting oxygen to water, the symbiotic bacteria would have protected their hosts from potentially toxic oxygen. This acquired immunity to oxygen poisoning would have enabled the early eukaryotes to inhabit the shallow waters where oxygen levels were highest, and so exploit the benefits of light — either by photosynthesis in the case of algae, or by grazing off the freshest pickings in the case of consumers. Over time, the success of this early pact would have encouraged an even closer union, in which the host cell spoon-fed its guests with nutrients, and they in return exported ATP into the cell.

The idea that cells might protect themselves against oxygen by associating with other cells is borne out at a looser level, which may have had even more profound consequences in the long run. When modern oxygen-hating eukaryotes such as the ciliate protozoa are placed in oxygenated water, their first impulse is to swim away to water with less oxygen. The more oxygen there is, the faster they swim. But what if there is no escape? When their surroundings are equally well oxygenated and flight is futile, the ciliates institute plan B: they clump together in a mass. Even anaerobic cells have some capacity to consume oxygen. When cells clump together in this way, each cell benefits from the oxygen consumption of its neighbours. Other communally living cells also seem to have benefited from spreading the burden in this way. For example, stromatolites, those great domed communities of cyanobacteria, are known to have contained many other types of cells, including anaerobic bacteria. Only the top few millimetres of most stromatolites are composed of oxygen-producing cyanobacteria, whereas the deeper levels are home to billions of anaerobic cells, despite high oxygen levels during the daylight hours. Again, each cell benefits from sharing the oxygen load.

Rising oxygen levels may therefore have favoured confederations of cells, from which grew the most efficient energy system for powering life — numerous mitochondria per cell[5] — and the first stirrings of multicellular organization. If so, it is quite possible that a tendency to huddle together as clumps of cells, to alleviate the toxicity of oxygen, was an impetus to the evolution of multicellular life. Certainly, it is a fact that all true multicellular organisms contain mitochondria. Of the thousand or so simple eukaryotes that lack mitochondria, not one is multicellular. People are thus confederates of cells and of cells within cells. We shall see in Chapter 8 that the design of the human body actually restricts oxygen delivery to individual cells: multicellular organization still serves the same purpose in us that it did for our single-celled ancestors.

The Precambrian is drawing to an end. We have travelled down 3 billion years. There has been little to see but much has changed. Without these changes, the explosion of multicellular life that is soon to follow would have been impossible. I have argued that the changes were linked with rises in atmospheric oxygen.

In summary: the first signs of life, the carbon signatures in the rocks of western Greenland, date back to 3.85 billion years ago. By 3.5 billion years ago we find microscopic fossils, resembling modern cyanobacteria, and large stromatolites. If appearances are not deceptive, these cyanobacteria were already producing oxygen. However, it is not until nearly a billion years later, 2.7 billion years ago, that we have the first definitive evidence of cyanobacteria, as well as the first signs of our own ancestors, the eukaryotes, in the form of tell-tale biochemical fingerprints in the rocks. These eukaryotes made sterols for their membranes, a task that requires oxygen. From the activity of sulphate-reducing bacteria we know that oxygen levels rose at this time, perhaps to around 1 per cent of present atmospheric levels. Another 500 million years later, 2.2 billion years ago, oxygen levels rose again, following hard on the heels of the snowball Earth. In the period of geological unrest that followed, huge banded iron formations precipitated from the oceans all around the world. Free oxygen was probably needed for the genesis of at least some of

[5] Just as a car might be 100 horse-power, so a eukaryotic cell with 100 mitochondria could be said to be 100 bacteria-power.

these formations. At the same time, around 2.1 billion years ago, we see the earliest fossils of eukaryotes. By 2 billion years ago, we have rock-hard evidence of oxygen accumulating in the air: fossil soils, continental red-beds and uranium reactors. Oxygen levels reached around 5 to 18 per cent of present atmospheric concentration. In the rocks, we see a sudden explosion of diversity in fossil eukaryotes. Many have mitochondria. All the elements of the modern world, bar true multicellular organisms, are in place.

Then little changed. For a billion years, oxygen levels remained steady at 5 to 18 per cent of present atmospheric levels. The prolonged period of tranquillity saw a number of quiet developments in the history of life — the flourishing of the eukaryotes, genetic diversification, colonization of new habitats, and, in the shape of the algae, the first tentative steps towards multicellular life. And yet: in the face of all these quiet advances, nothing more complicated than a few slimy green tendrils evolved in the course of a billion years. None of this prepares us for what happens next. In a geological blink of an eye, 543 million years ago, the whole of creation as we know it exploded into being. Whatever happened?

CHAPTER FOUR

Fuse to the Cambrian Explosion

Snowball Earth, Environmental Change and the First Animals

The Cambrian explosion — the eruption of multicellular life at the beginning of the Cambrian era — has taxed the finest minds in biology ever since Darwin himself. Why did it happen so suddenly? Did it really happen so suddenly? Darwin had assumed that natural selection should be a process of gradual, cumulative change, and was troubled by the abrupt appearance of fossilized animals in the rocks of the Cambrian era. He hoped, as many have since, that the Cambrian explosion would turn out to be an aberration of the fossil record. If this were the case, then the discovery of older fossils would one day prove that the Cambrian animals had evolved slowly after all — that there had, in reality, been a long Precambrian fuse to the Cambrian explosion. This position was not unreasonable, as most Cambrian fossils known at the time were hard calcified shells, with few remnants of the soft animals that once lived inside; small wonder then that the soft bodies of their unprotected predecessors had perished without fossil record. Perhaps the Cambrian explosion recorded no more than the evolution of shells.

The Burgess shale put paid to the idea that life had only invented shells at the beginning of the Cambrian. Discovered high in the Canadian Rockies by Charles Doolittle Walcott of the Smithsonian Institution in the early years of the twentieth century, this mid-Cambrian shale contains such an astonishing variety and preservation of soft body parts that it has acquired almost iconic status. Many of the fossils examined by Walcott were 'shoehorned', to borrow the late evolutionary biologist Stephen Jay Gould's phrase, into modern taxonomic groups. The story of their reclassification by Harry Whittington, Derek Briggs and Simon Conway Morris of Cambridge University, was the subject of Gould's book *Wonderful Life*, published in 1989. Under bright lights and operating microscopes, the Cambridge team reconstructed the anatomy of numerous strange bilaterally symmetrical creatures, placing one 'weird wonder' after another into taxonomic groups of their own. Their names spoke for themselves: *Hallucigenia, Anomalocaris, Odontogriphus* — each referred to creatures that seemed to correspond to nothing alive today; stalk-eyed, armour-plated, shutter-jawed monstrosities more reminiscent of cartoon Martians than sensible Earthly animals.

In celebrating their strangeness, Gould dwelt on both the sudden appearance of this wealth of biological variety and its eclipse over subsequent geological time. No fundamental body plans have been added to the collection that had evolved by the end of the Cambrian (all insects, for example, have three body segments and six legs), and many variants that existed then have since disappeared without trace. Then, ungratefully soon after Gould had published *Wonderful Life*, two well-preserved fossil beds from the same period were discovered in Greenland and China, and the strangeness of the Cambrian fauna came to be seen in a more conventional light. Some of the weird wonders turned out to have been interpreted upside down, or to have had the parts of other animals mistakenly grafted onto them. Conway Morris, now one of the world's leading authorities on Cambrian biology, has remarked that the real marvel is how familiar so many of these animals seem. The deep similarities between many of the Cambrian animals were first proved statistically in 1989, by Richard Fortey of the Natural History Museum, London and Derek Briggs, then at the University of Bristol, and have since been confirmed by other workers.[1] But if the strange variety of Cambrian fauna is no longer contentious, the roots of the explosion are still fiercely debated. The question remains surprisingly similar to that which troubled Darwin: was the Cambrian explosion really a sudden event, or had there been a slow-burning fuse stretching back into the Precambrian?

[1] Their approach is known as cladistics. Essentially, rather than seeking differences, cladistic analyses enumerate the fundamental similarities between different species to draw a web of inter-relatedness.

We do have more to go on than did Darwin — a century of searching late Precambrian rocks for signs of life has duly turned up a few examples. The most famous are the so-called Ediacaran fauna, a group of radially symmetrical animals, along the lines of jellyfish: pads and pillows of amorphous protoplasm. Some reached a considerable size, measuring a metre [3 feet] or so across. Originally named after their place of discovery, the Ediacara Hills in Australia, similar fossils have since cropped up across all six continents, and date to the Vendian period, 25 million years before the Cambrian. Their discovery, however, did not so much dispel the enigma of the Cambrian as deepen it. Dolf Seilacher, a German palaeobiologist now at Yale University, claims that these stuffed bags of protoplasm, the gentle vegan Vendobionts (as he affectionately calls them), were far from being the ancestors of the bilaterally symmetrical, armour-plated Cambrian animals, but were instead a doomed early experiment in multicellular life that either fell extinct before the beginning of the Cambrian period or got eaten by the shutter-jawed Cambrian predators. While Seilacher's view has been vigorously contested by many palaeontologists, who claim that at least some Vendobionts survived into the Cambrian, few dispute that these strange floating bags do not fit comfortably into modern taxonomic groups.

But the Vendobionts were not the only inhabitants of the Vendian period. Small worms (perhaps several centimetres [an inch or so] in length) burrowed through the mud of the sea floor. Their tracks are preserved, amazingly, in the sandstones of Namibia and elsewhere. These signs of animal movement in the bottom sediments are the first in the long course of the Precambrian, and from then on, similar tracks were left throughout the modern age. The worms live on and leave them still today.

No creature is as synonymous with lowliness as the worm, but its humbleness belies quite a complex design. To burrow through mud requires muscles, and in order to contract, these must be opposed by some form of 'skeleton' — in a worm's case a body cavity filled with fluid. Muscular contraction demands oxygen, and as this cannot diffuse through more than a millimetre or so of tissue, the early worm-like animals must also have had a circulatory system and a mechanism for pumping oxygenated fluid, such as a primitive heart. To move forward at all, the body segments of the worm must have contracted in a coordinated sequence, and this in turn implies at the least a simple nervous system. To displace sediment while worming forwards demands a mouth, a gut and an anus; and indeed some fossilized trails do contain pellets that are interpreted as of faecal origin. Some worms may have been predators of a sort, and to hunt might have been equipped with eyes, or light-sensitive spots, as are their descendants. In short, at a rudimentary level, these primitive worms must already have evolved many of the features that are needed by large animals that can move around. The worms were also bilaterally symmetrical (the same on either side) and segmented — two central features of the later Cambrian animals. It seems likely, then, that our earliest animal ancestors were an approximation to a worm, as Darwin's critics saw only too well when they satirized his view of Man's descent.

For all its lowliness, a worm is much too complicated a creature to have arisen overnight; other, earlier, fossils have indeed been found, dating back to about 600 million years ago — nearly 60 million years before the Cambrian explosion, a period of time as long as that from the extinction of the dinosaurs to the present. Most of the fossils of these earliest multicellular animals are equivocal to say the least: faint, circular impressions, sometimes a centimetre [about ½ an inch] across, but unrecognizable as animals in any conventional sense. Beyond this, nothing. If there were indeed any animals large enough to be visible to the naked eye before about 600 million years ago, they must have had an uncanny knack of avoiding fossilization. The existence of a longer Precambrian fuse can only be deduced from the evidence of molecular 'clocks', perhaps the most powerful and controversial of tools available to the molecular palaeontologist. Molecular clocks imply that the evolution of multicellular animals — the metazoans — may have stretched back to at least 700 million years ago, and possibly to more than a billion years ago.

Molecular clocks make use of the genetic differences between present-day species to predict the time since their divergence from a common ancestor. When an ancestral species splits off new species, these new species and their descendants all gradually accumulate different genetic changes — mutations in the DNA — over time, that eventually make them very different from each other. As different, for example, as humans are now from fruit flies. The basic assumption is that species drift away from one other, in terms of their shared genetic inheritance, at a steady rate. At face value, this assumption is of course nonsense — we have crossed a lot more evolutionary space over the last 600 million years than have worms, for example. The difficulty is to specify a distance across evolutionary space on the basis of averaging the rate of evolution in different species. Luckily, a few simple tricks can be applied, allowing a more robust guess to be made. Two of the most essential factors are calibration of the molecular clock using reliably dated fossils and the use of an average rate of genetic drift obtained by determining the changes in a large number of different genes in a wide variety of species. The evolutionary biologist Richard Fortey provides a nice analogy in his delightful book *Trilobite!* He compares molecular clocks with an old-fashioned horologist's shop, in which hundreds of clocks beat to their own music of time. Some have stopped completely, others tell wildly different times, but the majority indicate that the time is around two-thirty in the afternoon. While the onlooker may doubt the exact time, he will probably be satisfied that it is mid-afternoon. Similarly, the results of molecular clock calculations sometimes vary by hundreds of millions of years, according to the genes and species studied, but all indicate a substantial fuse of animal evolution during the Precambrian. The overall weight of evidence implies that the fuse lasted at least 100 million years, and possibly as much as 500 or 600 million. If this is true, the earliest animals must have been too small to leave visible fossils, so the search is on for tiny impressions measuring less than a millimetre [1/16 in] across.

The genetic studies reveal more than a Precambrian fuse, however. They also indicate that an ancient set of genes, which controls the embryological development of all animals today, were already fully operational in the earliest Cambrian animals. These genes are known as the Hox genes. They are remarkable in two ways. First, there are relatively few of them: just a handful of genes control many of the steps in the early development of all animal embryos — from flies to mice and men. Second, the Hox genes of different species have very similar coding sequences. Even distinct groups, such as the arthropods and the chordates — the group to which we and other vertebrates belong — share sets of very similar Hox genes. Let us consider the implications of these two points in turn.

How is it that so few genes can control embryological development? The Hox genes function as master switches along the length of the body, switching on or off the hundreds of other genes required to make, say, a leg or an eye, depending on the position in the body. They behave like opinionated newspaper proprietors, who influence the tone or coverage of their papers on particular issues, such as politics or European union. If the proprietor buys another newspaper, with a different political affiliation, he might bring about a shift in their political reporting to reflect his own views. A single rogue proprietor is enough to make the paper transfer its affiliation from right-wing to left-wing overnight. In the same way, if a master-switch Hox gene responsible for growing an eye in a fruit fly is switched on, by mistake or design, in a body segment further back, it alters the way other genes are switched on or off in that segment, causing bizarre developmental errors, such as the growth of an eye on a leg. Normal development thus requires a set of master-switch Hox genes plus a regulatory framework that ties the action of a given Hox gene at a particular position in the body to a particular effect on the genes under its influence.

Why are the Hox genes so similar in different species? The fact that animal groups that were already distinct in the Cambrian (such as the arthropods and chordates) share very similar Hox genes implies that all inherited them from a common Precambrian ancestor. This is purely logical reasoning. It is highly unlikely that all the Cambrian species evolved exactly the same genes independently. We might as well suggest that the physical characteristics we share with our brothers and sisters — light hair, blue eyes, white skin, or brown eyes, dark hair and black skin — have nothing at all to do with inheritance and everything to do with a common environment. It is conceivable that Cambrian species transferred genes to each other by lateral transfer, by sex, but for this to happen, utterly different species would have to exchange genes in a way that could not be imagined today. If a lobster were to copulate with a jellyfish, it is hard to imagine a successful outcome. It is more reasonable to assume that the Hox genes were in fact inherited from a common ancestor shared by all Cambrian animal species. If that was indeed the case, then the Hox genes — the basic genetic tool-kit needed to produce segmented body parts, such as a head with feelers and eyes on either side — must have evolved before the Cambrian explosion. Along with the fossil findings, this genetic evidence constrains the significance of the Cambrian explosion. It was not the evolutionary diversification of the first multicellular animals, which probably happened more than 600 million years ago; nor was it a radiation of relatively large animals, which took place among the stuffed bags of protoplasm, the Vendobionts, 570 million years ago. No, the Cambrian explosion was above all a diversification of segmented bilateral animals similar to modern-day crustaceans.

According to the Harvard University palaeobiologist Andrew Knoll and his University of Wisconsin colleague, the molecular biologist Sean Carroll, the Cambrian explosion was probably driven by a rewiring of the regulatory loops between the master Hox genes and the genes under their control. Shuffling and duplication of Hox genes allowed existing genes to take on new responsibilities. The number of Hox genes correlates roughly with morphological complexity. Thus, nematodes have one cluster of four Hox genes (and are simple in structure), whereas mammals have 38 Hox genes arranged in four clusters. Goldfish, rather surprisingly, have 48 Hox genes in seven clusters; biology never stoops to a perfect correlation. In essence, though, duplication of Hox genes allows the replication and subsequent evolutionary modification of repetitive body parts. Having extra, dispensable, body parts makes specialization and complexity easier to achieve. For example, in the ancestors of the arthropods, the large group to which modern insects and crustaceans belong, a small change in the workings of a Hox gene could cause new legs to sprout on previously bare segments, and these then evolved into antennae, jaws, feeding appendages and even sexual organs.[2] While many of the fine genetic details are being worked out step by laborious step, the outstanding question has shifted from 'how' to 'why now?', or rather, 'why then?' The broad answer offered by Knoll and Carroll is that the Cambrian explosion was the historical product of an interplay between genetic possibility and environmental opportunity. The most likely environmental gate-keeper was oxygen.

The long equanimity of the earth, which had persisted since the upheavals of around 2.3 to 2 billion years ago, was shattered for a second time by another series of snowball Earths, starting about 750 million years ago. This time the cataclysm was not a singular event, caused by the exhaustion of a greenhouse gas such as methane, but a 160-million-year roller-coaster ride, comprising possibly as many as four great ice ages, two of which, the Sturtian (at around 750 million years ago) and the Varanger (at around 600 million years ago) were arguably the most severe in Earth's history.

We don't know exactly what triggered this dramatic sequence. The most plausible explanation argues that the tectonic meanderings of the continents happened to bring about their freak assembly around the

[2] One reason why the segmented bilateral body plan is so pregnant with genetic potential is that small changes in *Hox* genes, shifting their zone of responsibility, can lead to sudden and dramatic changes in morphology – a purely Darwinian stepwise process that is easily mistaken for a giant leap over genetic space.

Equator for a time.[3] This would have meant that all the land masses on Earth were free of ice. To understand why this should matter, we must look at what happens when rock is exposed to the air, or to warm oceans with plentiful carbon dioxide. Rock can be eroded by dissolved carbon dioxide, which is weakly acidic. As a result of this reaction, carbon dioxide is lost from the air and becomes petrified in carbonates. But when glaciers form over land, the underlying rock becomes insulated from the air by the thick layer of ice. This means that the rate of rock erosion by carbon dioxide is cut to a fraction and the carbon dioxide stays in the air. In fact, in such a situation, carbon dioxide actually builds up in the air, because it is also emitted more or less continuously from active volcanoes.

Over aeons of time, such a build-up can make a very considerable difference, unless it is offset by erosion of rocks. Because carbon dioxide is a greenhouse gas, this build-up produces an increased greenhouse effect. The surface of the Earth gets warmer. Global warming ultimately halts the spread of the glaciers from the poles. So in today's world, where there are large land masses at or around the poles, any spread of the polar glaciers towards the Equator is offset by a greenhouse effect that gets stronger whenever the glaciers advance, and weaker whenever they retreat.

Now consider what happens if polar ice forms over the oceans instead of the continents. This is what may have taken place to form the late Precambrian snowball Earth. Because the continents were clustered together in the tropics, glaciers at the poles formed over sea only. These polar glaciers could not affect the rate of rock weathering on the continents. The rocks kept on drawing down carbon dioxide from the air. Atmospheric carbon dioxide levels began to fall. The gradual draw-down of carbon dioxide had an anti-greenhouse effect, encouraging the spread of the glaciers. There was nothing to stop the advance: the equatorial continents kept sucking up more and more carbon dioxide. Worse still, as the glaciers marched on towards the Equator, they reflected back the Sun's light and heat, cooling the planet still further, sending the Earth into a vicious spiral of cooling. Eventually, the whole Earth was covered in ice. The ice reflected back so much of the Sun's heat that the planet was in danger of turning into an eternal snowball. Yet Earth is not a snowball today. Somehow, the spell was broken; what happened?

When the equatorial continents were finally sealed beneath the ice, the continuous draw-down of carbon dioxide by rock weathering ceased.

[3] Palaeomagnetic studies do in fact support this arrangement, albeit with a wide margin of error.

With no liquid water exposed to the atmosphere, there was no evaporation, no rain. Any carbon dioxide in the air stayed in the air. All climatic traffic between the air and the frozen seas and the buried rocks came to an end. Deep beneath the surface of the Earth, however, the forces of vulcanism were oblivious to the icy crust. Active volcanoes burst through the ice, spewing volcanic gases into the air, among them carbon dioxide. Over millions of years, carbon dioxide accumulated in the air again, re-warming the Earth. Finally the glaciers began to melt. As this happened, more of the Sun's warmth was retained, less reflected back. The vicious circle of reflectance went into reverse. But there was a diabolical catch. The juxtaposition of the continents around the Equator continued to set the same snare: the whole crazy snowballing and melting repeated itself as many as four times before the continents were finally dispersed to the four corners of the Earth by the forces of plate tectonics.

This story is admittedly hypothetical, but Joseph Kirschvink (whom we met in Chapter 3) and others have dispelled any doubt that glaciers did encroach to within a few degrees of the Equator at that time. Their grand synthesis is supported by the presence of so-called cap carbonates in the rocks of Namibia and elsewhere. Cap carbonates are exactly what their name implies: belts of limestone, at times hundreds of metres thick, which cap the glacial deposits laid down during and immediately after the ice age. For many years, their intimate relationship with the glacial deposits seemed a paradox, as carbonate rocks normally form only in warm oceans and in the presence of plentiful carbon dioxide; and neither of these circumstances are compatible with ice-age conditions. A solution to the conundrum was put forward by the Harvard University geologists Paul Hoffman and Dan Schrag in 1998. They argued that a build-up of perhaps 350 times the current levels of carbon dioxide would be required to melt the ice. Once the reflectance of the snowball Earth had been overcome in this way, however, the extreme levels of carbon dioxide would swing the global climate from an ice-box to an oven in a matter of a few hundred years. Searing temperatures, tropical storms and torrential rain would scrub carbon dioxide from the skies, turning the oceans into an acid bath. The only way to regain a normal chemical balance would be to drop carbonates out of the oceans, straight on top of the glacial debris, so forming the cap carbonates. Thus Hoffman and Schrag use the cap carbonates themselves as proof of the plausibility of a snowball Earth.

Geologists continue to squabble over the plausibility of a snowball Earth. How did life survive? Could bacterial production of methane — another greenhouse gas — have helped to melt the ice before carbon dioxide levels became so high? Were the oceans completely sealed from the air under the deep ice? Or was the Snowball Earth perhaps more of a Slushball Earth, in which the oceans never completely froze, and icebergs floated on open water in equatorial regions.

Even though we don't yet know how severe the snowball glaciations really were, geochemists can add up the consequences, as recorded in the isotope signatures of the rocks. These signatures tell a fascinating story of their own. In particular, the ratio of carbon-12 to carbon-13 (see Chapter 3, page 34) in the cap carbonates and other rock formations veers from background volcanic levels to the highest levels of carbon-13 in the entire Precambrian (Figure 3). For carbon isotopes to plateau at background volcanic levels, there can have been almost no burial of organic matter, as burial of organic matter always disturbs the natural equilibrium left behind in the oceans. If there was no organic matter buried, then there must have been next to no organic matter produced — in other words, no biological activity. This stark conclusion is the geological equivalent of a flat-line cardiac trace, and is interpreted as the near-extinction of *all* living things either during or immediately after each ice age, when carbon dioxide was being scrubbed from the skies and the oceans were an acid bath. Conversely, a swing to the highest levels of carbon-13 in the whole Precambrian implies a massive production and burial of organic carbon, (mostly derived from microplankton, algae and bacteria), which left behind an excess of carbon-13 in the oceans, to form the next layer of carbonate rocks. At such peak times — after the Sturtian glaciation some 700 million years ago, for example (see Figure 3) — life was flourishing as never before.

This dramatic picture sounds plausible. If ice really did cover the whole Earth, then it is quite probable that only a few cells or tiny animals would survive, scratching a living in hot springs, or beneath translucent or thin ice, through which sunlight could penetrate.[4] No doubt life was

[4] Life would have survived comfortably in the hot springs and black smokers at the bottom of the oceans. Some say this is where life originated in any case; others say that the world was repopulated after the last snowball Earth by hydrothermal bacteria. It is therefore possible that evidence we think dates back to the origins of the life actually only dates back to the *repopulation* of the world following the snowball Earth bottleneck. I doubt that this view is true: the cyanobacteria are too distinct from the hydrothermal bacteria to have evolved from them so recently; and there is evidence of cyanobacteria from before all the snowball Earths. Somehow, then, the cyanobacteria survived all the snowball Earths, probably under thin ice near the Equator or in thermal springs on the Earth's surface.

lucky to keep a tenuous grip in the hellish acid bath that followed. No wonder there was so little burial of organic matter. After scourging itself so thoroughly, the Earth regained a climatic equilibrium. Now the survivors had a whole planet to themselves. They must have multiplied like mad. In this they were aided by high levels of minerals and nutrients, eroded by glaciers worldwide and swept into the oceans by the deluge. All these nutrients, all this empty space, must have stimulated the greatest

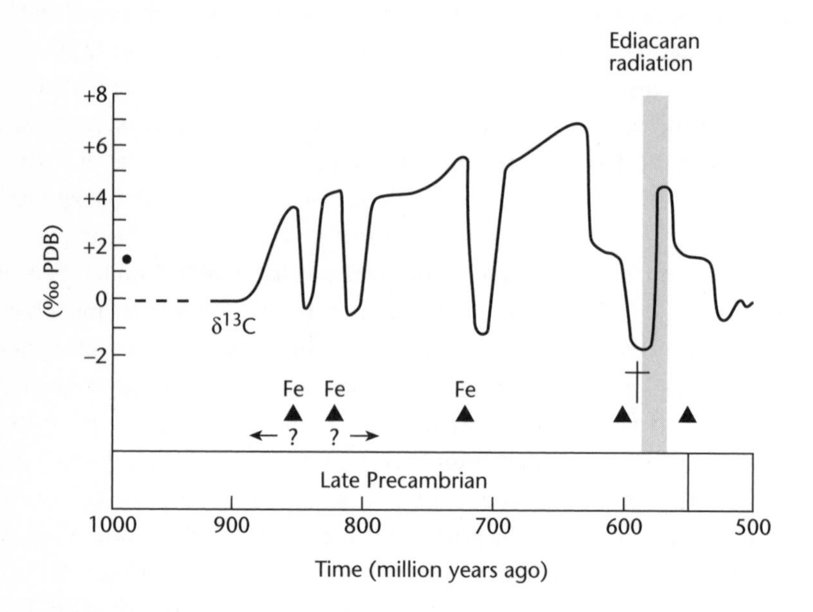

Figure 3: Changes in carbon isotope ratios during the late Precambrian and early Cambrian periods. The changes are given in parts per thousand (0/00) relative to the PDB standard. (The PDB standard is the carbon-13 level found in belemnite from the Pee Dee formation in South Carolina; belemnite is a form of limestone made from the calcification of an extinct order of molluscs related to squid, which were widespread in the Jurassic and Cretaceous periods.) The asterisk at the left-hand side shows the average present-day carbon-13 value. The peaks of carbon-13 (positive anomalies) indicate a substantial increase in the amount of organic-carbon burial (and therefore probably rising oxygen levels), whereas the troughs (negative anomalies) indicate virtually no organic-carbon burial. The negative troughs correspond to possible ice-ages or snowball Earths, of which the two most important were the Sturtian (750–730 million years ago) and the Varanger (610–590 million years ago). Fe indicates the presence of banded-iron formations. The cross marks a major extinction of microplankton that immediately predated the appearance of the Ediacaran fauna — the cushioned Vendobionts and the first worms. Adapted with permission from Knoll and Holland, and the *National Academy of Sciences*.

blooming of cyanobacteria and algae the world has ever seen: a world of blue-green ocean. These blooms must have produced prodigious amounts of free oxygen in a short period, oxygenating the surface oceans and air between each of the ice ages.

All this extra oxygen could only persist in the air if it was not consumed by the respiration of other bacteria, or by reaction with rocks, minerals and gases. When oxidized, a single atom of iron loses one electron to oxygen to form rust. In contrast, each atom of organic carbon gives up as many as four electrons to form carbon dioxide. A single atom of organic carbon therefore consumes four times as much oxygen from the air as does a single atom of iron. By far the best way of preventing the complete re-uptake of atmospheric oxygen is to prevent it from reacting with organic matter; and the easiest way of doing that is to bury the organic matter rapidly.

The essential difference between conditions today, in comparison with those that followed hard on the heels of the snowball Earths, is the rate of rock erosion, which is slower today than it was then. Slow erosion normally equates with slow burial of organic matter, as it takes longer for organic matter settling on the ocean floor to be buried under new sediments originating from rock erosion and organic matter. This leaves time for bacteria to break down the organic matter produced, for example, by algal blooms, consuming oxygen in the process. In present-day conditions, therefore, bacteria breaking down organic matter maintain the *status quo*. In contrast, high rates of erosion in the wake of glaciation lead to high rates of sedimentation and burial. Some organic carbon inevitably gets mixed up in this overall flux. In the aftermath of a snowball Earth, then, high rates of erosion *ought* to have led to high rates of carbon burial, and so to persistent oxygenation.

The theory makes sense, but is there any evidence that the rate of erosion was high after a snowball Earth? Did this really lead to a rise in free oxygen? Think about that for a moment. How do we even start to answer questions like these? How can we possibly know what the rate of erosion was, 590 million years ago? Where should we look for evidence that free oxygen increased at this time? This is the very stuff of science, and the conclusions that can be drawn from clever reasoning backed up by precise measurements never cease to amaze me. There is indeed evidence that the rate of erosion increased after the snowball Earth, and that this was coupled to an accumulation of free oxygen. Each piece of evidence, by itself, leaves room for doubt, but taken together, I find the central assertion convincing — there was a rise in free oxygen in the immediate aftermath of the snowball Earth. This rise corresponded in time with the evolution of the first large animals — the Vendobionts. Here's a brief resumé of the evidence so that you can make up your own mind (or marvel at the ingenuity of the human mind).

We start with another set of isotope signatures. The rate of erosion in the distant past can be estimated by measuring the ratio of strontium isotopes in marine carbonates. Two stable isotopes of strontium — strontium-86 and strontium-87 — differ in their distribution between the Earth's crust and the mantle underneath it. The mantle is rich in strontium-86, whereas the crust is more richly endowed with strontium-87. The major source of strontium-86 in the oceans is the igneous rock basalt. This rock is extruded continuously from the mantle at the mid-ocean ridges, from where it spreads slowly across the ocean floor before diving back into the mantle beneath the ocean trenches. A little strontium dissolves from the basalt into seawater. The speed of dissolution is more or less constant. The gradual build-up of dissolved strontium-86 in the oceans is balanced by a steady uptake of strontium by marine carbonates, such as limestone (calcium carbonate). This is because strontium can displace its sister element, calcium, in the crystalline structure of limestone. As each of these processes takes place at a steady rate, we would not expect the relative amount of strontium-86 in limestone to fluctuate a great deal. In fact it varies quite a lot. Strontium-87 is to blame.

The quantity of strontium-87 in the oceans depends on the rate of erosion of the continental crust. Periods of glaciation and mountain-building intensify erosion and run-off into the rivers, delivering strontium-87 to the oceans. Like strontium-86, strontium-87 is incorporated into marine limestones. The ratio of strontium-86 to strontium-87 incorporated depends on their relative concentration in sea water. In periods of high continental erosion, more strontium-87 gets into the oceans, so more is trapped in marine carbonates than in periods of low erosion. The ratio of the two strontium isotopes in limestone rocks of a certain age therefore gives an indication of the rate of erosion at the time that they were formed. According to Alan Kaufman of the University of Maryland and his Harvard University colleagues Stein Jacobsen and Andrew Knoll, the ratio of strontium-87 to strontium-86 in marine carbonates rose steadily in the aftermath of the snowball Earth, indicating a high rate of rock erosion. Not only this, but the correlation between carbon-isotope ratios (more carbon-12 buried) and strontium-isotope ratios (more strontium-87 in the rocks) implies that a high rate of erosion did indeed go hand in hand with a fast rate of carbon burial. This should have led to a rise in free oxygen.

Two independent methods corroborate such a rise in oxygen. The first method hinges on the ratio of sulphur isotopes in iron pyrites, which are iron sulphides (FeS_2), and was reported by Donald Canfield in *Nature* in 1996. We first met Canfield in Chapter 3, along with his ingeniously tangential conclusions based on the behaviour of sulphate-reducing bacteria, which reduce sulphate to hydrogen sulphide. Here he does it again for a later period. Working this time with Andreas Teske, at the Max Planck Institute for Marine Microbiology in Bremen, Canfield demonstrated an ecological turning point in the way that bacteria handled sulphur, starting soon after the meltdown of the last snowball Earth. For a full 2 billion years before this, the sulphide-producing activities of sulphate-reducing bacteria caused the sedimentary iron sulphides to become enriched in sulphur-32 by about 3 per cent compared with background ratios. Then, suddenly, after the last snowball Earth, around 590 million years ago, the sulphides in sediments became enriched by about 5 per cent. They have been enriched by about 5 per cent ever since, so this figure is virtually diagnostic of modern ecosystems. What had happened?

The figure of 3 per cent is easy to explain. Sulphate-reducing bacteria rely on a one-step conversion of sulphate into hydrogen sulphide. This simple process enriches the sulphur-32 in hydrogen sulphide by about 3 per cent. The enriched hydrogen sulphide is then free to react with iron to make iron pyrites. The trouble is, a 5 per cent enrichment cannot be achieved by a one-step bacterial process. It can only happen in an ecosystem that recycles its raw materials, in the same way that that we can concentrate carbon dioxide from our breath by repeatedly breathing in and out of a plastic bag.

In the case of hydrogen sulphide, the recycling needs oxygen. The system works as follows. Sulphate-reducing bacteria thrive in stagnant muds at the bottom of the sea. The hydrogen sulphide they produce percolates up the water column and reacts with oxygen filtering down. A mixing zone develops between the stagnant conditions deep down and the aerobic conditions higher up. Today, this zone is inhabited by numerous inventive bacteria that live on sulphur. Some of these oxidize hydrogen sulphide, producing the element sulphur, while others convert the elemental sulphur back into a mixture of sulphate and hydrogen sulphide. Because this sulphate is regenerated biologically, it is itself enriched in sulphur-32. The sulphate-reducing bacteria take this biologically generated sulphate and convert it back to hydrogen sulphide. Each cycle enriches the sulphates and the sulphides a little bit more in sulphur-32. Finally, an average of about 5 per cent enrichment is achieved. This value of 5 per cent is just an equilibrium point, at which hydrogen sulphide is likely to react with iron to produce iron pyrites. Once formed, the heavy iron pyrites settle out into the bottom sediments, preserving the equilibrium for posterity.

What Canfield and Teske propose, then, is that 'modern' types of ecosystems, requiring modern levels of oxygen, began to develop soon after the end of the last snowball Earth. They back their conclusions by molecular-clock calculations, which confirm an increase in the number of species of sulphur-metabolizing bacteria at this time. Thus, Canfield and Teske project a rise in atmospheric oxygen to nearly modern levels in the final years of the Precambrian.

The second method that points to a rise in free oxygen is the pattern of so-called rare-earth elements. The relative amounts of these trace elements, such as cerium, in marine carbonates depends on their abundance in sea water at the time; and this depends on their solubility. The solubility of many elements differs according to the level of oxygen. We have already seen that iron becomes less soluble in the presence of oxygen, whereas uranium becomes more soluble. If we see a shift in the relative concentrations of different elements in the rocks (some becoming more abundant, some less so) we get an indication of the degree of oxygenation of the oceans at the time of their formation. According to Graham Shields, at the University of Ottawa in Canada, and Martin Brazier, at the University of Oxford, the marine carbonates that formed in Western Mongolia during and after the snowball Earth period record a shift in the pattern of rare-Earth elements, indicating a rise in the oxygenation of the oceans.

Uniquely in the history of our planet, all these factors — the carbon isotopes, sulphur isotopes, strontium isotopes and rare-earth elements — simultaneously point to a rise in free oxygen. Indeed, the wild swings in environmental conditions during the 160-million-year snowball Earth period may have pushed atmospheric oxygen up to nearly modern levels. At the same time, however, there was a re-emergence of banded-iron formation, after a hiatus of nearly a billion years, suggesting that the deep oceans still contained large amounts of dissolved iron. If so, there can have been little oxygen in the ocean depths.

We emerge blinking, then, from the great Varanger ice age — the last snowball Earth — which ended some 590 million years ago, into a world in which the surface oceans and the air are well oxygenated — well enough for us to breathe — but the deep oceans are still stagnant, like the Black Sea today, saturated in hydrogen sulphide. Then suddenly, within a few million years of the dawn of this new and better world, we find the first large animals, the strange bags of protoplasm known as Vendobionts, floating in the shallow waters, and worms wending their way through the muddy bottoms of the continental shelves, all across the world. This was an age bursting with potential. Strangely, the fulfilment of the potential may have spelled its early demise.

The philosopher Nietzsche once remarked that mankind will never mistake himself for a god so long as he retains an alimentary tract; the need to defecate is our most unflattering quality. In a clever paper published in *Nature* in 1995, Graham Logan and his colleagues, then at Indiana University, contradicted Nietzsche, arguing, in effect, that we owe our most god-like qualities, indeed our very existence, to the primal need for defecation. Faecal pellets from the first large animals, they say, cleansed the oceans, paving the way for the Cambrian explosion. Few theories of environmental change in the terminal Precambrian are quite so down to earth (or at least, seafloor).

Basing their arguments on a detailed study of carbon isotopes in molecular fossils, Logan's group found that virtually all the organic carbon produced during the long period of environmental stasis from 1.8 billion to 750 million years ago was *not* buried in sediments, but was instead broken down again and reused by bacteria lower in the water column. The dead remains of the tiny, almost weightless, bacteria sank very slowly into deeper waters, giving consumers plenty of time to reuse any available carbon. As most carbon was reused, the rate of carbon burial was low. Because oxygen only accumulates when carbon is buried, there can have been very little long-term accumulation of oxygen in the air, and little stimulus for change. Worse than this, any oxygen percolating down the water column was neutralized by hydrogen sulphide rising up — a situation that could sustain the stagnant conditions indefinitely. In the wake of the very first snowball Earth (2.3 billion years ago), high rates of erosion and carbon burial brought about a sudden change, but the debris was ultimately depleted and the *status quo* — tortuously slow organic burial — was restored. The restoration of this *status quo* after the snowball Earth seems to account for the fact that oxygen levels did not rise above 5 to 18 per cent for a billion years. Left to their own devices, then, bacteria may never have broken out of an endless equilibrium.

Logan argues that the *status quo* was finally broken for ever by the evolution of animals with guts — a leap that could only be achieved in shallow waters with the aid of oxygen (only oxygen-requiring respiration is efficient enough to support the evolution of multicellular animals with guts). The relatively heavy faecal pellets of these animals must have sunk rapidly to the ocean bed, cutting swathes through the heaving population of oxygen-hating sulphate-reducing bacteria. Peppering the bottom sediments with nutrients, the faecal pellets were buried in turn under more sediments, so depriving the sulphate-reducing bacteria of their organic nutrients and, through their burial, contributing to the oxygenation of overlying waters. The lean pickings for the sulphate-reducing bacteria, combined with better oxygenation of deep waters, must have hastened the retreat of these oxygen-haters to the anoxic bottom sediments.

Suddenly, at a time when the huge genetic potential implicit in the segmented bilateral structure of the Cambrian animals was poised, just waiting for an opportunity, a vast, new, well-oxygenated ecosystem opened up like the promised land. The expansion of these motile, predatory, genetically loaded animals into the vacant eco-space must have given the gentle floating Vendobionts no chance. They must have been shredded like swollen plastic bags in a combine harvester.

The fall of the Vendobionts to predators is speculative, but there can be no doubt that rising atmospheric oxygen did correlate with radiations in biological diversity in the Precambrian era. In Chapter 3, we noted a link between oxygen and the rise of the eukaryotes, and following that, a link between oxygen and the first stirrings of multicellular life. Now we must consider the links between oxygen and the appearance of the first animals of any size, the Vendobionts, and the bilateral segmented Cambrian animals in turn (see Figure 4). While these links are beyond question in themselves, it is an embarrassingly common mistake to confound a correlation for a causal relationship. You may recall that Preston Cloud's third criterion for linking oxygen to evolution was a good biological basis for

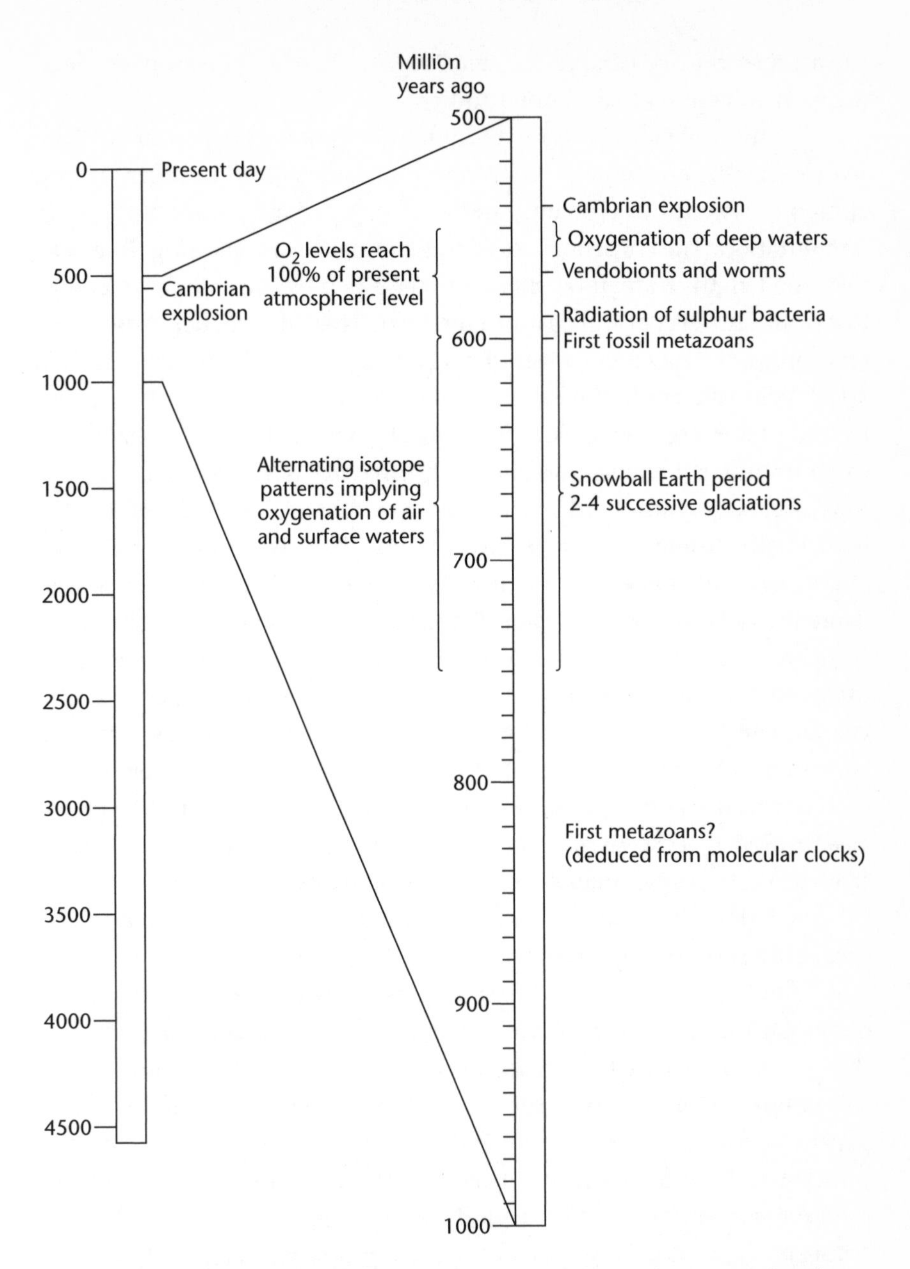

Figure 4: Geological timeline expanding the late Precambrian period and Cambrian explosion. Exactly when the first metazoans evolved is not known. Calculations from molecular clocks suggest a date of somewhere between 700 and 1200 million years ago. The succession of snowball Earths appears to have driven atmospheric oxygen up to atmospheric levels similar to those today.

the association (see Chapter 2, page 28). Are there good reasons for linking oxygen with biological opportunity?

The most obvious basis for a causal link is energy production: oxygen releases much more energy from food than do sulphur, nitrogen or iron compounds acting as oxidants, and is an order of magnitude better than fermentation. The consequences of this simple fact are startling. In particular, the length of any food chain is determined by the amount of energy lost from one level of the chain to the next. This, in turn, depends on the efficiency of energy metabolism. Energy metabolism is generally less than 10 per cent efficient in the absence of oxygen (that is, less than 10 per cent of the total energy available in the food is extracted). If this organism is eaten in turn, the energy available to the predator is less than 1 per cent of that originally synthesized by the primary producer. This is the end of the food chain: below a 1 per cent threshold, there is simply not enough energy available to eke out a living. As a result, food chains must be very short in the absence of oxygen. Bacteria usually become specialized, or compete with each other for scarce resources, rather than 'eating' each other. In contrast, oxygen-powered respiration is about 40 per cent efficient in energy extraction. This means that the 1 per cent energy threshold is crossed only at the sixth level of the food chain. Suddenly carnivorous food chains pay and the predator is born. The dominant position of predators in modern ecosystems is not possible without oxygen. It is no fluke that the Cambrian animals were the Earth's first real predators.

Predation is a powerful stimulus to weight gain in both predators and prey, either to eat larger prey or to avoid being eaten. Large size requires structural support. The two most important structural components of plants and animals, lignin and collagen, require oxygen for their synthesis. Lignin is best known as the cement that binds cellulose into a strong but flexible matrix in the wood of trees. Because the production of paper necessitates the expensive and time-consuming chore of removing most of the lignin, there has been commercial interest in genetically modifying plantation trees to produce less lignin. If nothing else, the failure of these efforts gives an idea of the importance of lignin: the paltry, stunted, lignin-depleted trees are swept to the ground and crushed in the lightest of winds. Lignin is produced by the reaction of phenols with oxygen. (Phenols, or phenolic antioxidants, are found at high levels in red grapes, and have been linked to the benefits of the Mediterranean diet.) Once formed, it is the most refractory biological polymer known: even bacteria cannot easily break it down for energy.

Collagen is the animal world's answer to lignin. It is a protein and an essential part of the supporting connective tissues in flesh, the skin, around organs and the tendons at joints. Before construction work can begin, additional oxygen atoms must be incorporated into the protein chains of collagen, cross-linking them together to form triple-chain molecules, entwined like rope. As animals get older, more collagen cross-links form, which is why meat from older animals is tougher than that from younger animals. Even tiny errors in collagen synthesis can bring about a pathological bendiness of joints and fragility of skin, such as in Ehlers-Danlos syndrome. The 'India-Rubber Man' of circus fame, for example, is said to have had Ehlers-Danlos syndrome. Given the universal importance of lignin and collagen, it is hard to see how large plants and animals could have borne their own weight without oxygen.

A final factor that is often mentioned in relation to rising levels of oxygen is the formation of an effective ozone screen. Ozone (O_3) is formed from the action of ultraviolet radiation on molecular oxygen in the upper atmosphere. Ozone is good at absorbing ultraviolet rays, so once a thick ozone layer has built up, the penetration of damaging ultraviolet rays into the lower atmosphere is cut by a factor of 30 or more. Much early work focused on the importance of the formation of an ozone layer in permitting the colonization of the land, though this work has been questioned in recent years. James Kasting, for example, has argued that only 10 per cent of the present atmospheric levels of oxygen is required to produce an effective ozone shield. This level may have been attained as long as 2.2 billion years ago, nearly 2 billion years before the invasion of the land took place.

James Lovelock argues that the living world is much more robust than we credit, and tells an amusing tale of his early work at the Institute of Medical Research in Mill Hill, London, where he and his collaborators tried to sterilize hospital air using high-intensity ultraviolet radiation — with a singular lack of success. The bacteria protected themselves under a layer of mucus, and were only destroyed if stripped of their mucus before being irradiated. In such circumstances, high levels of ultraviolet radiation, in the days before the ozone layer had formed, cannot have presented much of an obstacle to bacterial colonization of lakes and shallow oceans. Desiccation on land may have been a more intractable problem; but there is no reason why desiccated bacterial spores should not have survived prolonged dryness then, as today.

In fact, rather than protection by an ozone layer, size and structural support probably rank as the most important factors behind the colonization of the land by animals and plants. True land-lubbers, which are more or less independent of liquid water, must avoid desiccation. Resisting desiccation while remaining active requires special adaptations that are only possible in larger organisms, such as a waterproof skin combined, in animals, with internal lungs for maximum oxygen uptake and minimum water loss. And large size, as we have seen, is not likely without oxygen.

We may reasonably conclude that oxygen was a cornerstone of Precambrian evolution. While nobody would propose that oxygen itself stimulates evolution there can be little doubt that rising oxygen levels opened new horizons for Precambrian life. Not one important evolutionary step took place without an associated rise in oxygen, and no rise in oxygen was divorced from a rapid increase in biodiversity and in the complexity of life. Curiously, though, the major injections of oxygen were not brought about by biological innovations, as had been tacitly assumed for many years (with the exception of guts), but by non-biological factors, such as glaciers and plate tectonics.

Left to its own devices, life on Earth dawdled for billions of years. If the stimuli for change and evolution were little more than accidents of tectonics and glaciation, a quiet world untroubled by geological strife would almost certainly fail to accumulate much free oxygen. The Earth stagnated for two prolonged periods, which between them account for half its history. From 3.5 to 2.3 billion years ago, the world was dominated by bacteria. Then, after the violent upheavals of 2.3 to 2.0 billion years ago, another equilibrium was established, in which the oxygen levels remained between 5 and 18 per cent of present atmospheric levels. This new equilibrium stimulated a blossoming of genetic diversity among the early eukaryotes, but could not provide enough energy for the evolution of large animals. At such low oxygen levels, life is denied size and complexity; and without these, a brain is unthinkable.

The deadlock was broken by a second series of snowball Earths, which started 750 million years ago and catapulted oxygen to modern levels. Now the evolution of large animals was only a matter of time, and it didn't take long. The Vendobionts, the Cambrians, the whole plethora of modern life, exploded into being in a period less than that taken up by the preceding glaciations. If nothing else, this relationship between life and environmental conditions should sound a note of caution to those who seek intelligent life elsewhere in the Universe. We must look beyond the mere presence of water to the presence of volcanoes, plate tectonics and oxygen. Perhaps, if life once existed on Mars, it died out as the fires of vulcanism faded within.

Whether oxygen was linked with opportunities or extinctions in the modern age of plants and animals, the Phanerozoic, is a question for the next chapter. I can find no evidence to support the idea that free oxygen caused a global holocaust in the Precambrian, but there is a big difference between our modern oxygen level of about 21 per cent, and the postulated Carboniferous high point of 35 per cent, around 300 million years ago. Our experience with diving gas mixtures alone suggests that prolonged exposure to high levels of oxygen can provoke lung damage, convulsions and sudden death, to say nothing of the raging infernos and stunted plant growth predicted by most biologists. Did oxygen really reach fever pitch? If so, how did life cope? And if life flourished, what does this say about our health today as we pop another multi-antioxidant pill to stem the ravages of ageing?

CHAPTER FIVE

The Bolsover Dragonfly

Oxygen and the Rise of the Giants

THE SMALL ENGLISH MINING TOWN OF BOLSOVER in Derbyshire enjoyed an unexpected 15 minutes of fame in 1979. While working a coal seam 500 metres [1640 ft] beneath the surface, local miners dislodged a gigantic fossilized dragonfly with a wing-span of half a metre [20 in], rivalling that of a seagull. Experts from the Natural History Museum in London confirmed that the fossil dated to the Carboniferous period, about 300 million years ago. The giant was dubbed the Bolsover dragonfly, but although one of the oldest and most beautifully preserved of fossil insects, it was far from unique. Similar fossils from the coal measures of Commentry in south-east France had been described by the French palaeontologist Charles Brongniart as long ago as 1885, and giant dragonflies had since been unearthed in North America, Russia and Australia. Gigantism was unusually common in the Carboniferous.

The Bolsover dragonfly belongs to an extinct group of giant predatory flying insects, thought to have sprung from the same stock as the modern dragonflies (*Odonata*) and known as the *Protodonata*. Like their modern counterparts, the *Protodonata* had long narrow bodies, huge eyes, strong jaws and spiny legs for grasping prey. Pride of place went to the largest insect that ever lived, the colossal *Meganeura*, which had a wing-span of up to 75 centimetres [30 in] and a diameter across the upper body — the thorax — of nearly 3 centimetres [just over an inch]. For comparison, the largest modern dragonfly has a wingspan of about 10 centimetres [4 in] and a thoracic diameter of about 1 centimetre [1/3 in]). The prototype giant dragonfly differed mostly from its living relatives in the structure of its wings, which were primitive in the number and pattern of veins. The giant size and primitive wing structure led the French scientists Harlé and Harlé to propose in 1911 that *Meganeura* could never have managed to fly in our thin modern atmosphere. They argued that such a giant could only have found the power to fly in a hyperdense atmosphere containing higher levels of oxygen than the present 21 per cent. (If the extra oxygen was added to a constant amount of nitrogen, the air as a whole would be more dense.) This startling claim echoed down the corridors of twentieth-century science, to be repeatedly and vigorously rejected by the palaeobiological establishment. In 1966, the Dutch geologist M. G. Rutten could write, in a charmingly antiquated style that has passed forever from the scientific journals:

> Insects reached sizes of well over a metre during the Upper Carboniferous. In view of their primitive means of breathing, by way of trachea through the external skeleton, it is felt that these could only survive in an atmosphere with a higher O_2 level. As a geologist, the author is quite satisfied with this line of evidence, but other geologists are not. And there is no way of convincing one's opponent.

Insect flight mechanics are notoriously complex. A famous, albeit spurious, tale from the 1930s tells of an unnamed Swiss aerodynamicist who was said to have proved, on the basis of calculated flight mechanics, that the bumblebee cannot fly (in fact, he proved that the bumblebee cannot glide, which is quite true). We should not smile too superciliously, though; we have not progressed that far since then. In a detailed 1998 review of dragonfly flight, J. M. Wakeling and C. P. Ellington concluded that our grasp of dragonfly aerodynamics is limited by a poor understanding of the interactions between the two sets of wings, and admitted that we are unable to model their aerial performance with any confidence. In the face of such sweeping ignorance, we can hardly reach any firm conclusions about the composition of the ancient atmosphere on the basis of theoretical flight mechanics alone.

Even so, the idea that giant insects may have required hyperdense, oxygen-rich air to fly was never quite discredited, and stubbornly refused to go away. We shall see that empirical measurements may yet succeed where theory has failed. Other factors imply that oxygen levels fluctuated during the modern era of plants and animals — the Phanerozoic (see Figure 1, page 27). Unequivocal geological evidence shows that the deep oceans contained little dissolved oxygen for at least a short spell, corresponding to the mass extinction at the end of the Permian period (250 million years ago); and for this to have happened we can only presume that atmospheric oxygen levels fell, at least slightly. Conversely, if we are to believe the principle of mass balance (see Chapter 3, page 34), the vast amount of coal — which is essentially organic matter — buried in the Carboniferous and early Permian period must surely have forced the oxygen levels to rise. The question is, by how much?[1]

The chief difficulty in calculating changes in the air is to identify which factors control the composition of the atmosphere over geological time, and which are relatively trivial. Early attempts to model atmospheric evolution would have had us believe that oxygen levels swung from less than zero to several times the present value. If nothing else, these studies drew attention to our surprising ignorance of the factors that actually do control oxygen levels in the atmosphere. The difficulties in modelling atmospheric change may of course reflect no more than an erroneous starting assumption: that changes took place when they didn't. However, before dismissing the problem as one of our own making, we should note that the same difficulty applies to steady-state models, in which oxygen levels remain constant. We do not know how an unchanging oxygen level is maintained in the face of other environmental changes, which are known to have happened.

Take fire as an example. Because fires consume oxygen they are assumed to limit the accumulation of oxygen in the atmosphere. In the absence of human meddling, fires are typically ignited by lightning strikes. Under present conditions, most lightning strikes do not start fires because forest vegetation is damp, especially when electrical storms are accom-

[1] Rutten mentions the old argument that, because the bulk of photosynthesis is thought to come from oceanic plankton, it is not certain whether the flora of the Carboniferous really produced a measurable excess of oxygen. This argument is wrong: the important parameter is not absolute productivity, but *burial*. The complete decomposition of plankton means that far less carbon is buried at sea compared with the situation on land, where plants are more refractory to decomposition.

panied by torrential rain. But if wet organic matter burns freely in air containing more than 25 per cent oxygen, as we are told, then, given an atmosphere with such levels, lightning could trigger conflagrations even in rain forests. The higher the oxygen level, the greater chance of fire; and as the fires rage they use up excess oxygen. If oxygen levels rise too high, fire would restore the balance.

This simple scenario tends to be accepted uncritically, but is in fact quite misguided. Only if the forests are *vaporized* will the balance be maintained (just as we vaporize food when we burn it for energy during respiration, giving off carbon dioxide gas and water vapour in our breath). Anyone who has seen the gutted remains of a forest after a fire knows that a large amount of charcoal is formed. Charcoal is virtually indestructible by living organisms, including bacteria. No form of organic carbon is more likely to be buried intact.

We have already seen that oxygen can accumulate in the air only if there is an imbalance between the amount of oxygen produced by photosynthesis and the amount consumed by respiration, rocks and volcanic gases. Permanent burial of organic matter is the most important way of disrupting this balance, because it prevents the consumption of oxygen by respiration. Organic remains that are buried are not oxidized to carbon dioxide, so the oxygen is left over in the air. As charcoal is more likely to be buried intact than normal decaying plant matter, the net result of a forest fire is to increase carbon burial, and thus to raise atmospheric oxygen. This in turn makes fire more likely and pushes up oxygen levels until finally life on land is destroyed. Only then, when all organic production and photosynthesis on land has ceased, can oxygen levels dwindle slowly, as the gas is removed by reaction with eroded minerals and volcanic gases. If perhaps a spore survived, life can strike up again; but if it does so, the cycle of flames and destruction will be repeated endlessly. Fire is a very poor control of atmospheric oxygen.

Such catastrophic scenarios are only too familiar to the environmental scientists who try to model changes in atmospheric composition, but there are difficulties with more subtle forms of negative feedback too. One mechanism, proposed in the late 1970s by Andrew Watson, Lovelock and Margulis as part of the Gaia hypothesis, suggested that bacterial methane production might stabilize oxygen levels.

Methane-producing bacteria thrive in stagnant swamps where oxygen levels are very low, and they cannot tolerate higher levels. They gain their energy by breaking down organic remains in the swamps to release methane gas. This is no trivial process. Lovelock estimates that some 400 million tons of methane are emitted into the atmosphere each year by swamp bacteria (industrial pollution, farming and landfill sites have now more than doubled this figure, contributing to global warming, among other things). The theory goes that if the burial of organic matter in swamps were to increase, pushing up oxygen levels, new colonies of methane bacteria would thrive on the putrefying detritus, ultimately belching out excess methane into the surrounding air. Methane escaping from the swamps reacts with oxygen over a period of several years to form carbon dioxide, thereby lowering oxygen again. Conversely, a lower burial rate would support a smaller population of methane bacteria, which would emit less methane and so encourage oxygen levels to rise again.

The continuous feedback operating in this cycle might prevent large fluctuations in oxygen levels. The difficulty with the theory is that it predicts that methane bacteria should maintain *permanent* carbon burial at a roughly constant rate, as the bacteria regulate oxygen levels by breaking down organic matter that would otherwise be permanently buried. In fact, there are periods when the geological record shows that it varied substantially. The massive amount of coal formed in the Carboniferous and early Permian was clearly not broken down by methane bacteria in coal swamps, so we can only conclude that there were times when the methane cycle was insufficient to regulate atmospheric oxygen.[2]

More convincing as a biological feedback mechanism for regulating oxygen is a curious phenomenon that affects plants, suppressing their growth and productivity; in some circumstances, plant growth can be halted completely. The phenomenon is known as *photorespiration* and, unlike the plant's normal mitochondrial respiration, takes place only in sunlight. Its purpose is a mystery. The net effect is that the plant takes up oxygen and releases carbon dioxide, which parallels normal respiration (hence the name) but fails to generate any energy. Also unlike normal respiration, photorespiration competes with photosynthesis for the use of an enzyme known by the sonorous acronym of Rubisco (which stands for ribulose-1,5-bisphosphate carboxylase/oxygenase). This contest for the

[2] Availability of nutrients such as phosphate has also been proposed to limit the burial rate, but since the ratio of phosphorus to carbon is much lower in land plants than in marine algae and plankton, more carbon can be buried on land per unit of phosphorus. Phosphate abundance is therefore less likely to affect burial rates in terrestrial, compared with marine, environments.

affections of Rubisco undermines the efficiency of photosynthesis and so reduces plant growth.

Rubisco is the enzyme that binds carbon dioxide and incorporates it into carbohydrate in photosynthesis. It is often justifiably claimed to be the most important enzyme in the world. Certainly, weight for weight, it is the most abundant enzyme on Earth. Without Rubisco, photosynthesis as we know it could not take place. With it, we have a different problem. By enzyme standards, Rubisco is not at all discriminating. A molecular two-timer, it binds oxygen almost as eagerly as it does carbon dioxide. When Rubisco binds carbon dioxide, its lawfully wedded wife, the plant uses the carbon constructively to make sugars, fats and proteins. If, however, Rubisco binds mistress oxygen, then a large number of enzymes start catalysing a useless chain of biochemical reactions, the result of which is a return to square one. This energy-sapping chain of reactions slows down plant growth as surely as a politician's misdemeanours retard his rise to power.

The rate of photorespiration increases with both the temperature and the oxygen level. This means that plant growth grinds to a halt in hot, oxygen-rich conditions. Even in normal air, the apparently pointless squandering of resources can reduce growth by as much as 40 per cent in tropical areas. There is a corresponding drain on agricultural productivity, though this may be masked to the casual gaze by ameliorating factors such as rainfall, soil fertility and length of growing season.

Despite its apparent futility, photorespiration is almost universal among plants, although many have evolved ways of reducing its detrimental effects.[3] It has been maintained by evolution for some reason. In other words, it must be useful for something, otherwise it would have been unceremoniously dumped in the struggle for survival. This premise is backed by the failure of numerous commercial attempts to breed plants that do not photorespire, often motivated by an honest desire to increase productivity in developing countries. Curiously, none of these genetically modified plants could survive in normal air, yet most could thrive in air rich in carbon dioxide but low in oxygen, suggesting that photorespiration might protect against oxygen toxicity in some way. This would

[3] Photorespiration is a particular problem for so-called C3 plants, which include most trees and shrubs. Grasses are mostly C4 plants, and escape the worst excesses of photorespiration by compartmentalizing their photosynthetic machinery. They capture carbon dioxide and then release it in large amounts into the cellular compartment containing Rubisco. In these circumstances CO_2 out-competes O_2 for the services of Rubisco.

explain why photorespiration is less necessary at low oxygen levels, and more necessary at normal or high oxygen levels. Whatever the reason, the upshot is that photorespiration stunts plant growth if atmospheric oxygen levels are high.

The potential magnitude of photorespiration fits it as a plausible mechanism for stabilizing atmospheric oxygen. If oxygen levels were to rise, the rate of photorespiration would rise in tandem and plant growth would falter. Stunted plants would produce less oxygen by photosynthesis, thereby adjusting atmospheric oxygen back to previous levels. An elegant feature of this hypothesis is that it does not imply that carbon burial should be constant. Quite the contrary. The rate of burial should in principle vary with the exuberance of plant growth: if there is no growth there can be no burial, and vice versa. The big question is empirical — can photorespiration really account for both oxygen regulation and variations in carbon burial?

The answer is by no means certain, but the question is at least open to experiment. Some studies suggest that, while surely playing a role, photorespiration alone cannot maintain atmospheric oxygen at a constant level all the time. In coming to this conclusion, in an important 1998 paper in the *Philosophical Transactions of the Royal Society*, David Beerling and his colleagues at the University of Sheffield measured the growth of a variety of plants at incremental oxygen levels from 21 to 35 per cent. At 25°C, overall productivity was down by 18 per cent in oxygen-rich air compared with normal air, confirming an effect on growth rates. However, the magnitude of this effect was not the same for all plants: evolutionarily ancient groups of plants fared much better than their modern cousins. Plants that had evolved during the Carboniferous, such as ferns, gingko and cycads (palm-like evergreens with cones rather than coconuts) were less sensitive to increased oxygen than were plants that had evolved more recently, such as the angiosperms — the largest group of plants today, which include deciduous trees and shrubs, our staple crops, and all other herbaceous crops and flowers. The more ancient groups of plants were also more likely to adapt to the new conditions by changing the structure of their leaves. In particular, these plants increased the number of stomata (the pores through which gases enter and leave the leaf), allowing more carbon dioxide to accumulate inside the leaf.

Interestingly, if carbon dioxide levels in the air were doubled, from 300 to 600 parts per million in these experiments, plant growth was not diminished at all, and indeed productivity sometimes rose. Although carbon dioxide levels generally fall as oxygen rises, most geologists agree that carbon dioxide fell from a high point of about 3000 parts per million in the Devonian (385 million years ago) to a low point of 300 parts per million by the end of the Permian (245 million years ago) (Figure 5). Carbon dioxide levels may therefore have been higher in the Carboniferous than today. All in all, the Sheffield team concluded that high oxygen levels during the Carboniferous and early Permian could have done little more than thin plant growth in tropical areas.

Swamp bacteria, nutrients and photorespiration may well help to regulate oxygen levels under normal conditions, but could at best only blunt the

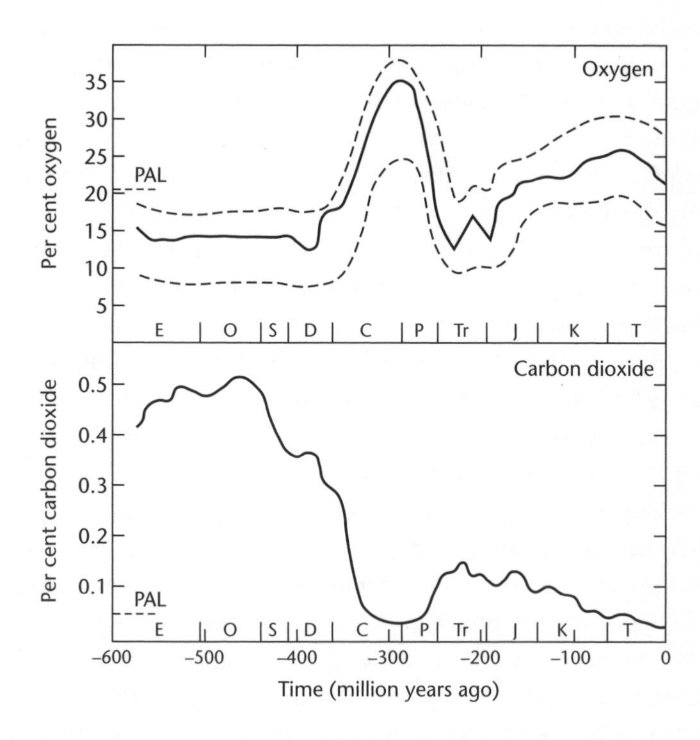

Figure 5: Changes in atmospheric composition over the Phanerozoic period, from 600 million years ago, based on the models of Robert Berner. Oxygen levels (top graph) reached a peak of 35% in the late Carboniferous and early Permian, before falling to 15% in the late Permian. Oxygen levels peaked a second time at 25 to 30% in the late Cretaceous (K) before falling to present atmospheric levels in the Tertiary (T). Carbon dioxide levels (bottom graph) fell from 0.5% in the Silurian (S) to around 0.03% by the end of the Carboniferous. Reproduced with permission from Graham *et al.*, and *Nature*.

large changes in atmospheric oxygen predicted by the high rate of carbon burial in the late Carboniferous and the early Permian. Perhaps it is time to look a little more closely at the riddle posed by this 70-million-year period, which lasted from around 330 to 260 million years ago. Ninety per cent of the world's coal reserves date to a period that accounts for less than 2 per cent of the Earth's history. The rate of coal burial was therefore 600 times faster than the average for the rest of geological time. Most organic matter is not buried as coal, of course (see Chapter 2), but systematic analyses of the organic content of sedimentary rocks all over the world confirm that the total amount of organic matter buried during the Carboniferous and early Permian was much greater than during any other period, including the present day.[4]

Unique events are best explained by singular circumstances. The most believable explanation for the high rate of carbon burial during the Carboniferous and early Permian invokes an accidental alliance of geology, climate and biology. Two factors in particular were probably important. First, the continents had recently converged to form a low-lying supercontinent called Pangaea, and the wet climate and vast flood plains could hardly have provided a better nursery for coal swamps. Second, the rise of large woody plants, the first trees, about 375 million years ago, brought about the verdant colonization of upland areas, as well as the swamps and seashores. Woody plants depend on lignin for structural support. Even today, bacteria have difficulty digesting lignin, but during the Carboniferous and early Permian there must have been a huge discrepancy between the amount of lignin formed by woody plants and the amount broken down by swamp bacteria.

The unparallelled rate of coal formation in the Carboniferous and early Permian is explained, then, by an exceptionally high rate of lignin production, an exceptionally low rate of lignin breakdown, and nearly perfect conditions for preserving organic matter on an unprecedented scale. We know of no negative feedback mechanism that could have constrained atmospheric oxygen under these conditions, so we can only conclude that oxygen levels must have gone up, and probably quite sub-

[4] In addition to carbon compounds, we must also consider the burial of iron pyrites, or fool's gold. When hydrogen sulphide is formed by sulphate-reducing bacteria, it can either react with oxygen to form sulphates, or if no oxygen is present it may react with dissolved iron to form iron pyrites. If iron pyrites are formed and then buried, some oxygen that would otherwise have been consumed by hydrogen sulphide is instead retained in the atmosphere. Simultaneously high rates of carbon and pyrite burial therefore translate into the highest rate of oxygen formation.

stantially. Like M. G. Rutten, I'm quite satisfied by this line of evidence; but the question remains, how much is a lot?

The balance sheet for photosynthetic oxygen production shows that a fixed amount of oxygen is left in the air for each equivalent unit of organic carbon buried (see Chapter 2, page 24). In principle, to calculate oxygen levels all we need to know is how much organic matter was buried in the past. From this figure we must subtract the amount of buried matter that was later exposed by erosion and returned to the atmosphere as carbon dioxide. On our balance sheet, carbon returned to the air through erosion is no different from carbon burnt for energy and returned immediately to the air as carbon dioxide. The difference in rate, however, is critical. Coal that was buried in the Carboniferous, and is today dug out and burnt, was nonetheless buried for 300 million years. Its burial helped raise atmospheric oxygen levels throughout this time, just as burning it is lowering them again today (albeit by a matter of 2 parts per million per year, against a background level of 210 000 parts per million).

To put numbers on rates of carbon burial and erosion at times in the distant past might seem a reckless intellectual escapade, but with a little hedging the Yale University geochemist Robert Berner and his former doctoral student Donald Canfield have succeeded in assigning some reasonably sensible parameters. They argued that, because the great bulk of organic matter is buried in coal seams, silting river estuaries and shallow continental shelves, we do not need to worry too much about rocks that formed at the bottom of the deep oceans. It is then a straightforward, if soul-destroying, task to determine the relative abundance of the different sorts of continental sedimentary rocks, as a quick glance at any detailed geological map will testify. The organic content of these rocks can be measured directly. The real difficulty comes in calculating differential rates of erosion. We can assume that older rocks are more likely to have been completely lost by erosion or metamorphism, whereas younger rocks, buried closer to the surface, are more readily exposed and eroded. Another factor to take into account is whether burial originally occurred in places with good potential for preservation, such as coal swamps (as occurred widely in the Carboniferous), compared with burial at sites subject to high rates of erosion, such as alluvial plains (which was more common in the Permian).

By estimating rates for carbon burial and rock erosion on the basis of the evidence available, Berner and Canfield calculated the apparent changes in oxygen levels over the past 600 million years. They came up with a graph that sent shock waves through the geological establishment. Oxygen levels, they said, rose to 35 per cent during the late Carboniferous and early Permian, then fell to 15 per cent in the late Permian, to cause the worst mass extinction ever recorded. Later, during the Cretaceous (the final age of the dinosaurs), oxygen levels crept up again, this time to around 25 or 30 per cent (see Figure 5).

However impeccable the logic, numbers such as these strain credibility, and stand opposed to most people's intuitive feeling. Perhaps for this reason, conclusions about ancient atmospheres based on computer modelling meet continuing resistance. Most scientists are distrustful of mathematical or philosophical reasoning unsupported by empirical observation — memorably dismissed as 'fact-free science' by the guru of evolutionary biologists, John Maynard Smith. A famous example of fatuous logic is the conundrum posed by an ancient Greek, Zeno of Elea, which troubled logicians for centuries but can rarely have cost a scientist a night's sleep: movement is impossible, said Zeno, because to complete a single step one must first complete half a step, then half of the remainder, and so on in exponential fashion. Just as an exponential curve does not touch the baseline until infinity, so the infinite number of half steps precludes the possibility of ever taking a whole step. Berner's and Canfield's models may be far removed from the perverseness of Zeno's paradox, but even though their calculations are based on empirical data, such an apparently improbable outcome will always invite the reply that an important factor has simply been overlooked.

The only really satisfying way of confirming or rejecting the hypothesis that oxygen levels once reached 35 per cent is to measure them: could there be, somewhere, a pocket of ancient air, miraculously undisturbed for hundreds of millions of years? The idea seems outrageous, but polar scientists have been drilling cores of ice from deep within the Arctic and Antarctic ice caps over many years, in an effort to read the preserved record of environmental change. The results have revealed much about the speed and magnitude of climatic changes in the past, as well as the extent of industrial pollution in Roman times and more recently. Unfortunately, ice-core data can take us back only about 200 000 years, before which the ice runs out. We have gone just 0.0007 per cent of the distance.

The situation seemed hopeless until the mid 1980s, when Gary Landis, a geochemist at the US Geological Survey in Denver, had a clever idea. Tiny bubbles trapped in amber might conceivably contain ancient air, which had once dissolved in the resin of trees and later formed pressured bubbles, as the resin hardened to form amber. As luck would have it, Landis had the right equipment for the job: a quadrupole mass spectrometer, recently designed by the US Geological Survey to analyse the amounts and chemical identities of gases in tiny samples. The instrument is sensitive enough to detect gases at concentrations as low as 8 parts in a billion, and quick enough to analyse samples released in milliseconds from bubbles as small as a hundredth of a millimetre (10 micrometres) in diameter.

Amber jewellery, containing embalmed insects and spores, has been highly prized ever since Neolithic times. Trade in Baltic amber is known to stretch back at least 5000 years. The amber itself ranges in age from Carboniferous (300 million years old) to Pleistocene (the period of the last major ice age) so many trapped insects are in fact very ancient. The reputation of amber as a time capsule reached its apotheosis in the notion that dinosaur genes might survive intact in the abdomens of blood-sucking insects that were engulfed in resin soon after their meal, an idea made famous by the novelist Michael Crichton in *Jurassic Park*. Although controversial, the idea did have sufficient scientific merit to attract serious attention, and ancient DNA has indeed been isolated from insects in amber dating from the Cretaceous (140 million years ago). If flimsy DNA molecules could survive, why not air?

Working with Robert Berner, Landis used his quadrupole mass spectrometer to detect the gases trapped in air bubbles by mechanically crushing the amber in a vacuum. To piece together a time line, Berner and Landis ground up amber samples that dated to successive geological periods, from the Cretaceous (140 million years ago) to modern times. There was, of course, a danger that the different types of amber would produce a range of values simply because they *were* different types of amber. In an attempt to exclude the possibility that the measurements reflected only local conditions, Berner and Landis used amber from a variety of sources, from the Baltic States to the Dominican Republic, and from open beaches to buried strata.

Their results were published in the journal *Science* in March 1988 and caused an immediate stir. The data implied that oxygen levels had been higher than 30 per cent during the Cretaceous, falling to our modern level of 21 per cent around 65 million years ago, a time that corresponded a little too closely for coincidence to the mass extinction of the dinosaurs. Might it be the case, Berner and Landis wondered, that, like the dragon-flies, the dinosaurs needed high oxygen levels to achieve their giant size, and could not survive in our thin modern atmospheres? By August of that year, the letters pages of *Science* were filled with detailed technical criticisms.

There are few better forums for a display of erudition than letters pages, and on a good day the top scientific journals extend this tradition into a marvellous parade of eclectic learning. No academic exchange is more chastening to an erring author, or more entertaining to an inter-ested onlooker. The amber results were subjected to scrutiny from just about every angle, from the diffusion constants and solubility ratios of matrix gases to the unusual chemistry of fractured polymers and the geometrical behaviour of pressurized bubbles. Curt Beck of the Amber Research Laboratory in New York, for example, drew attention to the methods by which the Romans restored translucence to milky (or bone) amber. Bone amber is rendered opaque by the presence of microscopic air bubbles, but can be made transparent, as well as dyed, by heating it in oil. The Romans apparently used the fat of suckling pigs, whereas the great nineteenth-century German authority, Dahms, recommended rapeseed oil. The method works because the air bubbles become filled with the oil, which has the same refractive index as the amber. This means that the oil can penetrate the amber matrix completely. That is, the bubbles are not in fact entirely isolated from the environment, and so the air inside them can exchange with air from outside, and thus would not accurately reflect the composition of the air at the time they were formed.

The overall consensus, which has now persisted for more than a decade, was that the air trapped in amber could not have been ancient, and that the results of Berner and Landis must be artefacts of inappro-priate methodology. While Landis claims to have refuted his critics in a new round of experiments, his compendious data are not yet published in any detail, and so have not been widely accepted by the geological com-munity. Berner himself acknowledges that their preliminary data were flawed, and concedes, with a rather charming candour, that Landis has so far failed to persuade him that their critics were wrong.

In rushing to denounce the amber results, one or two critics claimed that the old *status quo* had been restored, and that the "previously held views of palaeontology, geology and atmospheric science are still intact". Berner and Landis retaliated that these 'previously held views' were not at all intact, but were being revised independently of the amber studies. This brisk exchange seems to me to stand at the heart of the debate, and reveals quite a lot about the nature of scientific enquiry, which is very rarely the kind of dispassionate induction process that philosophers like to call the 'scientific method'. Most scientists instead cling to their favourite stories, or hypotheses, until they are either proved, or discredited to the point that even an obstinate old professor must concede their fallacy. The two sides here were entrenching themselves for battle, with a broad stretch of no-man's land separating the advocates of an unchanging atmosphere from the campaigners for high-oxygen air — a situation reminiscent of the squabbles that once took place in physics between the exponents of a steady-state universe and the big bang theory. Clearly, Berner's and Canfield's atmospheric model would convince some geologists but not others, and only fresh data could point to a way out of the impasse. If amber does not contain ancient air, it is hard to think of anything else that might. Is there another way of corroborating the model's results?

Well, ye-ess, is the best answer: there is another way, but it has a few problems of its own. As we saw in Chapters 3 and 4, carbon-isotope ratios can be used to measure changes in oxygen levels. The idea is more or less the reverse side of the coin to measuring carbon burial directly, and the problem has always been to make the opposite sides of the coin correspond to each other.

The isotope method hinges on the fact that life prefers the light carbon isotope, carbon-12. Organic matter therefore becomes enriched in carbon-12. When organic matter is buried, relatively more carbon-12 is buried and more carbon-13 is left behind in the air as carbon dioxide. The carbon dioxide in the air exchanges freely with carbonates in the oceans, as well as those in swamps, lakes and rivers; the whole world is interconnected. In shallow seas, the dissolved carbonates may in turn precipitate as carbonate rocks — marine limestones. Because of the equilibrium between carbonates in the seas and swamps and carbon dioxide in the air, periods of high carbon burial on land leave a strong carbon-13 signature in marine limestones. These signatures can be extrapolated backwards to calculate how much organic carbon must have been buried.[5] The advantage of the isotope method is that it paints a picture of organic burial based on changes in the ocean pool. This gives an idea of the total *global* rate of burial, as, like salt, carbonate is dispersed by the tides and currents at a roughly equal concentration throughout the oceans.

There can be little doubt from carbon-isotope measurements that the rate of burial of organic matter did vary substantially over geological time, with a large peak during the Carboniferous and early Permian. The problem now, as before, is how to constrain the predicted changes in oxygen levels within reasonable limits. The difficulty with estimates based on isotope evidence is the extreme sensitivity of the predicted atmospheric oxygen levels to small changes in the predicted rate of burial. We noted in Chapter 2 that the amount of buried organic matter dwarfs the organic content of the living world, by a factor of 26 000 according to Robert Berner. This means that a very small change in the calculated rate of burial (from isotope ratios), sustained over a period of millions of years, can result in extreme variations in calculated oxygen levels — shifts that are incompatible with life, and so cannot really have happened. Some kind of feedback mechanism must have constrained these variations — the trouble is, we don't know what it is.

After testing numerous possibilities in his atmospheric models, with what must have been a dispiriting lack of success, Berner finally had an ingenious idea that was open to experimental verification. What if life's preference for carbon-12 itself varied with oxygen levels? In other words, what if the degree of carbon-12 enrichment in organic matter varied with the amount of oxygen in the air? When we measure carbon burial, we assume that a fixed proportion was buried as carbon-12. If we then detect more carbon-12 in buried carbon (or more carbon-13 in limestone), we take this to mean that there was an increase in the rate of burial. However, it could also mean that there was a rise in the *proportion* of carbon-12 buried. If this happened, the total amount of buried carbon might remain

[5] A similar argument applies to sulphate-reducing bacteria, which discriminate between the two sulphur isotopes, sulphur-32 and sulphur-34. Burial of iron pyrites resulting from the activities of sulphate-reducing bacteria raises oxygen levels because the organic matter used as a carbon source by the bacteria is not completely oxidized. Strong sulphur-34 signatures in sulphate evaporites, such as gypsum, correspond to a high burial of sulphur-32 in iron pyrites. The studies discussed in the main text do take sulphur isotopes into consideration but a detailed exposition here runs the risk of losing any remaining readers. Berner himself writes clearly and engagingly on the subject, so for those who want more detail, I recommend his primary papers, listed in Further Reading.

constant, but the amount of carbon-12 measured in it could still increase. If we then misguidedly applied the law of fixed proportion, we would have an exaggerated impression of the total amount of carbon that had actually been buried. Any mechanism that makes plants use *more* carbon-12 than normal could have this effect. Correcting for the distortion would decrease the amount of buried carbon estimated by the model and so blunt the predicted variations in oxygen. In other words, the impossibly extreme fluctuations in atmospheric oxygen predicted by the unmodified isotope method could be made more realistic by taking into consideration the selectivity of plants for carbon-12 at different oxygen levels. If plants became more selective for carbon-12 at high oxygen levels, and less selective at low oxygen levels, the predicted fluctuations in atmospheric oxygen would be constrained within more reasonable bounds.

Convinced? Perhaps not, but in theory our old friend photorespiration could have exactly this effect. The carbon dioxide produced by photorespiration is released into the leaf spaces and may then either diffuse out of the leaf, or be taken up a second time by Rubisco and converted back into sugars, proteins and fats. Because the carbon dioxide released by photorespiration is formed from organic matter, it is already enriched in carbon-12. When subjected to a second round of discrimination, the newly made organic matter will be even more enriched in carbon-12. We noted a similar effect with sulphate bacteria in Chapter 4, analogous to re-breathing air in a plastic bag. The rate at which this super-enrichment occurs depends on the rate of photorespiration — and this, as we have seen, increases with rising oxygen levels. In theory, then, high oxygen levels should increase life's preference for carbon-12 and distort our calculations of atmospheric oxygen. The theory seems to fit; but do the numbers add up?

Berner teamed up with specialists from the University of Sheffield in England, including David Beerling, and the University of Hawaii. They chose photosynthesizers from diverse evolutionary groups including angiosperms, cycads and marine single-celled algae, and grew them in the laboratory under different oxygen concentrations. Their findings were published in *Science* in March 2000 and are an amazingly close fit to the wildest dreams of the theorists. All the species they tested responded to increasing oxygen levels by becoming more selective for carbon-12. Plants that had evolved during the Carboniferous or early Permian, however, showed the greatest selectivity. Cycads, for example, left behind an excess of 17.9 parts per thousand carbon-13 in the air at 21 per cent oxygen, but when oxygen levels were raised to 35 per cent, this rose to 21.1 parts per thousand. This is a relative increase in carbon-12 enrichment of 18 per cent. With fewer stomata in their leaves, the plastic-bag effect was enhanced. When this figure is taken into consideration in the isotope calculations, we find a very close match between the two sides of the coin: between the calculations based on carbon burial directly (mass balance), and those based on carbon isotopes. Both independent calculations indicate that oxygen levels reached 35 per cent during the Carboniferous. Berner had finally balanced his books.

These results do not prove conclusively that the air in Carboniferous and early Permian times was rich in oxygen, but they do place the boot firmly on the other foot: it is now up to those who disbelieve that such changes could have taken place to produce strong empirical evidence supporting their case. In the meantime, an expanding party of researchers from other fields have taken the high-oxygen case at face value and started to ask pertinent questions about the likely implications. Many of these implications can be examined directly by a fresh look at the fossil record, or by measurements of physiological performance in high-oxygen atmospheres, such as dragonfly flight. But first, we still have a paradox on our hands: fire. Shouldn't everything burst spontaneously into flames at such high oxygen levels? What about the runaway catastrophe of burning forests that we discussed earlier?

One of the difficulties that scientists face today is the sheer breadth of knowledge. Even within a particular field, such as medical research, it is hard to keep abreast of new research, while spending most working hours at the bench or in the clinic. Researchers typically have a minutely detailed knowledge of their own immediate discipline, for example population genetics, while maintaining enough of a broad understanding to weigh up the importance of developments in related subjects, such as molecular biology. Once we get further afield, however, scientists, like everyone else, just have to accept a lot on trust. Fire gives an idea of just how deeply a concept can become embedded in scientific lore without anybody questioning its experimental basis.

During the 1970s, Lovelock and Watson argued that "above 25 per cent oxygen very little of our present land vegetation could survive the raging conflagration which would destroy tropical rain forest and arctic tundra alike", and that even wet vegetation would have "a significant probability of ignition... a fire could be kindled even during a rainstorm." To back such strong statements, we may be forgiven for assuming that experiments must have been done, and the data filed away long ago as established fact; this, at least, was my assumption, until I became sufficiently troubled by the story to look up the original experiments. The work had indeed been done: Andrew Watson, then a graduate student of Lovelock's, devoted his 1978 PhD thesis to a detailed examination of burning in different oxygen atmospheres. Unfortunately, somewhere along the way his conclusions were stripped of their context.

To control the parameters of his experiments, allowing him to compare like with like, Watson worked mostly with paper strips. These he moistened to varying degrees and ignited. He went on to stage hundreds of small burns under conditions of controlled oxygen and moisture content, and drew curves for the probability of ignition by electrical discharge, the rate of fire spread, and the amount of water needed to extinguish the flames. The results confirmed our intuitive judgement that high oxygen intensifies fire and counteracts the dampening effect of moisture.

There is nothing wrong with these results. The problem is what they left unsaid, as Watson himself acknowledges. Critically, paper is a poor surrogate for the biosphere, as anyone who has attempted to light a wood fire using newspaper knows. As we noted in Chapter 4, paper manufacture involves the removal of most of the lignin, and this increases its flammability. Lignin does not burn well, but instead tends to smoulder gently. Trees with a high lignin content in their bark are relatively fire resistant. Nor can paper retain water by osmosis as living cells do. The moisture content of comparably thin plant tissues, such as leaves, is therefore much higher than paper. While Watson measured the flammability of paper up to a maximum of 80 per cent moisture saturation, some leaves approach an equivalent of 300 per cent saturation. Plants at high risk of fire may also accumulate high levels of fire retardants, such as silica. Some straws, for example, have an unusually high content of silica, which can make it hard to burn agricultural waste. Housewives apparently discovered this trick long ago: silicate paints were commonly applied to curtains during the Second World War to retard the spread of fire in the aftermath of bombing.

Perhaps the most surprising feature of all this is that we simply do not know the extent to which high atmospheric oxygen affects the spread of fire in real ecosystems. I understand that trash cans stuffed with wet organic matter are being detonated in high-oxygen atmospheres as I write — science marches ever on — but from work published to date we can come to no firm conclusions about whether fire really would have posed an insurmountable problem in the hypothetical Carboniferous atmosphere. Given the devastation caused by forest fires today, it is hard to imagine that so much extra oxygen would not have risked global conflagration, but we must bear in mind two factors. First, most fires today are ignited, either accidentally or intentionally, by people. There would be far fewer fires today if they were ignited only by lightning. If, in the past, the risk of fire was greater, this extra risk might have been counterbalanced by a much lower rate of ignition: fires may have been no more common then than they are today. Second, plants have an extraordinary capacity to adapt to regular devastation by fire.

Our knowledge of the adaptations of modern plants to fire allows us to scrutinize the fossil record for evidence of similar adaptations in Carboniferous or early Permian times. These issues have been examined in an illuminating and unsurpassed 1989 review by Jennifer Robinson, working at the time at Pennsylvania State University. She argued that if oxygen levels were high during the Carboniferous, we should expect to find adaptations to fire in fossil plants. Conversely, failing to find them would be a good case against a high-oxygen atmosphere. Going one step further, Robinson argued that, while adaptations to fire do not prove the case for high oxygen, a stronger case could be made if even swamp plants adapted to fire in the Carboniferous. This would indeed be curious. Today, most swamp plants have no need to adapt to fire, because the fire risk in waterlogged environments at present oxygen levels is virtually zero.

In her survey of Carboniferous swamp plants, Robinson came to the tentative conclusion that they really had adapted to fire. I say tentative, because there are some difficulties of interpretation. Succulent leaves, for example, might retard the spread of fire, but might also be an adaptation to watery surroundings, or at least an expression of plenty. Deep tubers (along the lines of potatoes) may store enough energy to fuel regeneration of the plant after destruction by fire, but may alternatively be forced on the plant by the depth of the bog. Morphological adaptations are even harder to interpret in plants that have since fallen extinct. Notwithstanding these caveats, however, the fossil record is consistent with the idea of fire resistance. Most large plants at the time had deep tubers, thick bark with a high lignin content, succulent leaves and branches high above the ground, out of reach of any fire that might have swept through the brushwood. And there were few hanging vines or fronds that would enable ground fire to move into the canopy.

The appearance of the giant lycopods, the dominant trees of the Carboniferous swamps, is reminiscent of palm trees, although they are not related. The beautiful geometric patterns of their thick, lignin-impregnated bark preserve well and are commemorated in some of the ornamental columns at the Natural History Museum in London. Whether or not the giant lycopods were specifically adapted to fire, they would certainly have been hard to burn. Smaller survivors from the era, such as ferns and the horsetail *Equisetum*, are less obviously fire-resistant, but can also be hard to burn as they contain high levels of fire retardant. In a wry aside, Robinson notes that "modern *Equisetum* is almost unburnable (personal observation), perhaps due to high silica content." I can't resist the image of Robinson as a petulant pyromaniac, stamping her foot in frustration as the horsetails fail to set alight. True science is born from this kind of passion.

Other aspects of the swamp environment also suggest periodic ravaging by fire, in particular the abundance and properties of fossil charcoal. Some coals contain over 15 per cent fossil charcoal by volume — an extraordinary amount if we consider that coal beds are formed in swamps, which under modern conditions virtually never catch fire. The closest modern equivalents to Carboniferous swamps, the swamps of Indonesia and Malaysia, are almost charcoal-free. The discrepancy led many scientists to question whether fossil charcoal was perhaps an impostor — another type of coal not formed by charring at all, but which just happened to look similar. Finally, however, Given, Binder and Hill demonstrated in 1966 that the questionable charcoal really had been exposed to temperatures of several hundred degrees: it really was charcoal and not some form of compressed, unburned coal. Today, few geologists dispute that wildfires once burned frequently in swamp environments. What is not agreed is why. It may have been a result of high oxygen, making fire in waterlogged surroundings more likely, or it may reflect no more than the local climate, and the frequency at which the swamps dried out.

Re-examining the fossil charcoal record from the perspective of variable oxygen throws new light on the conundrum. Coals that formed during periods of hypothetically high oxygen, such as the Carboniferous and Cretaceous, contain more than twice as much charcoal as the coals that formed during low-oxygen periods like the Eocene (54 to 38 million years ago). This implies that fires raged more frequently in times of high oxygen and were not related to climate alone. Some of the properties of the charcoal support this interpretation. The shininess of charcoal depends on the temperature at which it was baked. Charcoals formed at temperatures above about 400°C are shinier than those that cooked at lower temperatures, and so reflect back more of the light directed at them. The difference can be detected with great accuracy using a technique known as reflectance spectroscopy. The shininess of fossil charcoals from both the Carboniferous and Cretaceous implies that they formed at searing temperatures, almost certainly above 400°C and perhaps as high as 600°C, in fires of exceptional intensity. The temperature at which a fire burns, of course, depends on many factors, including the type of vegetation (modern conifers burn at much higher temperatures than deciduous trees), the thermal conductivity of the wood and the height of the water table; but an important factor is the level of oxygen. The simplest explanation for the twin peaks of charcoal reflectance in the Carboniferous and the Cretaceous is that oxygen levels were highest then.

The fireworks that brought the Cretaceous to an abrupt end support the idea of high oxygen levels. A catastrophic firestorm may have accompanied the extinction of the dinosaurs. One piece of evidence supporting the theory that a giant meteorite hit the earth 65 million years ago is a thin layer of rock rich in iridium that delineates the boundary between the Cretaceous and the Tertiary — the so-called K–T (Cretaceous–Tertiary) boundary. This thin band of iridium-rich rock has now been found at more than 100 sites worldwide. Iridium is rare on Earth, much rarer than gold, but is relatively common in meteorites.[6] In fact, the 2:1 ratio of iridium to gold in samples from the K–T boundary closely matches that found in meteorites. The presence of iridium in a thin band all around the world suggests that the meteorite shattered on impact, throwing fine dust high into the stratosphere, which later settled out to form the iridium layer.

[6] A competing theory argues that the iridium is derived from the giant volcanic eruptions of the Deccan traps in India. Either event might have produced a catastrophic fire, though it is hard to see how the Deccan traps theory could account for the evidence of extraterrestrial impact on the Yucatan peninsula in Mexico (the shallow seas retreated, leaving evidence of a crater filled in with sediments), or the proposed 'megawave' tsunami that followed.

In 1988, Wendy Wolbach, then working on her PhD thesis, and her colleagues at the University of Chicago, presented evidence in a paper to *Nature* that the iridium was mixed with soot at 12 sites from the United States, Europe, North Africa and New Zealand. She argued, on the basis of the isotopic uniformity of the carbon, that the soot had been deposited by a single global fire in the wake of the impact itself. Simple calculations suggested that about 25 per cent of terrestrial biomass of the planet had perished in the flames. The headlines at the time were colourful, if predictable, along the lines of 'Dinosaurs Barbecued in Giant Fireball!'

Wolbach's work has found support more recently in other evidence for a great conflagration. In 1994, Michael Kruge and his colleagues at the University of Southern Illinois described a belt of fossil charcoal 3 metres [10 feet] thick in Mimbral in northern Mexico, and argued, from its curious cocktail of terrestrial and marine sediments, that terrestrial plants must have charred in a firestorm (ignited by the passage of the meteorite through the skies), then drowned in the deep sea by the backwash from a giant tsunami, or megawave, caused by the impact of the meteorite in shallow tropical seas. While their interpretation has been queried, the evidence for an exceptional fire at the time is hard to ignore.

If indeed there was such a fire, then a high-oxygen atmosphere may have sealed the fate of the dinosaurs. This is believable, if only because other large meteorites have hit the Earth without causing mass extinctions. For example, a major impact formed the Ries crater in Germany 15 million years ago. The impact threw huge boulders more than 95 kilometres [60 miles] into Switzerland and the Czech Republic, and droplets of molten rock over several hundred miles, but not even the local mammal populations were affected. The Montagnais and Chesapeake Bay impacts formed craters 45 and 90 kilometres [28 and 56 miles] in diameter respectively, but neither brought about a mass extinction. It seems possible that the added zest needed for an explosive doom is a little extra oxygen.

We now have two independent empirical models — the mass balance and isotope models — which concur that oxygen levels reached 35 per cent during the Carboniferous and early Permian. Also, plants that evolved at the time are resistant to high levels of oxygen; their productivity is barely affected by photorespiration in high-oxygen atmospheres. Fire presents a serious risk in these conditions, but by no means rules out vegetative cover even in dry conditions. Jennifer Robinson, in another of her vivid personal observations, suggests that a modern analogy might be the thinned and locally denuded cover found on bombing ranges in seasonally dry climates. Swamps protect against fire, but even swamp plants of the time show morphological adaptations to fire, including thick, lignin-rich bark, deep tubers and high crowns. Among the survivors from the Carboniferous, some ferns and horsetails have a high content of fire-retardants such as silicate. The environment shows signs of regular wildfires in the form of abundant fossil charcoal, and this charcoal probably formed at the searing temperatures characteristic of high oxygen. There is some evidence that the Cretaceous ended with a catastrophic firestorm. Altogether the case is sufficiently strong to have convinced some scientists to return to the old bugbear of insect gigantism. Was that, too, related to high oxygen levels?

I cited the Dutch geologist M. G. Rutten at the beginning of this chapter. He argued that the primitive means by which insects breathe might limit their size and flight performance. Insects take in air by way of fine tubes or trachea that open directly to the air through pores in the external skeleton and then branch to penetrate every cell in the insect's body. The idea is that the size of flying insects is restricted by the need for oxygen to diffuse through the tracheal system. Any increase in insect size means that oxygen must diffuse over greater distances through the tracheal system, and so makes flight less efficient. The effective upper limit to passive diffusion down a blind-ending tube (at modern atmospheric levels of oxygen) is about 5 millimetres [1/5 inch]. According to Robert Dudley, a physiologist at the University of Texas, an increase in the oxygen content of the air to 35 per cent would increase the rate of oxygen diffusion by approximately 67 per cent, enabling it to diffuse over longer distances. In other words, air that contains more oxygen allows the minimum amount needed for respiration to reach further into the insect's trachea. This would improve the oxygenation of flight muscles, allowing thicker constructions and permitting insects to grow larger. While other selective pressures, such as predation, probably drive the actual tendency to get bigger, higher oxygen levels raise the physical barrier to greater size.

So far so good, but there is one problem with this line of reasoning: the tracheal system may be primitive, but it is far from inefficient — with it, flying insects achieve the highest metabolic rates in the whole of the animal kingdom. Almost without exception, insect flight is totally aerobic, which means that their energy production is dependent entirely on oxygen. In spite of our well-ventilated lungs, powerful hearts, elaborate circulatory systems and red blood cells packed with the oxygen-carrier haemoglobin, we are less efficient. Sprinters cannot breathe in enough oxygen to power their efforts and their muscle cells must instead resort to the less efficient process of energy production by anaerobic glucose breakdown, or glycolysis, which produces a mild poison, lactic acid, as a by-product. The longer we persist in violent exercise, the more lactic acid builds up, until finally we are left half paralysed, even if we are running for our lives. Heavy-legged exhaustion is the product of a respiratory failure that does not trouble insects. If you ever thought that a housefly never grows tired of buzzing, you were probably right: unfortunately for us, it does not poison itself with lactic acid.

The limits of insect flight are not at all easy to define. In a handful of rather quirky experiments dating back as far as the 1940s, experimenters have tried tethering insects, attaching tiny weights, cutting oxygen levels to a fraction of normal air, and replacing nitrogen with light-weight helium mixtures. All went to show the surprisingly wide safety margins of insect flight. Some insects are even able to fly in low-density helium mixtures with an oxygen content of just 5 per cent. In most experiments, insects gained no apparent benefit if oxygen levels were increased to 35 per cent. The broad conclusion was that insect flight is not limited by tracheal diffusion, so oxygen cannot act as a spur to greater size. This is still the opinion of many entomologists, but the tide is beginning to turn.

The reason the tracheal system is so efficient is that oxygen remains in the gas phase, where it can diffuse rapidly, and need not pass into solution until the last possible moment, as it enters the flight muscle cells themselves. As a result, the ability of the tracheal system to deliver oxygen typically exceeds the capacity of the tissues to consume it. The only real inefficiency is the blind endings of the trachea, which branch into fine tubules in much the same way as the blind bronchioles in our own lungs. Just as we suffocate if we cannot physically draw breath, so too insects are limited by the diffusion of gases in the blind alleys of the tracheal system. Most insects get around this difficulty, as we do ourselves, by actively ventilating their trachea.

For insects, there are two ways of ventilating the trachea, known as abdominal pumping and autoconvective ventilation. Most 'modern' insects, including wasps, honeybees and houseflies, rely on abdominal pumping, in which the insects contract their abdomens rhythmically to squeeze air through the tracheal network. The rate of pumping changes in response to the amount of oxygen available. If honeybees are placed in low-oxygen air, for example, their metabolic rate remains constant — they continue to get through the same amount of oxygen as they fly — but the rate of water loss by evaporation may increase by as much as 40 per cent, implying that the bees compensate for the low oxygen by pumping their abdomens more vigorously, thereby increasing the rate of tracheal ventilation, and so evaporation. The efficiency of this process allows most insects to keep an even keel in changeable conditions.

Dragonflies, locusts and some beetles rely on the second, more primitive, means of ventilation — autoconvective ventilation. This is a splendidly opaque way of saying that they create draughts when they flap their wings. Insects that depend on autoventilation can increase airflow in their trachea by increasing the frequency or amplitude of wing beats — they flap their wings harder. There is of course a catch here: beating wings demands energy, and the harder they beat the more energy is needed. Abdominal pumping requires little energy in comparison. As energy production requires oxygen, and the availability of oxygen can only be increased by beating, which consumes the extra oxygen, dragonflies and other autoventilating insects may be uniquely susceptible to fluxes in oxygen levels.

In principle, a rise in oxygen levels should enable dragonflies to beat their wings less actively to achieve the same flight performance; or, for a constant rate of beating, the body size might be increased. In a detailed study published in the *Journal of Experimental Biology* in 1998, Jon Harrison of Arizona State University and John Lighton of the University of Utah put these ideas to the test, and finally produced solid evidence that dragonfly flight metabolism is sensitive to oxygen. They measured carbon dioxide production, oxygen consumption and the thoracic temperature of free-flying dragonflies kept in sealed respiratory chambers. Raising the oxygen content from 21 to 30 or even 50 per cent increased the metabolic rate. This means that, in today's atmosphere, dragonfly flight is limited by oxygen insufficiency. If dragonflies can fly better in high-oxygen air, then presumably larger dragonflies, which could not generate enough lift to become airborne at all in today's thin air, would have been able to fly in the postulated oxygen-rich mix of the Carboniferous.[7] It seems that the Bolsover dragonfly really was only able to fly, and so hunt its prey and survive, in an oxygen-rich atmosphere.

Dragonflies were not the only giants of the Carboniferous — many other creatures attained sizes never matched again. Some mayflies had wing-spans of nearly half a metre [19 inches], millipedes stretched for over a metre [39 inches], and the *Megaranea* spider, with a leg-span of nearly half a metre [18 inches], would have chilled the marrow of Indiana Jones. Even more terrifyingly, scorpions reached lengths of a metre, dwarfing their modern cousins, the largest of which barely manages a fifth of that length. Among the terrestrial vertebrates, amphibians grew from newt-like proportions to reach body lengths of 5 metres [16 feet]. They left some of the oldest footprints in England, at Howick in Northumberland — 18 centimetres [7 inches] long and 14 centimetres [5½ inches] across. In the plant world, ferns turned into trees, while the giant lycopods reached heights of nearly 50 metres [164 feet]. Their only survivors today are the diminutive herbaceous club-mosses, such as the ground pine (*Lycopodium obscurum*), which rarely grow higher than 30 centimetres [12 inches].

Was all this rampant gigantism related to oxygen? It is certainly possible. Like the dragonfly, each of these organisms depends on the passive diffusion of gases in one way or another. The size of amphibians, for instance, is restricted by their capacity to absorb oxygen by diffusion across the skin, whereas the height of plants depends on the thickness of their structural support, which in turn is limited by the need for gases to reach internal tissues. But while it is plausible that high oxygen levels might enable greater size, it is hard to back the claim with direct evolutionary evidence. There is one tantalizing suggestion from modern ecosystems, however, that this is indeed the case.

Tucked away in the 'Scientific Correspondence' section of *Nature* in May 1999 was a short paper on the size of crustaceans — the class that includes shrimps, crabs and lobsters — in polar regions. This paper solved a long-standing riddle rather neatly: the relationship between gigantism and oxygen availability. The authors, Gauthier Chapelle of the Royal

[7] Because the extra oxygen was added to the existing atmospheric gases, the overall density of the air would have increased. Higher density aids lift (by increasing the Reynold's number) and would have encouraged the evolution of flight. In a series of fascinating papers, Robert Dudley has connected the origins of flight in insects, birds and bats with changes in atmospheric density.

Institute of Natural Sciences in Belgium, and Lloyd Peck of the British Antarctic Survey, examined length data for nearly 2000 species of crustaceans from polar to tropical latitudes and from marine to freshwater environments. They focused on a single group, known as amphipods, which are cold-blooded, shrimp-like creatures, ranging in length from a couple of millimetres [1/25 inch] to about 9 centimetres [3½ inches]. The amphipods are not exclusively marine, and are best known to most of us as sand-hoppers, or the shiny brown animals that leap about when pot plants are moved in the garden.

The thousands of marine species of amphipod are a cornerstone of polar food chains, being the staple diet of juvenile cod, which are in turn preyed on by seals, and the seals by polar bears. In some bottom sediments, amphipods are found at an extraordinary density of 40 000 per square metre [4000 per square foot]. These tiny creatures offer even more of a square meal in polar waters: the largest Antarctic species are some five times larger than their tropical cousins — true giants by amphipod standards. In this respect, amphipods are not alone. For the past hundred years or so, scientists have catalogued numerous giant species in polar seas. Although polar gigantism is usually ascribed to the low temperatures and the reduced metabolic rates of cold-blooded animals, the relationship is not straightforward. Surprisingly, polar gigantism had never been satisfactorily explained. The trouble is that the inverse correlation between size and temperature is curved rather than linear, and has a number of puzzling exceptions. In particular, many species achieve far greater sizes in freshwater environments than they ought to on the basis of temperature alone. Freshwater amphipods from Lake Baikal in Russia, for example, are twice as large as those in the sea at the same temperature.

Then Chapelle and Peck had a clever idea and applied it to their amphipod data. What if the true correlation was not with water temperature at all, but with the dissolved oxygen concentration? Oxygen dissolves better in colder water and is nearly twice as soluble in polar seas than in tropical waters. The salt content also affects the solubility of oxygen, which dissolves 25 per cent better in fresh water than in saline. The highest oxygen saturation is therefore in large freshwater lakes verging on the Arctic tundra, such as Lake Baikal — and this is where the largest crustaceans are to be found. When Chapelle and Peck re-plotted their length data against the oxygen saturation of the water, they got a nearly perfect fit (Figure 6). While it is true that a correlation says nothing about mechanism, it seems likely that inadequate oxygen availability

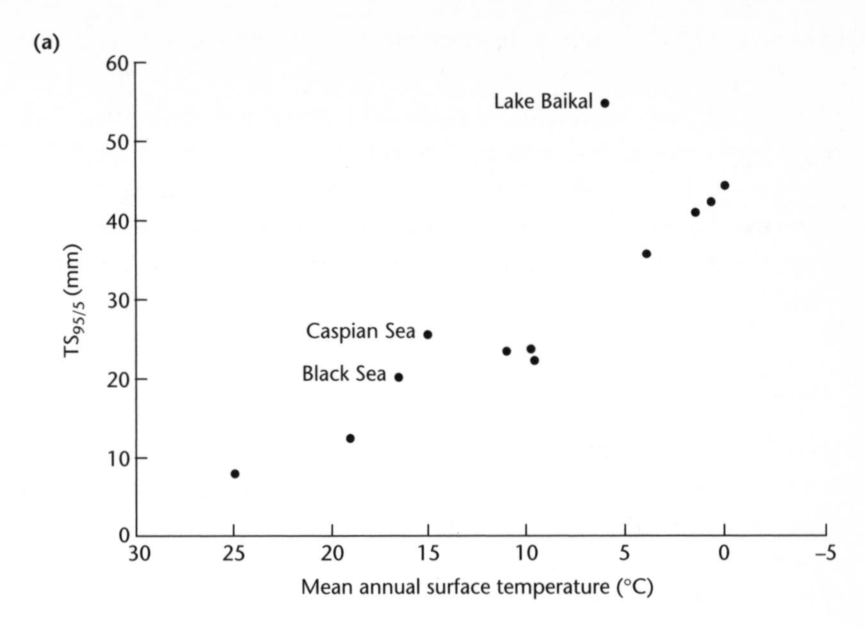

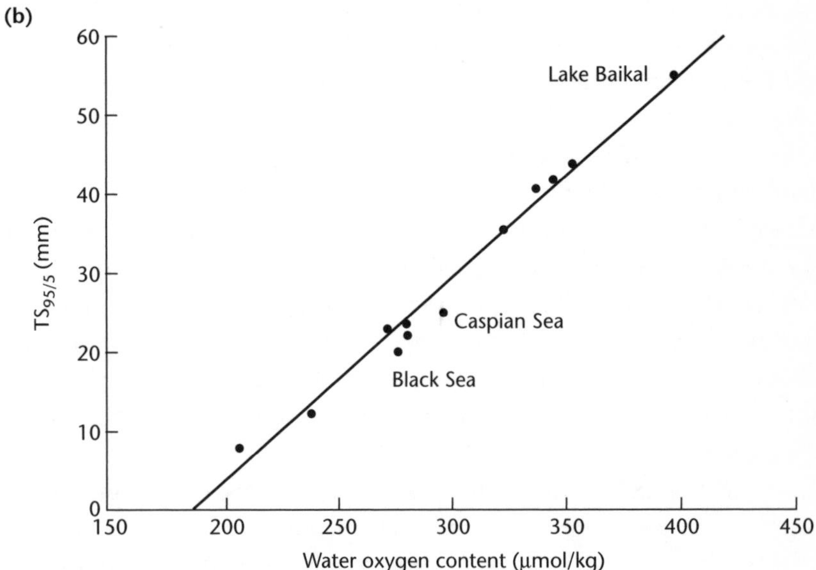

Figure 6: Correlation between body length of amphipods in millimetres (given as an index of average length, $TS_{95/5}$) and **(a)** temperature and **(b)** oxygen concentration. Lake Baikal, the Caspian Sea and the Black Sea are outliers from the temperature curve because they are brackish or freshwater, rather than saline. Oxygen dissolves more readily in freshwater, so the correlation is restored by plotting the length of amphipods against the dissolved oxygen concentration. Reproduced with permission from Chapelle & Peck, and *Nature*.

limits size in many species, or conversely, that high oxygen raises the barrier to gigantism.

Of course the dependence of giants on high oxygen means that they are perilously susceptible to falling oxygen levels. In a stark closing line, Chapelle and Peck predict that giant amphipods will be among the first species to disappear if global temperatures rise, or if oxygen levels decline. We can hardly begin to imagine what effect this might have on the rest of the food chain.

The case for changeable oxygen levels in the atmosphere is not easy to dismiss. This conclusion stands in opposition to Lovelock's Gaia theory, which argues that the living biosphere has regulated the levels of oxygen in the atmosphere over the past 500 million years. While this may be true for much of the time, there have surely been periods when the biosphere lost control over oxygen levels. If anything, the possibility that Gaia cannot always maintain the physiological balance that Lovelock ascribes to her only strengthens his concerns about our impact on the planet. Given the unequivocal evidence of global glaciation punctuating the Earth's history, it is plain that Gaia does not have a tight control over temperature. Something similar seems to be true of oxygen. We have a tenuous grasp of the factors that control oxygen or carbon dioxide levels, but the fact that there have been times when they have tipped out of balance means that it can happen again, perhaps with our help. The feedback mechanisms postulated by Lovelock and others have the power to resist change for a period. If we are to judge from the case of oxygen, they do not have indefinite flexibility and cannot resist catastrophic change. We should beware.

With the exception of fire, there is remarkably little evidence that high oxygen levels are in any way detrimental to life. On the contrary, high oxygen may have opened evolutionary doors that are closed to us today. Falling oxygen closes these doors, and the species left outside are unlikely to survive. Most of the giants of the Carboniferous, for example, failed to survive until the end of the Permian period, when Robert Berner's calculations suggest that oxygen levels plummeted to 15 per cent as the climate became cooler and drier.

We must conclude that high oxygen is good, low oxygen is bad. But we saw in Chapter 1 that high levels of oxygen are toxic, causing lung damage, convulsions, coma and death, and we are told that oxygen free radicals are at the root of ageing and disease. What is going on: is oxygen toxic or not? This paradox did not escape the notice of Barry Halliwell and John Gutteridge, authors of *Free Radicals in Biology and Medicine*, the standard text on the subject, who remarked laconically that "the plants and animals existing in the Carboniferous times must presumably have had enhanced antioxidant defences, which would be fascinating to study if these species could ever be resurrected." Yes indeed. How did they overcome oxygen toxicity? Might there be some way that we can imitate them to protect ourselves against free radicals in disease? It is time to look in a little more detail at the strange spectre of oxygen toxicity, and what life does about it.

CHAPTER SIX

Treachery in the Air

Oxygen Poisoning and X-Irradiation: A Mechanism in Common

IN 1891 A SHY 24-YEAR OLD POLISH GIRL named Marya Salomee Skłodowska, arrived in Paris with the dream of becoming a scientist. In the chauvinist academic culture of France at the time, such a dream stood little chance of fulfilment, but Marya possessed a brilliant mind and stubborn determination. Nor was she a stranger to hardship. The Poland of her youth was occupied by the Russians, and her mother had died when she was just four. The youngest of five children, she had been raised in poverty by an idealistic father, and educated in the Flying University, which met each week in different locations to avoid detection by the Russians — the Poles resisted oppression through education, and Polish culture flowered in underground centres of learning. Not surprisingly, the passion for learning that swept her homeland left an indelible mark on Marya.

When she was 18, Marya made a pact with her sister Bronya. Bronya would pursue her own dream of studying medicine in Paris, and Marya would support her by working as a private tutor in Warsaw; then Bronya would return the favour. Marya duly worked as a governess for six years, while continuing her underground studies in chemistry and mathematics and suffering an unhappy love affair. In the meantime, Bronya completed medical school and married a fellow medical student. So it was that Marya arrived as a mature student in Paris, changed her name to the French version, Marie, and enrolled at the Sorbonne. She passed a master's degree in physics with flying colours in 1893, and a second in mathematics in 1894. Then, while seeking extra laboratory space to conduct more elaborate experiments, she was introduced to an equally brilliant, introverted and free-thinking Frenchman, who had already made a reputation for his work on crystallography and magnetism. They quickly fell in love, and he wrote to her saying how nice it would be "to spend life side by side, in the sway of our dreams: your patriotic dream, our humanitarian dream and our scientific dream." Marie and Pierre married in 1895, going on a cycle tour around France for their honeymoon, and when Marie found fame as a scientist it was under her new name, Marie Curie.

In the next two years, Pierre gained a teaching position at a science college while Marie studied for a teaching certificate. In 1897 their first child, Irène, was born, and that same year Marie began work on her doctoral studies — another pioneering step for a woman at the time; she was to become the first woman in Europe to receive a doctorate in science.

Although both Marie and Pierre had been interested primarily in magnetism until then (and the field still pays homage to their name in the 'Curie point', the temperature at which materials lose their magnetism), the Curies had become close friends with another brilliant young French scientist, Henri Becquerel. Inheriting the large phosphorescent mineral collection of his scientist father, Becquerel had just discovered that if crystals of uranium sulphate were exposed to sunlight, then placed on photographic plates and wrapped in paper, an image of the crystals would form when the plates were developed. At first he assumed that the rays emitted by the crystals were a type of fluorescence derived from the sunlight, but this theory was confounded by the overcast skies of Paris that February. Becquerel returned his equipment to a drawer and waited for better weather, but after a few gloomy days he decided to develop his plates anyway, anticipating no more than faint images. To his surprise, the images turned out to be clear and strong, and Becquerel realized that the crystals must have emitted rays even without an external source of energy such as sunlight. He soon showed that the rays came from the small amounts of uranium in the crystals and that all substances containing uranium gave off similar rays. He even found that uranium causes the air around it to conduct electricity. His excitement transmitted to the Curies, and Marie decided to study the strange phenomenon, which she later termed radioactivity, for her doctorate.

Marie set to work on a uranium ore known as pitchblende. She and Pierre had realized that radioactivity could be measured by the strength of the electric field that it generates in the surrounding air, and Pierre invented an instrument that could detect the electric charge around mineral samples. Using this instrument, Marie discovered that the radioactivity of pitchblende was three times greater than that of uranium, and concluded that there must have been at least one unknown substance in pitchblende, with a much higher activity than uranium. By chemically separating the elements from uranium ore and measuring their radioactivity, the Curies discovered a new element that was 400 times more radioactive than uranium, which they named polonium after her native Poland. Later, Marie discovered tiny quantities of another element, this time a million times more radioactive than uranium, and called it radium. Pierre tested a tiny piece of radium on his skin, and found it caused a burn, which developed into a wound. The Curies recognised its potential as an anti-cancer treatment. Radium was first used for this purpose by S. W. Goldberg in St Petersburg, as early as 1903. Radium needles are still inserted into tumours as a cancer therapy today.

To study the properties of radium in detail, the Curies needed to isolate more, and to do this entailed working with tonnes of pitchblende to isolate just a few milligrams of radium. Radium is present in such small quantities that, even today, world production amounts to only a few hundred grams. The Curies worked in what must have been appalling conditions. Their lab was described by a contemporary chemist as looking more like a stable or a potato cellar. Refusing to patent radium, for humanitarian reasons, the Curies continued to struggle on. Despite their financial hardship and poor conditions, they took great pleasure in their work, especially at night, when they could see all around them "the luminous silhouettes of the beakers and capsules that contained our products."

For their work on natural radioactivity, the Curies and Becquerel received the Nobel Prize for physics in 1903. The year after that, Marie and Pierre had a second daughter, Eve. It must have been the best time of their lives. In 1906, Pierre, weakened by radiation, was killed in a road accident, his head crushed beneath the wheel of a horse-drawn cart. Traumatized, Marie began writing to him in a diary, which she kept for many years, but her scientific resolve did not falter, and she determined to complete alone the work they had undertaken together. She struggled against the French establishment for recognition, finally taking up her husband's old position at the Sorbonne in 1908 — the first woman in its

650-year history to be appointed professor there. In 1911 she received a second Nobel Prize, for the isolation of pure radium, and in 1914 she founded the Radium Institute, now renamed the Curie Institute, with its humanitarian goal of easing human suffering. Throughout the First World War she trained nurses to detect shrapnel and bullets lodged in wounds, using mobile x-ray vans, and after the war, with her daughter Irène alongside her, she pioneered the use of radium to treat cancer patients. Irène herself, with her husband Frédéric Joliot, went on to receive a Nobel Prize for the discovery of artificial radioactivity in 1935.

Marie did not live to see her daughter's Nobel laurels. She died of leukaemia on 4 July 1934 at the age of 67, exhausted and almost blinded by cataracts, her fingers burnt and stigmatized by her beloved radium. She had not been the first to die of radiation poisoning, nor was she the last. During the 1920s, several workers at the Radium Institute had died of a cancer that other doctors attributed to radioactivity. Not believing the truth, Marie put it down to a lack of fresh air. Later, her daughter Irène also died of leukaemia.

Today, with the experience of Hiroshima and Chernobyl behind us, radiation is not seen in quite the same humanitarian light. High doses of radiation kill cancer cells, but also kill normal cells. Within weeks of the discovery of x-rays there had been reports of tissue damage among researchers who worked for many hours a day with x-ray-producing discharge tubes. Many of them lost their hair and developed skin irritations, which sometimes festered into severe burns. Lower doses of radiation were found to *increase* the risk of cancer. The signs were there even in Marie Curie's time. Forty per cent of the early researchers in radioactivity died of cancer. They were joined by others who worked with radioactive materials. In 1929, doctors in Germany and Czechoslovakia noticed that 50 per cent of the miners working in Europe's only uranium mine, in Bohemia in northern Czechoslovakia, had lung cancer, which was attributed to their inhalation of radon gas, a radioactive decay product of uranium (via radium), escaping from the ore. The incidence of lung cancer among uranium miners in the United States was also much higher than normal.

An awful fate befell many of the young women hired to paint radium onto the dials of watches, so that they would glow in the dark. The original luminous watches had been designed for soldiers fighting in the trenches during the First World War, but their novelty stimulated a consumer fad in the 1920s. To point the tips of their paint brushes, the girls were taught to moisten the bristles with their lips. At the time, radium was still hailed as a panacea and was sold for a variety of medical purposes, as elixirs, snake oils and aphrodisiacs. The girls were told that radium would put a glow in their cheeks and give them a smile that shone in the dark, and they would sometimes paint their nails, lips and teeth. Within a year their teeth began to fall out and their jaws disintegrated. When they began to sicken and die in large numbers, doctors found that their bodies, even their bones, contained large amounts of radon and other radioactive substances. Not surprisingly, the watch companies rejected the link and government regulators concluded that existing evidence did not warrant further investigation. An editorial in the New York World called a trial in 1926 "one of the most damnable travesties of justice that has ever come to our attention."

Although the watch companies eventually agreed to pay token financial compensation, they never admitted their guilt or submitted to formal regulation. One worker, Catherine Wolfe Donahue, sued the Radium Dial company in 1938. She testified in a Chicago courtroom that she and a co-worker had once asked their supervisor, Rufus Reed, why the company had not posted the results of the physical examinations that had been carried out during the 1920s. Reed had apparently responded: "My dear girls, if we were to give a medical report to you girls, there would be a riot in the place." The medical community finally instituted a dose limit for radon in 1941, but the confusion and vested interests had concealed the delayed effects of radiation, and few people, even within the Manhattan Project that built the first atomic bomb, predicted the full horror of nuclear fallout.

Nuclear fallout is the settling of unconsumed radioactive waste left over from the atomic blast, and can be dangerous for a long time. The explosion causes intense firestorms and whirlwinds stretching high into the air, and the atmospheric disruption often provokes rain. After both Nagasaki and Hiroshima, the air was so full of radioactive ash that the rain was dark and tarry — the infamous 'black rain'. In Hiroshima, the black rain fell over a wide area that stretched from the centre of the town to the surrounding countryside, polluting water and grass alike. Fish died in the rivers, cows died in the fields.

Tens of thousands of survivors of Hiroshima and Nagasaki, who were uninjured by the initial blast, found they had not escaped the bomb after all. Within days, their hair began to fall out and their gums began to bleed. The victims suffered from bouts of extreme fatigue and excruciating headaches. They were weakened by nausea, vomiting, anorexia and diarrhoea. Painful sores filled their throats and mouths. They bled from the mouth, nose and anus. Those with the most acute symptoms died in a few months. Others were blinded by cataracts in the space of two years. Many died from cancer years, or even decades, later. Leukaemia is the cancer most commonly linked with radiation poisoning. The characteristic 'blue stigmata' of radiation victims are a sign of leukaemia. The stigmata are formed by clumps of proliferating white blood cells. For 30 years after the nuclear bomb, the number of cases of leukaemia in Hiroshima remained 15 times higher than the rest of Japan. The incidence of other cancers with longer incubation periods, such as lung, breast and thyroid cancer, all began to rise after about 15 years.

As the threat of nuclear war has receded, safety concerns about nuclear power plants and other potential sources of radiation have sharpened in focus. Confidence in reactor safety was undermined by two serious accidents, one in 1979 at the Three Mile Island nuclear power plant in Pennsylvania and the other in 1986 at Chernobyl in the Ukraine. Chernobyl was the worst reactor accident in history, with 31 people dying of direct radiation poisoning, and thousands more exposed to high doses of radiation. Even without accidents, fears of leakage and contamination are increasing. In England, reprocessing of nuclear waste at Sellafield has raised legitimate concerns about the high incidence of leukaemia in nearby villages. Other groups exposed to above-normal levels of radiation are also at higher risk of leukaemia. The so-called Balkan War syndrome (a form of leukaemia) among troops who had been stationed in Kosovo, and among potentially thousands of local people, is attributed to the use of armour-piercing depleted-uranium shells. Even commercial aircrews may have a relatively high risk of leukaemia, as they are subjected to higher levels of cosmic radiation at flight altitudes.

With such a history, it is not surprising that even medical x-rays and radiotherapy generate fears, sometimes hysteria, about radiation poisoning. No nuclear power stations have been built in the United States since the late 1970s. The existence of a 'safe' radiation dose has been debated for decades without consensus. As one expert puts it, the most practical approach is to limit human exposure to ionizing radiation and hope for the best.

What has all this to do with oxygen, you may be wondering? The answer is that radiation exerts its biological effects through a mechanism that is very similar to the effects of oxygen poisoning. The mechanism hinges on an invisible thread of reactions, linking oxygen to water. The lethal effects of radiation and oxygen poisoning are both mediated by exactly the same fleeting intermediates along this pathway. These intermediates can be produced from either oxygen or water (Figure 7). In radiation poisoning, they are produced from water, in oxygen poisoning from oxygen. However, normal respiration also produces the same reactive intermediates from oxygen. Respiration can therefore be seen as a very slow form of oxygen poisoning. We shall see that both ageing and the diseases of old age are caused essentially by slow oxygen poisoning.

The fleeting intermediates produced by radiation and respiration are called *free radicals*. We discussed them briefly in Chapter 1. Later in the book, we will refer to free radicals many times. I use the term rather loosely for convenience. Not all of these fleeting intermediates are free radicals within the usual definition of the term. Applying the correct terminology, however, is cumbersome. Another umbrella term, 'reactive oxygen species', is even more cumbersome and also untrue — not all are especially reactive and some, such as nitric oxide (NO) are technically reactive nitrogen species. A third possible term, oxidants, is also incorrect:

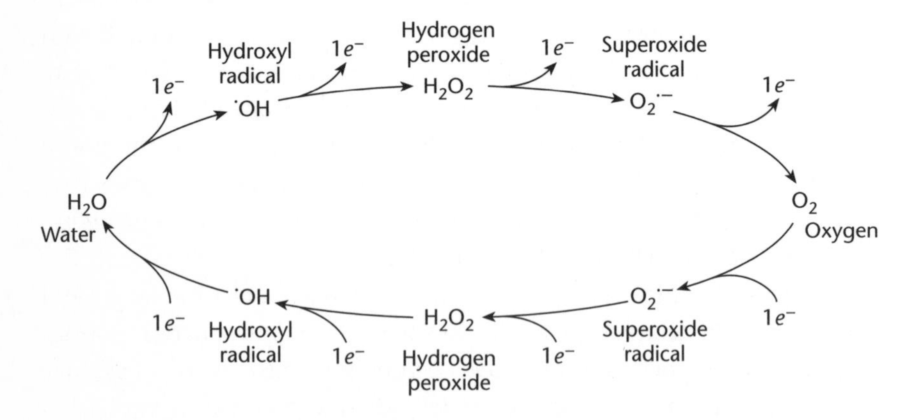

Figure 7: Schematic representation of the intermediates between water and oxygen. Only changes in the number of electrons (e^-) are shown in each direction. The reactions also depend on the availability of protons, although this is not shown for simplicity. Because protons are positively charged, electron rearrangements tend to lead to compensatory proton rearrangements.

the superoxide radical, for example, is more likely to act in the opposite way (as a reductant). Given these difficulties with definitions, I will stick with the name free radicals.

To follow the argument in the rest of the book, all you really need to know is that free radicals are reactive forms of oxygen, produced continuously at low levels by respiration. However, this definition is over simple and inexact. In the rest of this chapter, then, we will take a closer look at what free radicals are, and how and why they are formed.

The splitting of water by radiation was first described by Becquerel, who began experimenting with radium soon after Marie Curie had isolated workable quantities. In the late 1890s, Becquerel had classified the known radioactive emanations according to their penetrating power. Emissions that are stopped by a sheet of paper were termed *alpha rays* (they are in fact helium nuclei); those stopped by a millimetre-thin sheet of metal were called *beta rays* (now known to be fast-moving electrons); and those penetrating a centimetre [2/5 inch] of metal were called *gamma rays* (electromagnetic rays, analogous to x-rays). All three types of radiation displace electrons from atoms, which gives the atoms an electric charge. This is why the Curies could detect an electric field in the air around pitchblende. The loss or gain of electrons, giving a substance an electric charge, is called *ionization*, hence the term ionizing radiation. Radiation also produces many other effects, including heat generation, electron excitation, breaking of chemical bonds and nuclear reactions, such as nuclear fission, as we saw in Chapter 3.

Becquerel discovered that radium emits alpha rays and gamma rays. These decompose water into hydrogen and oxygen. The decomposition of water was not in itself unexpected, as water had been shown to consist of a mixture of hydrogen and oxygen by Laplace and Lavoisier in the 1770s. However, radiation cannot dissociate water directly into hydrogen and oxygen gases (which are made up of molecules of hydrogen and oxygen — H_2 and O_2 — each containing two atoms), because the ratio of hydrogen to oxygen atoms in water (H_2O) is wrong:

$$H_2O \rightarrow H_2 + O_2$$

Most people will remember having to balance chemical equations at school. The equation here will not do. We have two atoms of oxygen on the right side of the arrow and just one on the left. The immediate solution we might leap to is to double up the water side of the equation, so that everything balances:

$$2H_2O \rightarrow 2H_2 + O_2$$

But when we're dealing with radiation, this will not do either. The problem is that we are not dealing with a chemical reaction between two molecules, but with an interaction between radiation energy and a *single* water molecule. Ionizing radiation always interacts with matter at the level of individual atoms, and it cannot produce hydrogen and oxygen molecules 'just like that'. What it does produce stimulated debate throughout the twentieth century, as the products are normally so short-lived. Even the modern consensus is rather uneasy. A first step might be:

$$H_2O \rightarrow H^+ + e^- + {}^{\bullet}OH$$

Where H^+ is a proton (a hydrogen atom that has lost an electron), e^- is a dissolved, or solvated, electron and ${}^{\bullet}OH$ is a free radical called the hydroxyl radical — a ferocious molecule that is among the most reactive substances known.

A free radical is loosely defined as any molecule capable of independent existence that has an unpaired electron. This tends to be an unstable electronic configuration. An unstable molecule in search of stability is quick to react with other molecules. Many free radicals are, accordingly, very reactive. Nonetheless, we should not assume that *all* free radicals are reactive. Molecular oxygen, for example, contains two unpaired electrons, and so can be classed as a free radical by some definitions. The fact that everything does not burst spontaneously into flame shows that not all free radicals are immediately reactive. We shall see why later in this chapter.

In the reaction above, the oxygen atom has lost a single electron, but we are still a long way from generating oxygen gas, or O_2. In fact, to produce oxygen gas from water, a total of four electrons must be *removed* from two oxygen atoms. To reverse this process, to produce water from oxygen, as occurs during respiration, requires the *addition* of four electrons. These electrons must be added or lost one at a time, to produce a series of three possible intermediates: the hydroxyl radical (${}^{\bullet}OH$), hydrogen peroxide (H_2O_2); and the superoxide radical (denoted $O_2^{\bullet-}$).[1] These

[1] There are many more possible intermediates, but these three are the most important. Their stability and reactivity depends partly on their protonation (whether they have hydrogen ions attached, as in hydrogen peroxide H_2O_2), and this depends on the conditions. The superoxide radical ($O_2^{\bullet-}$), for example, is much more reactive when protonated ($HO_2^{\bullet}$).

intermediates are formed regardless of whether we are going from water to oxygen or vice versa (see Figure 6). They are responsible for more than 90 per cent of the biological damage caused by some forms of radiation.

Radiation can interact directly with all kinds of molecule, but in our bodies it is most likely to interact with water. This is largely a matter of probability. Our bodies contain between 45 and 75 per cent water, depending on age and body fat content. Children have the highest water content, up to 75 per cent of their body weight, while an adult man consists of about 60 per cent water. Adult women, who on average carry more subcutaneous fat than men, are about 55 per cent water. In addition to probability, the odds of an interaction between radiation energy and water is skewed by molecular factors. Some forms of radiation, such as gamma rays and x-rays, interact more readily with the bonds in water than they do with carbon bonds in organic molecules. This means that a fat elderly woman (with a low body-water content) is more likely to survive irradiation with x-rays than a young child.

The three intermediates formed by irradiating water, the hydroxyl radicals, hydrogen peroxide and superoxide radicals, react in very different ways. However, because all three are linked and can be formed from each other, they might be considered equally dangerous. Indeed, the three actually work together as part of an insidious catalytic system. We will consider each in turn, in the order that they are produced by radiation on route from water to oxygen.

Hydroxyl radicals ($^{\bullet}OH$) are the first to be formed. These are extremely reactive fragments, the molecular equivalents of random muggers. They can react with all biological molecules at speeds approaching their rate of diffusion. This means that they react with the first molecules in their path and it is virtually impossible to stop them from doing so. They cause damage even before leaving the barrel of the gun. If you ever hear someone talking about antioxidants that 'scavenge' hydroxyl radicals in the body, they won't know what they're talking about. Hydroxyl radicals react so quickly that they attack the first molecule they meet, regardless of whether it is a 'scavenger' or any other molecule. To scavenge hydroxyl radicals in the body, the scavenger would need to be present at a higher concentration than all other substances put together, to give it a higher chance of being in the way. Such a high level of any substance, even if benign, would kill you by interfering with the normal function of the cell.

Once hydroxyl radicals are formed, trouble escalates. When a hydroxyl radical reacts with a protein, lipid or DNA molecule, it snatches an electron to itself and then sinks back into the sublime chemical stability of water. But of course the act of snatching an electron leaves the reactant short of an electron, just as a mugger leaves the victim short of a handbag. In the case of the hydroxyl radical, another radical is formed, this time part of the protein, lipid or DNA. It is as if having your handbag snatched deranges your mind and turns you into a mugger yourself, restless until you have snatched someone else's bag. This is a fundamental feature of all free radical reactions — one radical always begets another, and if this radical is also reactive, then a chain reaction will ensue. Thus the cardinal feature of a free radical is an unpaired electron, while the cardinal feature of free-radical chemistry is the chain reaction.

We are all familiar with free-radical chain reactions when they happen in fatty foods such as butter: they are responsible for rancidity. The fats in the butter oxidize and taste disgusting. The same type of reaction also takes place in cell membranes, which are made mostly of lipids. The process is then called lipid peroxidation. Trying to stop lipid peroxidation happening has caused much wailing and gnashing of teeth among researchers. Free-radical damage is less obvious when it affects proteins or DNA, but free-radical damage to DNA is one of the main causes of genetic mutation, and accounts for the high rates of cancer suffered by radiation victims.

A dramatic non-biological example of the power of free-radical chain reactions is the hole in the ozone layer. The devastation that can be caused by chlorofluorocarbons (CFCs) such as freon is a result of the formation of free radicals in the upper atmosphere. CFCs are quite robust molecules that survive buffeting by the weather in the lower atmosphere. However, they are shredded by ultraviolet rays in the upper atmosphere, and disintegrate to release chlorine atoms. Being one electron short of a full pack, chlorine atoms are dangerously reactive free radicals. They can steal electrons from almost anything. Just one chlorine atom can set in motion a chain reaction that might destroy 100 000 ozone molecules. According to the US Environmental Protection Agency, a single gram of freon will often destroy as much as 70 kilograms of ozone.

There are only two ways for a free-radical chain reaction to end: when two radicals react with each other, and their unpaired electrons conjoin in blissful chemical union; or when the free radical product is so feebly reactive that the chain reactions fizzle out, like handbag thieves overcome with remorse. Some well-known antioxidants, such as vitamin C and vitamin E, act in this way. Although the products of their reactions with a free radical are themselves free radicals, they are so poorly reactive that the chain reactions squib out before too much damage is done.

If radiation strips a second electron from water, the next fleeting intermediate is hydrogen peroxide (H_2O_2) — whose bleaching properties gave its name to the peroxide blonde. Bleaching is caused by the oxidation of organic pigments as the hydrogen peroxide strips electrons from them. The oxidizing properties of hydrogen peroxide can kill bacteria, and are in part responsible for the mildly antiseptic properties of honey, which has been used to treat wounds since ancient times. Most industrial uses of hydrogen peroxide also draw on its power as an oxidizing agent. For example, hydrogen peroxide is used to oxidize pollutants in water and industrial waste, to bleach textiles and paper products, and to process foods, minerals, petrochemicals and detergents.

Despite its widespread use as an oxidizing agent, hydrogen peroxide is unusual in that it lies chemically exactly half way between oxygen and water. This gives the molecule something of a split personality. Like a would-be reformed mugger, whose instinct is pitted against his judgement, it can go either way in its reactions (losing or gaining electrons) depending on the chemical company it keeps. It can even go both ways at once, when reacting with another hydrogen peroxide molecule. In this case, one of the molecules gains two electrons to become water, while the other loses two electrons to become oxygen. The decomposition of hydrogen peroxide in this way is partly responsible for the generation of oxygen from water by radiation:

$$2H_2O_2 \rightarrow 2H_2O + O_2$$

A far more dangerous and significant reaction, however, takes place in the presence of iron, which can pass electrons one at a time to hydrogen peroxide to generate hydroxyl radicals. If dissolved iron is present, hydrogen peroxide is a real hazard. Organisms go to great lengths to avoid contamination with dissolved iron. The reaction between hydrogen peroxide and iron is called the Fenton reaction, after the Cambridge chemist Henry Fenton, who first discovered it in 1894:

$$H_2O_2 + Fe^{2+} \rightarrow OH^- + {}^{\bullet}OH + Fe^{3+}$$

He later showed that the reaction could damage almost any organic molecule. Thus, the main reason that hydrogen peroxide is toxic is because it produces hydroxyl radicals in the presence of dissolved iron. Ironically, the greatest danger lies in its *slow* reactivity in the absence of iron. It has time to diffuse throughout the cell. Hydrogen peroxide may diffuse into the cell nucleus, for example, and there mix with the DNA, before it encounters iron, which transforms it into a brutish hydroxyl radical.[2] The insidious infiltration of hydrogen peroxide means that it is more dangerous than the hydroxyl radicals produced outside the nucleus. Some proteins, such as haemoglobin, also contain iron. If they happen to run into hydrogen peroxide they can be mutilated on the spot. Hydrogen peroxide is like a gangland thug. Normally quiet, posing little danger to casual passers-by, it turns violent on meeting a rival gang member. Damage to proteins containing embedded iron can be as swift and specific as a kneecapping operation.

We have now met two out of the three intermediates between water and oxygen. The first, the hydroxyl radical, is one of the most reactive substances known. It reacts with all biological molecules within billiseconds, initiating chain reactions that spread damage and devastation. The second of the intermediates, hydrogen peroxide, is much less reactive, almost inert, until it meets iron (regardless of whether the iron is in solution or embedded in a protein). Hydrogen peroxide reacts quickly with iron to generate hydroxyl radicals, taking us back to square one. What, then, of the third of our intermediates, the superoxide radical ($O_2^{\bullet-}$)? Like hydrogen peroxide, the superoxide radical is not terribly reactive.[3] However, it too has an affinity for iron, dissolving it from proteins and storage depots. To understand why this is harmful, we need to think again about the Fenton reaction.

The Fenton reaction is dangerous because it produces hydroxyl radicals, but it grinds to a halt when all the accessible iron is used up. Any chemical that regenerates dissolved iron is capable of re-starting the reaction. Because the superoxide radical is one electron away from molecular

[2] Iron is normally tightly bound within proteins inside the cell, but some pathological conditions may bring about its release, allowing iron to be found even in the nucleus. Here, the positively charged iron binds to negatively charged DNA, exacerbating the damage to DNA.

[3] Despite the heroic-sounding name, superoxide radicals are not very reactive with lipids, proteins or DNA. Superoxide does react vigorously with other radicals such as nitric oxide, however, and this may cause cellular damage. It is also reactive in slightly acidic conditions, which occur in the vicinity of cell membranes, so superoxide may damage membranes directly.

oxygen, it is more likely to lose that electron to form oxygen than it is to gain three electrons to form water. Only a few molecules are able to *accept* a single electron, however. One of the best places for the superoxide radical to jettison its spare electron is iron. This converts iron back into the form that can participate in the Fenton reaction:

$$O_2^{\bullet-} + Fe^{3+} \rightarrow O_2 + Fe^{2+}$$

In summary, then, the three intermediates between water and oxygen operate as an insidious catalytic system that damages biological molecules in the presence of iron. Superoxide radicals release iron from storage depots and convert it into the soluble form. Hydrogen peroxide reacts with soluble iron to generate hydroxyl radicals. Hydroxyl radicals attack all proteins, lipids and DNA indiscriminately, initiating destructive free-radical chain reactions that spread damage and destruction.

These same intermediates are also formed from the oxygen that we breathe. The parallel between radiation damage and oxygen toxicity was first described by Rebeca Gerschman in the early 1950s, while she was working at the University of Rochester in New York State — the centre selected to study the biological effects of radiation for the Manhattan Project. In a 1953 seminar, she caught the imagination of a young doctoral student with a background in muscle physiology named Daniel Gilbert, and together the pair pioneered the theory that oxygen free radicals are responsible for the lethal damage caused by both oxygen poisoning and radiation. Their findings were published in a seminal 1954 paper in *Science*, with the splendidly unambiguous title *Oxygen Poisoning and X-irradiation: A Mechanism in Common*, which I've taken as the subtitle to this chapter. Since the 1950s, research has confirmed that radiation damage and oxygen toxicity amount to much the same thing.

Oxygen is a paradox. From a theoretical point of view, it ought to be easier to add electrons to oxygen than it is to remove them from water. Water is chemically stable. Taking electrons from water requires a large input of energy, which can be provided by ionizing radiation, ultraviolet rays, or by sunlight in the case of photosynthesis. Oxygen, on the other hand, releases energy when it reacts — a sure sign of favourable energetics. Burning is the reaction of oxygen with carbon compounds; and the massive amount of heat released suggests that burning should proceed almost spontaneously. In terms of energetics, it does not matter if fuel is burnt rapidly, as in combustion, or slowly, as in respiration. Regardless of whether we metabolize it or burn it, 125 grams [4 oz] of sugar (the amount required to make a sponge cake) will produce 1790 kilojoules [428 kilocalories] of energy: enough to boil 3 litres [6.3 US pints] of water or light a 100-watt bulb for 5 hours.

With such favourable energetics, and with oxygen all around us, the fact that living things do not burst spontaneously into flame betrays an odd reluctance on the part of oxygen to react. The reason for its reticence is buried within the bonds of the oxygen molecule itself. Although slightly abstruse, the chemistry of oxygen explains not only why oxygen free radicals are formed inside us all the time, but also why we do not spontaneously combust. We will therefore consider it briefly.

One of the first signs that oxygen is a little odd was reported in 1891 by the great Scottish chemist Sir James Dewar, who found that oxygen is magnetic. His discovery came after a competitive and acrimonious race to liquefy oxygen, in which the Frenchman Louis Cailletet narrowly defeated his Swiss rival Raoul Pictet, when he succeeded in producing a few droplets of liquid oxygen just before Christmas in 1877. The following year, Dewar liquefied oxygen in a demonstration before the hawkish audience of the Royal Institution, in one of their celebrated Friday Evening Discourses. Dewar was a star turn on these occasions, which were traditionally held in an auditorium known, to the terror of many invited to perform there, as 'that semi-circular fountain of eloquence'. But Dewar was far more than a gifted performer: he was also one of the most brilliant practical scientists of his day. By the mid 1880s, Dewar had improved his methods and was able to produce liquid oxygen in large enough quantities to study its properties in detail. He soon found that liquid oxygen (and indeed ozone, O_3) was attracted to the poles of a magnet. In 1891, he demonstrated his findings with characteristic flamboyance at one of the Discourses, using a strong magnet and his newly devised vacuum flask, still known in laboratories across the world as the Dewar flask. His classic demonstration is repeated today in many university foundation courses (and video demonstrations can be found on the Internet). Liquid oxygen is poured from a Dewar flask between the poles of a powerful magnet. The cascading liquid halts in mid air and sticks to the magnet, forming a plug that hangs majestically between the magnetic poles, until evaporating away.

What is going on? In 1925, Robert Mulliken finally explained why oxygen is magnetic, using recently developed quantum theory. Magnetism results from the spin of unpaired electrons, and Mulliken showed that molecular oxygen normally has two unpaired electrons.[4] These electrons dominate the chemistry of oxygen and make it hard for the gas to receive a bonding pair of electrons; hence the reluctance of oxygen to react by forming new chemical bonds (Figure 8). There are only two ways out of this chemical cul-de-sac. First, oxygen can absorb energy from other molecules that have been excited by heat or light, and this can cause one of its unpaired electrons to flip its spin. Some excited pigment molecules have this effect, and are being put to medical purposes, as in photodynamic therapy in which a pigment is activated by light to destroy a tumour or other pathological tissue. Flipping the spin of an electron leaves one electron pair and one vacant bonding orbital, and so frees up oxygen to react. It is said to remove the 'spin-restriction'. This form of oxygen is called *singlet oxygen*. Unlike its spin-restricted sister, singlet oxygen is fast to react with organic molecules. If, by a chemical quirk of fate, singlet oxygen was the only form that existed, we could never have accumulated oxygen in the atmosphere, or crawled out of the oceans.

The second way of coaxing oxygen to react is to feed it with electrons one at a time, so that each of the unpaired electrons receives a suitable partner independently. Iron can do this, as it has its own unpaired electrons (which makes it magnetic too). Iron loses these electrons without becoming unstable because it has several different 'oxidation states', all of which are energetically stable under relatively normal conditions. (This is partly because the iron atom is large, and the electrons furthest from the nucleus are loosely bound to the atom.) The ability of iron to feed electrons one at a time explains its affinity for oxygen, and the tendency

[4] Rotating electrical charges generate magnetic fields. This applies to electrical current in a coil of wire or to a single spinning electron. In theory, all chemical (covalent) bonds are formed from pairs of electrons. The electrons within a bond usually spin in opposite directions, and their spin is said to be antiparallel. The opposition of spins in typical bonds cancels out, leaving the molecule as a whole with no net spin. Most compounds are therefore not magnetic. Atoms or molecules that are magnetic, such as iron and oxygen, must have at least one unpaired electron; and this, as we have seen, is not usually a stable electronic arrangement. In the case of oxygen, though, quantum mechanical considerations mean that unpaired electrons are actually more stable than the regular double bond structure that most of us were taught at school (see Figure 8). On the basis of bond strengths and magnetism, oxygen has three bonds rather than two: one bond has two electrons, while the other two bonds have three electrons, one of which is unpaired in each bond.

of iron minerals and cars to rust. It also explains our own need to lock iron away in molecular safe-houses in the body. Other metals that exist in two or more stable oxidation states, such as copper, can feed electrons to oxygen just as efficiently, and are equally dangerous unless well caged.

Life is not free from the restrictions imposed by the odd chemistry of oxygen, and we too must supply electrons one at a time in order to tap into its reactivity. Cells contrive to break down the oxidation of food into a series of tiny steps, each of which releases a manageable quantity of energy that can be stored in a chemical form as ATP (see Chapter 3). Unfortunately, at each of these steps there is the risk of single electrons

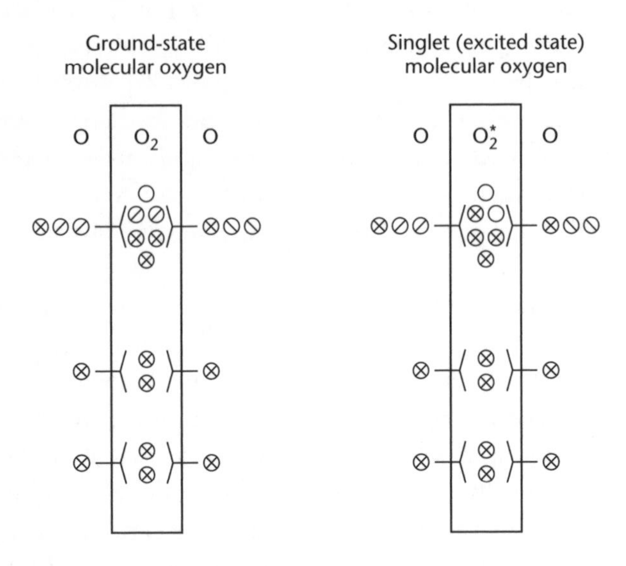

Figure 8: Electron orbitals for **(a)** ground-state molecular oxygen (normal O_2) and **(b)** singlet oxygen (the excited form of molecular oxygen). Each cross represents a pair of electrons, while a single diagonal line represents a single electron. The circles represent available electron orbitals; empty circles represent empty orbitals. According to Hund's rule, electron orbitals of the same energy (shown on the same level) must be filled one at a time, before pairing can occur. This is why atomic oxygen (shown to the side of the box in each case) contains two unpaired electrons. When the electron orbitals are merged in molecular oxygen, the same rule applies, leaving ground-state oxygen with two unpaired electrons, each with parallel spin. Oxygen can therefore only receive single electrons with anti-parallel spin to complete the electron pairings. In singlet oxygen, one of the parallel-spin electrons has flipped spin, enabling a pair to form, but vacating a lower energy orbital in violation of Hund's rule. This is why singlet oxygen is so reactive.

escaping from their shackles and joining with oxygen to form superoxide radicals. The continual production of superoxide radicals by cells means that, despite the emotive associations of radiation sickness, breathing oxygen carries a qualitatively similar risk.

Estimates suggest that, at rest, about 1 or 2 per cent of the total oxygen consumed by cells escapes as superoxide radicals, while during vigorous exercise this total might rise to as much as 10 per cent. Lest these figures sound trivial, we should remember that we consume a large volume of oxygen with each breath. An average adult, weighing 70 kilograms [154 pounds], gets through nearly a quarter of a litre of oxygen every minute. If only 1 per cent of this leaks away to form superoxide radicals, we would still produce 1.7 *kilograms* of superoxide each year. From superoxide we can go on to produce hydrogen peroxide and hydroxyl radicals by way of the reactions discussed earlier.

We *can* produce hydrogen peroxide and hydroxyl radicals, but do we really? Our bodies have evolved efficient mechanisms for eliminating both superoxide radicals and hydrogen peroxide before they can react with iron to produce hydroxyl radicals (we will look at these mechanisms in more detail in Chapter 10). Is there any way of working out how many hydroxyl radicals are actually formed in the body despite these mechanisms?

There are two possible approaches. First, from a theoretical point of view, we can calculate the rate of production of hydroxyl radicals on the basis of an *estimated* steady-state concentration of hydrogen peroxide and iron, and the *known* reaction kinetics. In the human body, it is likely that both iron and hydrogen peroxide are present at a steady-state concentration of about a millionth of a gram per kilogram body weight. If so, we would generate less than a million millionth of a gram of hydroxyl radicals per second per kilogram. Such small numbers are virtually incomprehensible. However, if we convert this figure from the number of grams to the number of molecules produced each second, using Avogadro's number, then we have a far more intelligible number. According to this calculation, we should produce around 50 hydroxyl radicals in each cell every second.[5] In a full day, each cell would generate 4 million hydroxyl

[5] Avogadro's number states that there are 6.023×10^{23} molecules in one mole of any substance. One mole of a substance is the molecular weight in grams. The molecular weight of a hydroxyl radical is 17, so one mole is 17 grams. This means that 17 grams of hydroxyl radicals would contain 6.023×10^{23} hydroxyl radicals. The volume of an average mammalian cell is approximately 10^{-9}–10^{-8} ml. The calculation is adapted from Halliwell and Gutteridge, *Free Radicals in Biology and Medicine* (see Further Reading).

radicals! Many of these would be neutralized in one way or another, and damaged proteins or DNA replaced, but over a lifetime, in a body composed of 15 million million cells, it adds up to a very great deal of wear and tear: more than enough to underpin a process such as ageing.

This is all very well, but somewhat theoretical. If so much damage is taking place, we ought to be able to measure it. This brings us to the second way of estimating the production of hydroxyl radicals: analysing the damage they cause. One possible marker for their effects, which was first proposed in the late 1980s by Bruce Ames and his team at Berkeley, is the rate of excretion of oxidized DNA building blocks in the urine. There are some obstacles to this approach — DNA is being constantly tinkered with by various enzymes in the normal processes of DNA replication and repair, and only some of the types of oxidized fragments produced as a result are *diagnostic* of attack by hydroxyl radicals. Other types are sometimes produced by hydroxyl radicals but can also be generated normally. This means that we need to know *which* fragments are diagnostic of hydroxyl radical attack, and what proportion of the overall molecular wreckage they constitute.

One modified DNA building block that probably does reflect hydroxyl radical attack is *8-hydroxydeoxyguanosine* (8-OHdG), a chemically modified form of deoxyguanosine, the G in the four-letter DNA code. Ames and his team measured the concentration of this molecule in the urine of rats, and then extrapolated from the results to calculate the likely number of hydroxyl radical 'hits' to DNA in each cell of the rat's body. They concluded that there are as many as 10 000 'hits' to the DNA of each cell every day, although most of these are presumably repaired immediately — hence the 8-hydroxyguanosine excreted in the urine. More recent studies have examined the equivalent rate in people. The rate seems to be lower than in rats, but may still approach several thousand hits per cell each day. This is several orders of magnitude lower than the projected 4 million hydroxyl radiacals estimated to be produced in a cell each day, but bear in mind that the figure of 10 000 represents hits to DNA alone, and does not include potentially damaging hits to lipids in cell membranes or to proteins, which make up a far greater part of the cell than does DNA.

Despite the wide margins for error, the fact that hydroxyl radicals are produced from both breathing and radiation allows a direct comparison to be made, at least in principle. James Lovelock used figures similar to those cited above to calculate the equivalent dose of radiation that we absorb by breathing. He estimates that the damage done by breathing for one year is equivalent to a whole-body radiation dose of 1 sievert (or 1 joule energy per kilogram). As a typical chest x-ray delivers 50 microsieverts of radiation, breathing for a year would seem to be 10 000 times more dangerous than a chest x-ray, or 50 times as dangerous as all the radiation that we normally receive from all sources in the course of an entire lifetime.

These figures, while impressive, are somewhat disingenuous. For a start, we do not know whether these 'hits' to DNA are to working genes, or to inactive genes, or to 'junk' DNA, which does not code for anything and, in fact, comprises the bulk of human DNA. In addition, there is one major difference between radiation and breathing — the starting point. Radiation produces reactive hydroxyl radicals from water immediately, and they have a random distribution throughout the cell. Because our normal exposure to radiation is low we have not evolved to deal with this pattern of distribution or immediate reactivity. Respiration, on the other hand, produces mainly superoxide radicals, which are altogether less reactive than hydroxyl radicals. The cell has much more time to dispose of them safely. Moreover, the superoxide radicals formed during respiration are produced at very particular locations within the cell, and we have evolved to deal with their normal production at these sites. There may also be a threshold of repair, linked to the rate of damage. In the course of normal respiration, where damage to DNA presumably accumulates slowly, virtually all damaged DNA will be repaired. Following serious radiation poisoning, when a massive amount of damage is produced very quickly, it is clearly not.

But despite all this, there is no *qualitative* difference between the kind of protection that is effective against radiation poisoning and that which protects against oxygen toxicity. This was understood by Gerschman and Gilbert, and other early workers in the field during the 1950s, who reported that several antioxidants helped to protect mice against the lethal effects of both x-irradiation and oxygen poisoning.

The point is driven home forcibly by the case of an unusual bacterium that is so resistant to radiation — 200 times more resistant than the common gut bacterium *Escherichia coli*, and perhaps 3000 times more resistant than we are — that the controversial astrophysicist Fred Hoyle proposed it must have evolved in outer space. Hoyle put forward this theory in the context of an argument for panspermia (meaning 'seeds everywhere') in his 1983 book *The Intelligent Universe*. Bacterial spores are so resistant to radiation that they can float about in space more or less unharmed, despite the intense cosmic radiation. It is therefore conceivable that life on Earth *could* have been seeded from space. Hoyle's ideas were updated by the cosmologist Paul Davies, in his book *The Fifth Miracle*, who agreed that such impressive radiation tolerance made little sense unless life was forced through a radiation bottleneck at some stage in the past.

The tiny monster that Hoyle and Davies describe is a red-pigmented bacterium called *Deinococcus radiodurans*, one of a tightly knit family of six similarly resistant cousins. *D. radiodurans* itself is among the most radiation-resistant organisms yet found on Earth. It was originally detected as a contaminant of irradiated canned meat, and has since cropped up in weathered granite rocks in a virtually sterile Antarctic valley, in medical equipment sterilized by radiation, and in many more normal environments, such as room dust and animal faeces. Its resistance to ionizing radiation is coupled with a more general resistance to other types of physical stress, including ultraviolet radiation, hydrogen peroxide, heat, desiccation and a variety of toxins. Such an enviable suite of qualities makes *D. radiodurans* an ideal candidate for the bioremediation of sites contaminated with radiation and toxic chemicals. These possible commercial applications have stimulated interest in its DNA genome — its total collection of genes — and in November 1999, Owen White and a large team, mostly from the Institute of Genomic Research in Rockville, Maryland, published the complete genomic DNA sequence, that is, the readout of its complete four-letter DNA code, in *Science*. We now have a much better idea of how it works.

The bacterium is a chimaera, a wonderful example of nature's ability to cobble together a hotchpotch solution and give it every semblance of preconceived design. There is no magic trick, and no need to invoke a stellar origin. Almost all the DNA-repair mechanisms present in *D. radiodurans* are also present in other bacteria, although usually not all together. The only one unique to *D. radiodurans* is an unusually efficient garbage-disposal system, which discards damaged molecular building blocks before they can be incorporated into DNA during DNA replication or repair. The amazing ability of the bacterium to survive insult and injury stems from its capacity to hoard multiple copies of its own genes, as well as other useful genes it acquired by transfer from other bacteria.[6] While most bacteria survive quite happily with a handful of protection devices, *D. radiodurans* uses the complete repertoire, and has multiple copies of each. This allows it to flourish in hostile environments, where it has little competition from less well-endowed brethren.

Far from the cosmic radiation bottleneck suggested by Davies, it seems likely that *Deinococcus* may have adapted relatively recently to radioactive environments. In their *Science* paper, White and his colleagues compared the genome sequence of *D. radiodurans* with the sequences of other bacteria, and argued that the closest living relative of the superbug is an extreme heat-loving bacterium by the name of *Thermus thermophilus*. Of the 175 known *T. thermophilus* genes, 143 have close counterparts in *D. radiodurans*, suggesting that its toughness may have originated through the modification of systems that evolved originally to provide resistance to heat.

There is an important general point to make here, which we will return to later in the book. The genes that protect against radiation are not only the same as those that protect against oxygen toxicity, but are also the same as many of those that protect against other types of physical stress such as heat, infection, heavy metals or toxins. In people, the genes activated by radiation also protect against oxygen poisoning, malaria or lead poisoning. The reason for this cross-protection is that many different physical stresses all funnel in to a single common damage-causing process in the cell, so all can be withstood through common protective mechanisms. This shared pathology is a rise in *oxidative stress*. Oxidative stress is defined as an imbalance between free-radical production and antioxidant protection. However, it is not just a pathological state, but also acts as a

[6] Bacteria can pass genes to each other by conjugation, which occurs between bacteria of the same species or related species, and has been likened to sex. Sometimes the genes transferred are part of the bacterial chromosome, sometimes they are part of rings of DNA known as plasmids. Most drug-resistance genes are carried on plasmids and this facilitates their spread through bacterial populations. Bacteria can also acquire new genes from distantly related bacteria in a variety of ways, including direct uptake of pieces of DNA from the environment. This type of transfer is often known as lateral or horizontal gene transfer. Gene transfer can complicate attempts to define the evolutionary lineage of bacterial species.

signal to the cell that it is under threat. Oxidative stress is therefore both a threat and a signal to resist threat. In the same way, the bombing of Pearl Harbor was both an act of aggression and a signal to the Americans to resist the Japanese threat.

The integration of protective mechanisms against oxidative stress raises the possibility that life might have evolved ways of dealing with oxygen toxicity long before there *was* any oxygen in the atmosphere — ionizing radiation alone might do the trick. We have already concluded that rising oxygen levels did not bring about a mass extinction during the Precambrian or afterwards (Chapters 2–5). As oxygen is undoubtedly toxic, life might somehow have adapted to the threat in advance. Could it be that life adapted to radiation early on, and that this formed the basis of its response to other kinds of threat? If so, then Fred Hoyle and Paul Davies may have been right in one sense. Life *was* forced through a radiation bottleneck, but it happened on Earth, not in space, and it happened not recently, but 4 billion years ago.

This possibility is supported by the conditions discovered by the Viking lander on the surface of Mars in 1976. The lander biology instrument had been designed to perform three experiments to determine whether life was present in the soils of Mars. The results of these experiments were not sufficiently clear-cut to support a definitive answer, and arguments about the correct interpretation of the data continue today. But one of the studies, the gas-exchange experiment, produced results that were, if not clear-cut, at least a total surprise. The experiment was intended to distinguish between gas emissions arising from microbial metabolism and those arising from purely chemical reactions. The instrument incubated Martian surface samples under dry, humid or wet conditions, and then measured any gases released. The samples were treated with a nutrient broth consisting of a mixture of organic compounds and inorganic salts, described by Gilbert Levin (one of the original Viking scientists and a steadfast proponent of life on Mars) as 'chicken soup'. The experiment was conducted in two stages. First the lid was taken off the broth, to allow the water vapour escaping from it to humidify the soil in the box; then a little soup was sprinkled onto the soil, to promote metabolism by any organisms present.

To the amazement of the team, removing the lid from the broth was enough to produce a great burst of oxygen from the Martian soil — 130 times more than they had anticipated. The investigators toyed with the idea that the broth might have stimulated photosynthesis, but the same reactions took place in the dark, and even after the samples were heated at 145°C for 3.5 hours to kill any microbes. If the samples were recharged with fresh nutrient following the initial burst of oxygen, however, no more oxygen was released, implying that the reactions had come to an end. Although these results did not strictly rule out life, the reactions were much better explained by chemistry than by biology. But if this was the case, the chemistry of the soil must have been potent, as even the simple expedient of adding water made it fizz. After some puzzlement, the Viking team eventually concluded that the soil samples must have contained superoxides and peroxides, generated by ultraviolet rays acting on the atmosphere or on the soil itself. This conjecture has been confirmed by other analyses of the chemical composition of the rocks.

What had happened? We can infer that hydroxyl radicals, hydrogen peroxide and superoxide radicals were formed over the aeons by ultraviolet rays splitting traces of water vapour in the Martian atmosphere or soil. Without surface water, these intermediates would have reacted with iron and other minerals in the soil to produce the rusty oxides that give the red planet its colour. Most of these metal oxides would be unstable on the Earth, but on Mars they remained petrified, precariously stable, as long as the dry and sterilizing conditions persisted. When the Viking scientists took the lid off their broth, the suspended chemical reactions could finally run through to completion. As the unstable iron oxides broke down, the petrified intermediates reacted by way of the pathways we have discussed in this chapter to conjure up oxygen and water from the very rocks themselves. Ironically, the science-fiction heroes struggling to detoxify the soil and replenish the air of Mars with oxygen may need no more than a little warm water to turn the Red Planet blue.

We must conclude that Mars is under serious oxidative stress. Although its thin atmosphere contains barely 0.15 per cent oxygen, any hypothetical life dwelling there would have to contend with oxygen toxicity generated by radiation, certainly as serious as anything faced by life on Earth today. If this is the case on Mars today, it must surely have been the case on the Earth 4 billion years ago. The Earth, after all, is closer to the Sun, and subject to more intense solar radiation. Before there was any oxygen, there was no ozone layer, and the cruel ultraviolet rays must have cut straight through to the ground. But the traditional view, that the continents and shallow seas would have been sterilized by the rays, is being turned on its head. New evidence suggests that resistance to oxygen and radiation alike was built into the very earliest organisms, rather than being built in later as an afterthought. The implications for evolution, and for our own make-up, run deep, as we shall see in the next few chapters.

CHAPTER SEVEN

Green Planet

Radiation and the Evolution of Photosynthesis

THE *HITCH-HIKER'S GUIDE TO THE GALAXY* describes the Earth as an utterly insignificant blue-green planet, orbiting a small and unregarded yellow sun in the uncharted backwaters of the western spiral arm of the galaxy. In deriding our anthropocentric view of the Universe, Douglas Adams left the Earth with just one claim to fame: photosynthesis. Blue symbolizes the oceans of water, the raw material of photosynthesis; green is for chlorophyll, the marvellous transducer that converts light energy into chemical energy in plants; and our little yellow sun provides all the solar energy we could wish for, except perhaps in England. Whether Adams intended it or not, his scalpel was sharp — photosynthesis defines our world. Without it, we would miss much more than the grass and trees. There would be no oxygen in the air, and without that, no land animals, no sex, no mind or consciousness; no gallivanting around the galaxy.

The world is so dominated by the green machinery of photosynthesis that it is easy to miss the wood for the trees — to overlook the conundrum at its heart. Photosynthesis uses light to split water, a trick that we have seen is neither easy nor safe: it amounts to the same thing as irradiation. A catalyst such as chlorophyll gives ordinary sunlight the destructive potency of x-rays. The waste product is oxygen, a toxic gas in its own right. So why split a molecule as robust as water to produce toxic waste if you can split something else much more easily, such as hydrogen sulphide or dissolved iron salts, to generate a less toxic product?

The immediate answer is easy. For living organisms, the pickings from water-splitting photosynthesis are much richer than those from hydrothermal activity, the major source of hydrogen sulphide and iron salts. Today, the total organic carbon production deriving from hydrothermal sources is estimated to be about 200 million tonnes (metric tons) each year. In contrast, the amount of carbon turned into sugars by photosynthesis by plants, algae and cyanobacteria is thought to be a million million tonnes a year — a 5000-fold difference. While volcanic activity was no doubt higher in the distant geological past, the invention of oxygenic (oxygen-producing) photosynthesis surely increased global organic productivity by two or three orders of magnitude. Once life had invented oxygenic photosynthesis, there was no looking back. But this is with hindsight. Darwinian selection, the driving force of evolution, notoriously has no foresight. The ultimate benefits of a particular adaptation are completely irrelevant if the interim steps confer no benefit. In the case of oxygenic photosynthesis, the intermediate steps require the evolution of powerful molecular machinery that can split water using the energy of sunlight. From a biological point of view, if you can split water, you can split anything. Such a powerful weapon must be caged in some way lest it run amok and attack other molecules in the cell. If, when the water-splitting device first evolved, it was not yet properly caged, as we might postulate for a blindly groping first step, then it is hard to see what advantage it could offer. And what of oxygen? Before cells could commit to oxygenic photosynthesis, they must have learnt to deal with its toxic waste, or they would surely have been killed, as modern anaerobes are today. But how could they adapt to oxygen if they were not yet producing it? An oxygen holocaust, followed by the emergence of a new world order, is the obvious answer; but we have seen that there is no geological evidence to favour such a catastrophic history (Chapters 3–5).

In terms of the traditional account of life on our planet, the difficulty and investment required to split water and produce oxygen is a Darwinian paradox. The usual solution presented is selective pressure. Perhaps, for example, the stocks of hydrogen sulphide and dissolved iron salts eventually became depleted, putting life under pressure to adapt to an alternative, such as water. Perhaps, but on the face of it there is a difficulty here — the argument is circular. For the large geochemical stocks of hydrogen sulphide and iron to have become depleted in this way, they must have been oxidized by something, and the most likely, if not the only, candidate for oxidation on this scale is oxygen itself. The trouble is that there *was* no free oxygen before photosynthesis. Only photosynthesis can produce free molecular oxygen (O_2) on the scale required. Thus, it seems that the only way to generate enough selective pressure for the evolution of photosynthesis is through the action of photosynthesis.

This argument is not just circular, it is also demonstrably false. We know from biomarkers diagnostic of cyanobacteria that oxygenic photosynthesis evolved more than 2.7 billion years ago. Despite this early evolution, we know that iron was still being precipitated from the oceans in vast banded iron formations a billion years later (Chapter 3). In no sense were oceanic iron salts depleted. Similarly, stagnant conditions, in which deep ocean waters are saturated with hydrogen sulphide, seem to have persisted until the time of the first large animals, the Vendobionts, and recur sporadically even today (Chapter 4). When these dates are taken together, we are forced to conclude that oxygenic photosynthesis evolved before the exhaustion of iron and hydrogen sulphide, at least on a *global* scale.

Why, and how, then, did oxygenic photosynthesis evolve? In the light of the last chapter, you may have guessed the answer already. There is good circumstantial evidence that oxidative stress, produced by solar radiation as on Mars (see Chapter 6, page 129), lies behind the evolution of photosynthesis on the Earth. The details are fascinating but also reveal just how deeply rooted is our resistance to oxygen toxicity: part and parcel, it seems, of the earliest known life on Earth. The earliest known bacteria did not produce oxygen by photosynthesis, but they could breathe oxygen — in other words they could apparently generate energy from oxygen-requiring respiration before there *was* any free oxygen in the air. To understand how this could be, and why it is relevant to our health today, we need to look first at how photosynthesis works, and how it came to evolve.

Of the different types of photosynthesis carried out by living organisms, only the familiar oxygenic form practised by plants, algae and cyanobacteria generates oxygen. All other forms (collectively known as *anoxygenic* photosynthesis) do not produce oxygen and pre-date the more sophisticated oxygenic form. Despite our anthropocentric interest in oxygen, plants are not much concerned with the gas — what they need from photosynthesis is energy and hydrogen atoms. The different forms of photosynthesis are united only in that they all use light energy to make chemical energy (in the form of ATP) needed to cobble hydrogen onto carbon dioxide to form sugars. They differ in the *source* of the hydrogen, which might come from water, hydrogen sulphide or iron salts, or indeed any other chemical with hydrogen attached.

Overall, plant photosynthesis converts carbon dioxide (CO_2) from the air into simple organic molecules such as sugars (general formula CH_2O). These are subsequently burnt by the plant in its mitochondria (see Chapter 3) to produce more ATP, and also converted into the wealth of carbohydrates, lipids, proteins and nucleic acids that make up life. We met the enzyme that cobbles hydrogen onto carbon dioxide in Chapter 5 — Rubisco, the most abundant enzyme on the planet. But Rubisco needs to be spoon-fed with its raw materials — hydrogen and carbon dioxide. Carbon dioxide comes from the air, or is dissolved in the oceans, so that is easy. Hydrogen, on the other hand, is not readily available — it reacts quickly (especially with oxygen to form water) and is so light that it can evaporate away into outer space. Hydrogen therefore needs a dedicated supply system of its own. This is, in fact, the key to photosynthesis, but for many years the lock resisted picking. Ironically, the mechanism only became clear when researchers finally understood where the oxygen waste came from.

In oxygenic photosynthesis, the hydrogen can only come from water, but the source of the oxygen is ambiguous. If we look at the overall chemical equation for photosynthesis, we see that it could come from either carbon dioxide or water:

$$CO_2 + 2H_2O \rightarrow (CH_2O) + H_2O + O_2$$

At first, scientists guessed that the oxygen came from carbon dioxide — a perfectly reasonable and intuitive assumption, but quite wrong as it turned out. The fallacy was first exposed in 1931, when Cornelis van Niel showed that a strain of photosynthetic bacteria used carbon dioxide and hydrogen sulphide (H_2S) to produce carbohydrate and sulphur in the presence of light— but did not give off oxygen:

$$CO_2 + 2H_2S \rightarrow (CH_2O) + H_2O + 2S$$

The chemical similarity between H_2S and H_2O led him to propose that in plants the oxygen might come not from carbon dioxide at all, but from water, and that the central trick of photosynthesis might be the same in both cases. The validity of this hypothesis was confirmed in 1937 by Robert Hill, who found that, if provided with iron ferricyanide (which does not contain oxygen) as an alternative to carbon dioxide, plants could continue to produce oxygen even if they could not actually grow. Finally, in 1941, when a heavy isotope of oxygen (^{18}O) became available, Samuel Ruben and Martin Kamen cultivated plants with water made with heavy oxygen. They found that the oxygen given off by the plants contained only the heavy isotope derived from water, proving conclusively that the oxygen came from water, not carbon dioxide.

In oxygenic photosynthesis, then, hydrogen atoms (or rather, the protons (H^+) and electrons (e^-) that constitute hydrogen atoms) are extracted from water, leaving the 'husk' — the oxygen — to be jettisoned into the air. The only advantage of water is its great abundance, for it is not easy to split in this way. The energy required to extract protons and electrons from water is much higher (nearly half as much again) than that needed to split hydrogen sulphide. Controlling this additional energy requires special 'high-voltage' molecular machinery, which had to evolve from the 'low-voltage' photosynthetic machinery previously used to split hydrogen sulphide. To understand how and why this voltage jump was made, we need to look at the structure and function of the machinery in a little more detail.

Whatever the source of hydrogen atoms — hydrogen sulphide or water — the energy for their extraction is supplied by the electromagnetic rays that we know as sunlight. All electromagnetic rays, including light, are packaged into discreet units called photons, each of which has a fixed quantity of energy. The energy of a photon is related to the wavelength of the light, which is measured in nanometres (a billionth of a metre). The shorter the wavelength, the greater the energy. This means that ultraviolet photons (wavelength less than 400 nanometres) have more energy than red photons (wavelength of 600 to 700 nanometres), which have more energy than infrared photons (wavelength above 800 nanometres).

The interaction of light with any molecule always takes place at the level of the photon. In photosynthesis, chlorophyll is the molecule that absorbs photons. It cannot absorb any photon — it is constrained by the structure of its bonds to absorb photons with very particular quantities of energy. Plant chlorophyll absorbs photons of red light, with a wavelength of 680 nanometres. In contrast, the anoxygenic purple photosynthetic bacterium *Rhodobacter sphaeroides* has a different type of chlorophyll, which absorbs less-energetic infrared rays with a wavelength of 870 nanometres.[1]

When chlorophyll absorbs a photon, its internal bonds are energized. The energetic vibrations force an electron from the molecule, leaving the chlorophyll short of one electron. Loss of an electron creates an unstable, reactive form of chlorophyll. However, the newly reactive molecule cannot simply take back its missing electron. That is snatched by a neighbouring protein and is whipped off down a chain of linked proteins, putting it beyond reach, like a rugby ball being passed across the field by a line of players.[2] On the way, its energy is used to power the synthesis of ATP in a manner exactly analogous to that in mitochondrial respiration.

The theft of an electron is half way to stealing an entire hydrogen atom, as hydrogen consists of a single proton and a single electron. Little extra work is needed to steal the proton. Electrostatic rearrangements draw a positively charged proton (from water in the case of oxygenic photosynthesis) after the negatively charged electron. The proton and the electron are eventually reunited by Rubisco as a hydrogen atom in a sugar molecule.

What happens to the chlorophyll? Having lost an electron, it becomes far more reactive, and will snatch an electron from the nearest suitable source. Reactive chlorophyll is constrained in the same way as a mediaeval dragon which is fed with virgins to stop it ravaging the neighbourhood. The source of suitable virgins — electrons in the case of chlorophyll — includes any plentiful sacrificial chemical, such as water, hydrogen sulphide or iron. Devouring an electron settles the chlorophyll back into its normal equable state, at least until another photon sets the whole cycle in motion again.

[1] Because plants absorb large amounts of red light, and reflect back more blue and yellow light, they appear green to us. In fact, chlorophylls are not the only light-absorbing molecules in plants. They are coupled with other pigments, such as carotenoids, which can absorb light of different wavelengths and transfer it to chlorophyll. It is the overall absorption spectrum of all these pigments operating together that gives plants their green colour.

[2] Electrons pass along a gradient of electrochemical potential, from compounds with a low demand for electrons (low redox potential) to compounds with a high demand for electrons (high redox potential). 'Electron-transport chains' comprise strings of proteins and other electron-transfer molecules linked together in the order of their electrochemical potential. Electrons usually pass smoothly down the chain from one end to the other, although sometimes they are poached by oxygen to produce superoxide radicals. At a certain step in the photosynthetic chain, the transfer of electrons from one molecule to another provides sufficient energy to power the manufacture of ATP.

Which electron donor is used in photosynthesis — hydrogen sulphide, iron or water — ultimately depends on the energy of the photons that are absorbed by the chlorophyll. In the case of purple bacteria, their chlorophyll can only absorb low-energy infrared rays. This provides enough energy to extract electrons from hydrogen sulphide and iron, but not from water. To extract electrons from water requires extra energy, which must be acquired from higher-energy photons. To do this requires a change in the structure of chlorophyll, so it can absorb red-light photons instead of infrared light.

The evolutionary question is this: why did the structure of chlorophyll change, allowing it to absorb red light and split water, when the existing chlorophyll of purple bacteria could already extract electrons from hydrogen sulphide and iron salts, which were plentiful in the ancient oceans? More specifically, what environmental pressure could have led to the evolution of a new and more potent chlorophyll, capable of oxidizing water and much else in the cell, when the old chlorophyll was less reactive and less dangerous — and yet still sufficiently strong to oxidize hydrogen sulphide?

Technically, the answers to these questions are surprisingly simple. According to Hyman Hartman, of the Institute for Advanced Studies in Biology at Berkeley, California, tiny changes in the structure of bacterial chlorophylls can lead to large shifts in their absorption properties. Two small changes to the structure of bacteriochlorophyll *a* (which absorbs at 870 nm) are all that it takes to generate chlorophyll *d*, which absorbs at 716 nanometres. In 1996, an article in *Nature* by Hideaki Miyashita and colleagues of the Marine Biotechnology Institute in Kamaishi, Japan, reported that chlorophyll *d* is the main photosynthetic pigment in a bacterium called *Acaryochloris marina*, which splits water to generate oxygen. Thus, an intermediate between bacteriochlorophyll and plant chlorophyll is not only plausible: it actually exists. From chlorophyll *d* another trifling change is all that is required to produce chlorophyll *a*, the principal pigment in plants, algae and cyanobacteria, which absorbs light at 680 nanometres.

Technically, then, the evolutionary steps required to get from bacteriochlorophyll to plant chlorophyll are simply achieved. The question remains, why? A chlorophyll that absorbs light at 680 nanometres is less good at absorbing light at 870 nanometres. It is therefore less efficient at splitting hydrogen sulphide, and so bacteria carrying it are at a competitive disadvantage compared with the bacteria that kept their original chlorophyll. Even worse, switching chlorophylls to split water poses the problem of what to do with the toxic oxygen waste, as well as any leaking free-radical intermediates — the same as those produced by radiation. Without foresight, how did life manage to cope with its dangerous new invention?

Chlorophyll extracts electrons from water one at a time. To generate oxygen from water, it must absorb four photons and lose four electrons in succession, each time drawing an electron from one of two water molecules.[3] The overall water-splitting reaction is:

$$2H_2O \rightarrow O_2 + 4H^+ + 4e^-$$

Only in the final stage is oxygen released. The rate at which chlorophyll extracts electrons depends on how quickly the photons are absorbed. As the successive steps cannot take place instantly, a series of potentially reactive free-radical intermediates must be produced, if only transiently.

For plants, this whole system is precarious in the extreme. Reactive oxygen intermediates are produced from water as it is stripped of electrons one by one to form oxygen. Some of these reactive intermediates might escape from the reaction site to devastate nearby molecules. Even if they don't escape, in the final step molecular oxygen is released into the cell in large quantities. Inside a modern plant leaf the oxygen concentration can reach three times atmospheric levels. Tiny cyanobacteria pollute themselves and their immediate surroundings in a similar fashion. This would have happened even in ancient times before there was any oxygen in the surrounding air. Some of this excess oxygen inevitably steals stray electrons to form superoxide radicals. The risks are huge. Chaos could break out at any moment. The closest analogy is a nuclear power station. If the reactors are sealed properly it is safe enough, but if a leak develops we face a disaster on the scale of Chernobyl. In both nuclear power and oxygenic photosynthesis the safety margins are slim, but the potential benefits — unlimited energy — are huge.

[3] As usual, it is a little more complicated than this. In fact, a second light-activated centre is required for oxygenic photosynthesis. Neither reaction centre alone can bridge the wide chemical gulf between stealing electrons from water and attaching them to carbon dioxide — so the two centres must work together. The centres are coupled, essentially in series, in what is known as a Z scheme, and each absorbs four photons in a single cycle. To produce one molecule of oxygen therefore requires eight photons of light.

If photosynthesis is to work at all, the reactive intermediates from water must be sealed inside a structure that immobilizes them, preventing them from escaping before oxygen is released. Needless to say, they *are* sealed in such a cage, this is how photosynthesis works. The cage is made of proteins and is called the *oxygen-evolving complex* (or sometimes the *water-splitting enzyme*). Water is bound tightly inside the protein cage while the electrons are extracted one at a time. But this is no ordinary cage. Its structure conceals a secret that is much older than the hills, which transports us back to the time before oxygenic photosynthesis evolved, to a time more than 2.7 billion years ago, before there was any oxygen in the atmosphere. This structure is the key to life on Earth, for without it the Earth would have remained as sterile as Mars.

The structure of the oxygen-evolving complex is very similar to that of an antioxidant enzyme called *catalase*. In fact, the oxygen-evolving complex looks as if it evolved from two catalase enzymes lashed together.[4] If so, then catalase must have evolved *before* the oxygen-evolving complex. If so, the chronology must be as follows. Catalase evolved on the early Earth, in an atmosphere devoid of oxygen. One day, two catalase molecules became bound together to form a cage that enabled the safe splitting of water: the oxygen-evolving complex. This cage allowed the evolution of oxygenic photosynthesis. As a result, the atmosphere filled with oxygen. Life was put under serious oxidative stress. Luckily it could cope: it already had at least one antioxidant enzyme that could to protect it — catalase. How convenient! But wait a moment. If catalase came before photosynthesis, then even if there was no atmospheric oxygen, there must have been oxidative stress. Is this plausible? To answer this question, we must take a look at how catalase works.

---

[4] The evidence is not compelling, but is certainly intriguing. First, there is a broad similarity in reaction mechanisms. Both catalase and the oxygen-evolving complex bind two identical molecules (either $2H_2O_2$ or $2H_2O$), which are then reacted together to generate oxygen, via a strikingly similar sequence of steps. Second, both contain clusters of manganese atoms at their core. Hyman Hartman and others have noted that the manganese core of catalase is structurally very similar to *half* that of the oxygen-evolving complex, implying that the latter may have evolved by the lashing together of two catalase units. However, it is possible that the structural similarities between catalase and the oxygen-evolving complex are no more than coincidence, or a case of convergent evolution towards a similar endpoint, like a stick insect's resemblance to a stick. Even if the similarity *is* authentic evidence of genetic relatedness, we cannot rule out the possibility that catalase evolved from the oxygen-evolving complex rather than the other way round.

Catalase is responsible for getting rid of hydrogen peroxide. This is a potential killer for bacteria as we saw in Chapter 6. Virtually all aerobic organisms possess a form of this enzyme, and even some anaerobic bacteria, which try to avoid oxygen like the plague, retain some catalase just in case. It works extraordinarily quickly. Without catalase, and in the absence of iron, hydrogen peroxide takes several weeks to break down into water and oxygen. Dissolved iron, of course, catalyses the breakdown of hydrogen peroxide into hydroxyl radicals, and eventually water, via the Fenton reaction (see Chapter 6, page 118). If iron is incorporated into a pigment molecule such as haem (the pigment in haemoglobin) the rate of decomposition is increased 1000-fold. If the haem pigment is embedded in a protein, as is the case with catalase, then hydrogen peroxide is broken down directly and safely into oxygen and water, at a rate that is estimated to be 100 million times faster than the rate in the presence of iron alone.

There are several different types of catalase. Most animal cells have a form that has four haem molecules embedded in its core. In contrast, some microbes have a different sort of catalase, which contains manganese instead of haem at its core. Despite their different structures, both enzymes are equally fast, and are correctly called catalase, in the sense that they work in the same way — they both catalyse the reaction of two molecules of hydrogen peroxide *with each other* to form oxygen and water:

$$2H_2O_2 \rightarrow 2H_2O + O_2$$

This simple reaction mechanism reveals a great deal about conditions on the Earth 3.5 billion years ago. It is the exact equivalent of the natural reaction between two molecules of hydrogen peroxide, but is speeded up 100 million times by the enzyme. The need for *two* molecules of hydrogen peroxide means that catalase is extremely effective at removing hydrogen peroxide when concentrations are high, when it is easy to bring two molecules together. It works less well at low concentrations of hydrogen peroxide, when it is harder to find two molecules close together. Catalase is thus swift to remove high concentrations of hydrogen peroxide, but is poor at mopping up trace amounts or at maintaining a stable low-level equilibrium.

Today, most aerobic organisms have a second group of enzymes, known as the *peroxidases*, which can dispose of trace amounts of hydrogen peroxide. These enzymes work better at low levels of hydrogen peroxide because they act in a fundamentally different way. Rather than bringing two molecules of hydrogen peroxide together, they use antioxidants such as vitamin C to convert a single molecule of hydrogen peroxide into two molecules of water, without generating any oxygen. Most aerobic cells have both sets of enzymes, and break down hydrogen peroxide using both mechanisms. Catalase is used for bulk removal, peroxidase for subtle adjustments.

We might infer that any cell using catalase would need to cope with large fluxes of hydrogen peroxide, at least occasionally. Catalase is highly specialized: it has no other known target and works at an extraordinary speed. Such tremendous efficiency does not appear out of the blue by chance: we might as well believe that the eighteenth-century theologian Tom Paley stumbled across a nuclear reactor, rather than his celebrated watch, and instead of inferring the hand of a designer, ascribed it to an accidental arrangement of the elements.

There is nothing accidental about catalase. If it was present on the early Earth, before photosynthesis, then there must have been hydrogen peroxide too, and in abundance. This is counter-intuitive, to say the least. Is it really credible that the early Earth could have been so rich in hydrogen peroxide that there was a selective pressure for the evolution of catalase?

As we have seen (Chapter 6), Mars is rich in iron peroxides; but their abundance in Martian soils tells us nothing about how quickly they were formed on the early Earth. While they were almost certainly formed on Earth (which is, after all, closer to the Sun, and so more drenched in ultraviolet rays), the abundance of hydrogen peroxide on Earth would have depended on its rate of formation and destruction — and these in turn are dependent on atmospheric and oceanic conditions. While the existence of catalase implies that hydrogen peroxide was indeed abundant, the story is suggestive but far from conclusive. Luckily, there are other ways to answer the question, and they not only support the notion that photosynthesis evolved in response to oxidative stress, but they also explain a few other long-standing paradoxes.

One of the most respected atmospheric scientists of recent decades is James Kasting, now at Pennsylvania State University, and during the 1980s at the NASA Ames Research Centre in California. In the mid-1980s, Kasting set out to answer the question of just how abundant hydrogen peroxide might have been on the early Earth. He was not really aiming to answer questions about the evolution of photosynthesis but rather to look at the time-line for oxygen production.

As we saw in Chapter 3, a surrogate measure for the accumulation of oxygen in the air is the extent of iron-leaching from fossil soils. Because iron is soluble in the absence of oxygen, it can be washed out of soils by rainfall on an oxygen-free planet. As oxygen builds up in the atmosphere, it reacts with iron in the soil to produce insoluble rusty iron deposits, which cannot be leached out by rainfall in the same way. In theory, then, fossil soils preserve a record of atmospheric oxygen levels in their iron content — the more oxygen there is in the air, the more iron is left in the soil. The trouble is that the fossil-soil record can be read to imply that oxygen began to accumulate in the air well over 3 billion years ago (long before the major rise 2 billion years ago). This early date does not tally with the sulphur-isotope measurements discussed in Chapter 3, or with the larger-scale deposits, such as banded iron formations, red-beds and uranium ores. Kasting was interested in the discrepancy.

Earlier studies of fossil soils had tacitly assumed that the most important oxidant dissolved in rainwater had always been oxygen itself. Kasting queried this assumption and set out to compute the possibility that hydrogen peroxide had been the most important oxidant in rainwater before the advent of atmospheric oxygen. In a detailed theoretical paper, Kasting, working with Heinrich Holland and Joseph Pinto at Harvard, calculated the rate at which water is split by ultraviolet rays under a variety of conditions. They then took into consideration the solubility of the degradation products (such as hydrogen peroxide) in rain droplets, to calculate their likely steady-state concentrations in rainwater and in lakes on the early Earth.

Under the most likely conditions 3.5 billion years ago — high carbon dioxide levels, a trace of oxygen (less than 0.1 per cent of present atmospheric levels) and virtually no ozone screen — Kasting calculated that there should have been a continuous flux (based on the rate of formation and removal by reaction or rainfall) of about 100 billion molecules of hydrogen peroxide per second per square centimetre. Although this number sounds fantastically big, we should bear in mind the inconceivably large number of molecules that make up matter. There are said to be more molecules in a single glass of water than there are glasses of water in all the oceans. We should not be too surprised to discover, then, that 100 billion molecules of hydrogen peroxide weigh about

56 thousand billionths of a gram.[5] To put these numbers into some sort of perspective, Kasting calculates that dissolved hydrogen peroxide, which is much more soluble than oxygen, accounts for between 1 and 6 per cent of the total oxidant concentration in rainwater today. There is no reason why the amount of hydrogen peroxide in rainwater should have been any less 3 billion years ago, and it may well have been higher, as the intensity of ultraviolet radiation was more than 30 times greater.

Such a large flux of hydrogen peroxide must have placed the first cells under oxidative stress. The level of stress would have been exacerbated by the reactivity of hydrogen peroxide in comparison with oxygen. In particular, hydrogen peroxide reacts quickly with dissolved iron, to produce hydroxyl radicals, whereas oxygen reacts much more slowly. In today's well-oxygenated oceans, the reactivity of hydrogen peroxide is limited by the low availability of dissolved iron (which long ago reacted with oxygen and precipitated out as banded iron formations), but during the early Precambrian, the oceans were so full of dissolved iron that hydrogen peroxide must have been continuously reacting with iron to produce hydroxyl radicals via the Fenton reaction. Thus, not only was there more hydrogen peroxide on the early Earth, it was also more likely to react to produce oxidative stress.

The effect that hydrogen peroxide had on the environment must have depended on the amount of iron available. In the deep oceans there was such a vast amount of dissolved iron that any hydrogen peroxide dissolved in rainwater could never have altered the overall chemical balance. In the shallow seas and freshwater lakes, however, there was much less iron. These low levels of iron could plausibly have been depleted, or exhausted, by a steady drizzle of hydrogen peroxide. With the loss of iron and hydrogen sulphide, such secluded environments would have grown steadily more oxidized. According to the mathematical models of Hyman Hartman and his colleague Chris McKay, at the NASA Ames Research Center, the sheltered lakes and sea basins may well have become oxidizing enough to stimulate the evolution of antioxidant enzymes such as catalase. Once this had happened, bacteria living in shallow-water environments would have been pre-conditioned to the appearance of free oxygen.

[5] This can be calculated from Avogadro's number, the number of molecules in one mole of any substance, which is 6.023×10^{23}. One mole of hydrogen peroxide weighs 34 grams. One gram of hydrogen peroxide therefore contains $1/34 \times 6.023 \times 10^{23}$ molecules, or about 177×10^{21}. A hundred billion molecules of hydrogen peroxide weighs 100 billion/177×10^{21} grams, or 56×10^{-12}. This is one 56 thousand billionths of a gram.

Thus, there are good grounds for thinking that there was indeed plentiful hydrogen peroxide on the early Earth, and that it built up in sheltered environments. The oxidation of such environments by hydrogen peroxide was probably a strong enough selective pressure to stimulate the evolution of the antioxidant enzyme catalase. Catalase itself seems to have been the basis of the oxygen-evolving complex, enabling the evolution of oxygenic photosynthesis. So far the story makes sense, but we are left with one difficult question. Why would oxygenic photosynthesis evolve from catalase?

Catalase would presumably have been present in the photosynthetic bacteria that generated energy by splitting hydrogen sulphide or iron salts in the era before oxygenic photosynthesis. In fact, hydrogen peroxide has some parallels with these early photosynthetic fuels. To remove electrons from hydrogen peroxide requires a similar input of energy to that required to remove electrons from hydrogen sulphide, and so could have been achieved using the same bacteriochlorophyll. Hydrogen peroxide would therefore have been a good source of hydrogen for photosynthesis. And, while far less plentiful than hydrogen sulphide and iron salts, it was nonetheless formed most readily in the surface waters, closest to the full power of the Sun. If this scenario is true, then catalase could have doubled as a photosynthetic enzyme. Because splitting hydrogen peroxide generates oxygen, this recruitment of catalase to photosynthesis also bridges the evolutionary gap between anoxygenic and oxygenic photosynthesis.

If catalase was acting as a photosynthetic enzyme, then it would be natural for a number of catalase molecules to cluster around the photosynthetic apparatus. In these circumstances, it would be simple for two catalase molecules to became associated as a complex: the prototype oxygen-evolving complex. At first it would have continued to use hydrogen peroxide as an electron donor, but given the right energy input, this complex could split water. We know that three small changes in the structure of bacteriochlorophyll can transform its properties, enabling it to absorb high-energy light at a wavelength of 680 nm. We now have a prototype oxygen-evolving complex (the nut-cracker that can physically split water) and a chlorophyll that can provide enough energy for it to do so (or the hand that presses the nut-cracker). Thus, with no foresight and no disadvantageous steps, we have taken a path leading from anoxygenic photosynthesis to oxygenic photosynthesis.

The evolution of oxygenic photosynthesis, then, seems practically inevitable, as long as three conditions are met: a selective pressure to use water; a mechanism for splitting water; and a tolerance to the oxygen waste. The selective pressure to use water was the loss of iron and hydrogen sulphide from sheltered environments. The mechanism for splitting water was a simple binding together of two catalase molecules. Tolerance for oxygen was imparted by catalase, and probably several other antioxidant enzymes which had evolved in response to oxidative stress from ultraviolet radiation.

These conditions could never have been fulfilled in the deeper oceans. They were full of iron and hydrogen sulphide, and shielded from the effects of ultraviolet radiation. Life there would have had no need to tolerate oxygen. In these places, even if given enough light, any mutations that produced chlorophyll from bacteriochlorophyll would have been eliminated by natural selection as worse than useless. They would have slashed the light-capturing capacity of bacteria without any gainful return.

The explanatory power of an 'oxidative stress before free oxygen' hypothesis is strong. If true this reverses received wisdom. The hypothesis implies that photosynthesis would not have been possible without the oxidative stress generated by ultraviolet radiation. Far from cowering away at the bottom of the oceans, in sulphurous hydrothermal vents (or black smokers), life embraced the surface oceans very early, and dealt with the conditions there through the evolution of potent antioxidant enzymes such as catalase. Without these radiation-scorched conditions, water-splitting photosynthesis could never have evolved. Even more significantly, the evolution of oxygenic photosynthesis hangs by a single thread: the accidental association of two catalase molecules.

If this hypothesis seems to be over-reliant on a single lucky chance, it is worth remembering that, unlike flight or vision, which evolved independently many times, oxygenic photosynthesis only ever evolved once. All algae, all plants, the entire green planet, use exactly the same system. All of them inherited it from the cyanobacteria, which invented it once, perhaps 3.5 billion years ago. No other cells on Earth ever learnt to split water. All known water-splitting complexes are related in structure, and all are similar to catalase. Perhaps life once existed on Mars, but found another way of dealing with the less-intense solar radiation. Catalase never evolved. Without catalase, oxygenic photosynthesis never evolved. Without photosynthesis, free oxygen never accumulated in the air. And without oxygen, there was no multicellular life, no little green Martians, no war of the worlds.

Convinced? Perhaps not, but there is more. To my mind, the most conclusive evidence comes not from atmospheric modelling, or structural and functional similarities, but from comparative genetics. Not the genetics of photosynthesis, which are still rather murky, but the genetics of respiration: how life came to use free oxygen as a means of extracting energy from food in the first place. Again, intuition is turned on its head. It seems impossible for oxygen-respiring organisms to have evolved before there was any free oxygen in the air. Surely they could not have evolved before oxygenic photosynthesis! We may have to think again. According to another iconoclastic viewpoint, put forward and backed with increasingly strong evidence by José Castresana and Matti Saraste of the European Molecular Biology Laboratory in Heidelberg, this is exactly what happened.[6] Respiration using oxygen evolved before photosynthesis, oxygen breathers before there was any free oxygen in the air. The arguments of Castresana and Saraste hinge on the identity of a single-celled creature named LUCA, the Last Universal Common Ancestor. We will find out who she was in the next chapter.

[6] Sadly, Matti Saraste died on the 20 May 2001 at the age of 52. His colleagues in Heidelberg let his own words speak as his testimonial, and there is no better tribute to his memory, or celebration of the fun and magic of biochemistry than these words. I hope they will inspire students in future generations to become biochemists. "The nicest aspect of biochemistry is the possibility to combine mental and practical work. One can even do these simultaneously. For me, controlling the practical work, the experiment, is extremely fun. An experiment is to approach the current scientific problem at the border of the known and unknown. It is as much fun to try to grasp the location of this magic border in your mind. To be a good biochemist, you do not have to be an egg-head or an absolute genius in maths or physics, but understanding the problems requires thinking, reading, experience and planning. On the other hand, you can keep your hands busy with minimal thinking: the bottleneck in research is often the experimental work." Matti Saraste, 1985.

CHAPTER EIGHT

Looking for LUCA

Last Ancestor in an Age Before Oxygen

LUCA, THE LAST UNIVERSAL COMMON ANCESTOR, was baptized in the luminous light of Provence, in the south of France, in 1996. She was named at a rare meeting, a gathering of the tribes. Attending were chemists who study the primal stirrings of life, molecular biologists who trace the origins and evolution of genetic replication, thermophilists who study the bacteria living in hydrothermal vents at searing temperatures, microbiologists who decipher the metabolic traits of primordial life, and geneticists who compare and contrast the complete genomes of living organisms to unravel their evolutionary relationships. The name LUCA was a happy compromise between LUA, the Last Universal Ancestor, and LCA, the Last Common Ancestor, and is less clumsy and ugly than the scientific terms 'cenancestor' or 'progenote'. LUCA conjures up images of Lucy, our own African forebear, and seems to encapsulate the trajectory of life on Earth. She (she has to be she) was a sympathetic entity who seeded life on our planet. She was not the first living thing, but rather the *last* ancestor common to all life known today, whether alive or extinct. Together with the bacteria, algae, fungi, fish, mammals, dinosaurs, grass and trees, we all owe our lives to LUCA.

Where or when LUCA might have lived is controversial, but most researchers in the field now agree on a date of about 3.5 to 4 billion years ago. She was a single-celled form of life — though some think she might not even have been as sophisticated as a cell in the way we know it today — and she probably lived in the ocean. Whether she lived a solitary life in an empty world, or mixed her genes with closely related cells, or fought a bitter evolutionary battle with quite different primordial forms of life, we do not know. If she did not live alone, then all her contemporaries have vanished without trace: LUCA seeded our planet as completely as biblical Eve seeded all humanity.

This means that LUCA's properties are the basis of all subsequent evolution on Earth. We may reasonably infer that any features that are shared by all living things were once features of LUCA herself, and were passed on with varying degrees of modification to all her descendants. Some attributes of present-day life, such as the ability to photosynthesize, are present only in some lines of descent, such as the purple bacteria, cyanobacteria and algae (which gave rise to plants), and so presumably evolved only in these lineages after the age of LUCA (unless they were independently lost by most of her descendants, which seems improbable but cannot be ruled out). The identity of LUCA may thus be pieced together, at least in theory, by comparing all the organisms that ever lived. The features that *all* organisms share were presumably inherited from LUCA herself.

Although comparing the properties of all living organisms might seem an impossible task, scientists have succeeded in defining a few of LUCA's attributes. At first sight, these attributes may seem even more improbable, but they have an internal logic of their own. Most importantly, they agree with the evidence discussed in the last chapter. LUCA probably could use oxygen to generate energy before there was free oxygen in the air. She could defend herself against oxidative stress generated by ultraviolet radiation. Her defences came before, and enabled, the evolution of oxygenic photosynthesis. Oxygen radicals were therefore the ultimate driving force behind all sophisticated life on Earth. In this chapter, we will look at what we can learn from this emerging portrait of LUCA.

The poet and polymath Goethe once said that nothing in Italy makes sense until one has been to Sicily. For biologists, nothing makes sense without the theory of evolution, which offers an intelligible framework for interpreting the overwhelming variety of biological detail. The veracity of the theory of evolution by natural selection has been confirmed not so much by a single awe-inspiring experiment as by the everyday observations of a million biologists worldwide. These innumerable observations and discoveries have invariably upheld the fundamental unity of life.

The kinship of all living things is less than obvious if we look idly around us — what do we have in common with a mulberry bush? Yet when we begin to probe beneath the surface, the similarities become more and more clear. We share an extraordinary 98.8 per cent of our total DNA sequence with chimpanzees, for example. But then, we are both bipeds with arms and legs, heads, eyes, noses, ears, brains, hearts, kidneys, a circulatory system — we even have the same number of fingers. Size apart, most people could not distinguish the kidneys of a human being from those of a chimpanzee. Our behaviour and mating rituals can be compared with some profit. Few people would maintain that such similarity is coincidental. But, we also have a remarkable number of things in common with fishes, or even with the earliest indisputable ancestor of modern animals, a lowly worm. A worm, after all, has bilateral symmetry (it is the same on both sides), a rudimentary heart, a circulatory system, a nervous system, eyes, a mouth and an anus. Unlike plants, it moves around and burrows holes in the sand.

As late as the 1950s, textbooks continued to cite the obvious differences between plants and animals, retaining Linnaeus's original subdivision of life into the two great kingdoms of Plantae and Animalia. But the dichotomy was breaking down, and was soon to be replaced by a new classification comprising five kingdoms, proposed by R. H. Whittaker in 1959: animals, plants, fungi, protists (a mixed bag including protozoa and algae) and bacteria. The new system had simplicity and convenience on its side, and for these reasons is still in common use today. Despite its honest virtues, though, it suffers a fundamental flaw. The problem is that the distinctions between the five kingdoms were based on morphology or behaviour rather than genetic ancestry. Holding a distorting mirror up to the problem, it is a little like classing a fly-eating plant and a woodpecker together on the grounds that both are multicellular and both devour insects. In reality, plants, animals, fungi and protists are much closer cousins than their visible appearance or their behaviour would have us believe. This kinship is at the cellular level, and only becomes evident under the microscope. When we consider the structure of their cells, these four kingdoms have far more in common with each other than any of them do with the fifth kingdom, the bacteria. The similarities are so fundamental that these four kingdoms are classed together as members of a single overarching taxonomic group, or domain, called the *eukaryotes*, from the Greek meaning 'with true nuclei'. All eukaryotes have a nucleus, the largest thing in the cell. It is a roughly spherical structure separated from the rest of the cell — the cytoplasm — by a double membrane. Most eukaryotic cells range from around a hundredth to a tenth of a millimetre (10–100 micrometres) in diameter, although there are exceptions, including our own nerve cells, which have fine projections that can be over a metre [several feet] long. The cytoplasm of a eukaryotic cell is crammed with assorted jumble — hundreds or even thousands of tiny specialized organs, called organelles, such as mitochondria (in virtually all eukaryotes) and chloroplasts (in algae and plants), mixed up with small sacs of membrane, stacks of folded membranes and a protein skeleton. This riot of compartmentalization gives eukaryotic cells the look of evolution by conglomeration, which is indeed what happened, as we saw in Chapter 3.

The nucleus contains the genetic heritage of eukaryotic cells — an oddly amorphous-looking material known as chromatin, made of DNA wrapped in proteins. When eukaryotic cells divide, the DNA is first replicated and then the chromatin condenses into tight coils, or chromosomes, which are eased apart on a protein framework to form two new nuclei. The detailed structure of eukaryotic genes came as one of the greatest surprises of molecular biology in the last quarter of the twentieth century. Far from being continuous compact coding sequences, neatly lined up like beads on a string, as we once imagined (and as bacterial genes had been shown to be), eukaryotic genes are discontinuous and comprise just a few per cent of the cell's total DNA. Most eukaryotes have 'genes in pieces', where the pieces of the gene coding for the protein product are interspersed with long tracts of apparently junk DNA, coding for nothing. Individual genes are also separated by large stretches of apparently junk DNA. Much of this extraordinary excess is thought to be 'selfish' DNA, which hitches a ride in the cell simply to get itself replicated, contributing nothing to the common good. Other stretches of junk DNA are the sunken wrecks of genes, holed by mutations below the waterline, such as our own derelict gene for producing vitamin C.[1] The eukary-

[1] The gene for making vitamin C still works in plants and in most animals, as we shall see in Chapter 9. Its sequence is comparable to the remains of our broken gene. We, or at least our ancestors, must once have eaten enough vitamin C from plants to satisfy our needs, since the loss of our gene for producing vitamin C was not penalised by natural selection, as it most certainly would have been had the deficit proved counter-adaptive. In general, a good guide to the importance of a gene is the degree of suffering caused by its loss. The loss of some genes is incompatible with reproductive vigour, or even with life. Most very ancient genes that are still in working order are very important — their loss has been punished repeatedly by the personal extinction of the bearer or their lineage.

otes, cells with 'true nuclei' hardly live up to their name. If we were to rename them today, any notion of 'truth' would surely be dismissed from their name. Eukaryotes are a tissue of lies, in the sense that they are not at all what they seem.

Bacteria are built on a totally different ground plan. Most important, they have no nucleus, and so are classified in a separate domain as *prokaryotes*, meaning simply 'without nuclei'. They are much smaller than eukaryotes, usually only a few micrometres in diameter, and are encased in a rigid cell wall, giving them the appearance of tiny capsules. The bacterial cell wall is composed of peptidoglycans, long chains of amino acids and sugars. Many types of eukaryotic cells also have cell walls, but none are made of peptidoglycans.

Bacterial genes are naked, in that the DNA thread is not wrapped around with proteins. Bacteria have a fraction of the number of genes of most eukaryotes — a few thousand, instead of tens of thousands. They organize these genes into groups with similar function, known as operons, and carry very little junk DNA. Nor are they encumbered with stacks of internal membranes, protein skeletons or organelles such as mitochondria. This lack of clutter allows them to divide at huge speeds simply by binary fission, or splitting in half. They can also recombine their genes with those of other bacteria by what amounts to copulation — the direct injection of genes into a neighbouring bacterium in a process known technically as conjugation. This allows genetic innovations, such as resistance to antibiotics, to spread rapidly through an entire bacterial population. In comparison with the eukaryotes, which lumber like battleships, bacteria have the evolutionary speed and agility of fighter aircraft.

The gulf between prokaryotes and eukaryotes is genuinely deep, but the two groups are clearly related at a fundamental level — the biochemistry of their cells. This is the kind of detail that persuades biologists that all living things on Earth are indeed related. As the proverb says, there are many ways to skin a cat. The fact that *all* life systematically follows the same way, step by step, suggests that all life has been following the same set of instructions from the very beginning. The genes in all cells, for example, are made from DNA (deoxyribonucleic acid). DNA constitutes a four-letter 'genetic code', which stipulates the sequences of the amino-acid building-blocks in proteins. The code is, to all intents and purposes, identical in all living things. Not only this, but the detailed mechanism by which proteins are built, following the instructions encoded in DNA, is the same in all cases. The sequence of letters in DNA is read off, or transcribed, into a rather similar molecule called messenger RNA (ribonucleic acid). This carries the instructions for building a particular protein from the gene to the protein-assembly stations, tiny molecular machines called ribosomes where all the action takes place. Here, on the ribosomes, the coded message is translated into a protein. This is done in the same way in all cells, using a repertoire of 'adaptor' molecules, also made from RNA, but this time with amino acids in tow. These adaptors are called transfer RNA. The molecules of transfer RNA recognize sequences of letters in messenger RNA and match them with the appropriate amino acid. Essentially the same process takes place in all living organisms, making use of the same code, the same messenger RNA and transfer RNA, the same ribosomes and the same amino acids. For Mother Nature, it seems, there is only one way to skin a cat.

The conclusion from all this — that all life on Earth has a common ancestor — is backed up, most convincingly of all, by the surprising unity of life at its most basic level: the 'handedness' of individual molecules. Many organic molecules, including amino acids and simple sugars, exist in two mirror-image versions, analogous to the left hand and right hand of a pair of gloves. Both hands are equally abundant in nature, and there is no reason why one should be used by living organisms instead of the other. Once a decision has been taken, though, it is hard to swap. A left-handed glove will never fit on the right hand. Similarly, an enzyme that catalyses the reactions of a left-handed molecule is useless when confronted with the right-handed version. Once that enzyme's sequence of amino acids has been encoded in the DNA, it is too late to swap. To make two mirror-image enzymes that work with opposite-handed molecules is a waste of resources: life must make a random choice and then stick to it. Given a random choice, we might expect that some species would have right-handed molecules, and others left-handed molecules (making full use of natural resources), but this is not the case. All life is right-handed in its molecular preference. The only sensible explanation for this extreme conservatism is that LUCA herself was right-handed — a historical accident — and that all her descendants have been obliged to follow suit.

When did LUCA give rise to her diverse offspring? Cells resembling modern prokaryotes date back 3.5 billion years, to the stromatolites in south-western Australia, as described in Chapter 3. The first signs of eukaryotic cells, the biomarkers of membrane sterols, date to about 2.7 billion years ago. The first unequivocal eukaryotic fossils are found in rocks dating to about 2.1 billion years ago. An explosion in the number and variety of eukaryotic cells took place around 1.8 billion years ago.

Eukaryotic cells share their fundamental biochemistry with prokaryotes, but are larger and more complex. It seems likely that only the innate complexity of the eukaryotic cell can support the added layers of organization required for the evolution of multicellular life. Certainly, all true multicellular organisms are eukaryotes. Taken together, these bare facts suggest that the prokaryotes were the first primitive cells (as suggested by their name), and that the more advanced eukaryotes evolved from them later, gradually accruing complexity.

Many features of the eukaryotes support this conclusion. During the mid-1880s, the German biologists Schmitz, Schimper and Meyer proposed that chloroplasts were derived from cyanobacteria. In 1910, the Russian biologist Konstantin Mereschovsky took this idea further, arguing that eukaryotic cells had evolved from a union of different types of bacteria. With only rudimentary microscopic techniques to back up his arguments, however, he failed to convince the biological establishment. His ideas stagnated for nearly 70 years until the late 1970s, when Lynn Margulis, at the University of Massachusetts at Amherst, championed the cause and marshalled evidence that organelles were once free-living bacteria, at a time when new molecular methods could prove the case.

It is now accepted, as one of the basic tenets of biology, that chloroplasts and mitochondria (the energy 'power-houses' of eukaryotic cells) were once free-living bacteria. Many details betray their former status. Both, for example, retain a genetic apparatus, including their own DNA, messenger RNA, transfer RNA and ribosomes. These bear witness to their bacterial origins. Mitochondrial DNA, for example, like bacterial DNA, comes packaged as a single circular chromosome, and is naked (not wrapped in proteins). The sequence of letters in its genes is closely related to the equivalent genes in a class of purple bacteria called the alpha-proteobacteria.[2] Mitochondrial ribosomes also resemble those of the proteobacteria in their size and detailed structure, as well as their sensitiv-

[2] Mitochondria are also purple, and are in fact one of the few coloured components of the cell. In a vivid aside in *The Energy of Life*, Guy Brown remarks that 'were it not for the melanin in our skin, myoglobin in our muscles and haemoglobin in our blood, we would be the colour of mitochondria. And, if this were so, we would change colour when we exercised or ran out of breath, so that you could tell how energized someone was from his or her colour.'

ity to antibiotics such as streptomycin. Again, like bacteria, mitochondria divide simply by splitting in half, usually at different times from each other and from the rest of the cell.

Despite these atavistic features, mitochondria have lost almost all their former independence. Two billion years of shared evolution have left the mitochondrial genome with little to call its own. Its closest bacterial relatives, the alpha-proteobacteria, have at least 1500 genes, whereas the mitochondria of most species have retained less than a hundred. As we saw in Chapter 3, evolution tends towards simplicity as readily as complexity. Any bacterial genes unnecessary for survival inside the eukaryotic cell would have been quickly lost, as genes in the nucleus could take over their role without competition or antagonism. Other mitochondrial genes have physically moved to the nucleus — 90 per cent of the genes that determine mitochondrial structure and function now reside in the cell nucleus. Why the remaining 10 per cent of genes stayed put in the mitochondria is something of an enigma, but their location probably confers some sort of advantage.

As far as our story of LUCA's identity is concerned, the movement of genes from free-living bacteria into eukaryotes has a profound impact on how we must view the web of genetic relations between living things. Clearly, the nuclei of eukaryotic cells contain bacterial genes abstracted from mitochondria. Any attempt to trace the earliest genetic heritage of eukaryotes on the basis of these genes would be misleading: they are a late graft rather than an ancestral trait of the eukaryotes. But in many respects the mitochondrial genes are easy to track. At least we know their context and their function. What we don't know is how many of the rest of the genes in the eukaryotic nucleus were once subsumed in this manner; or indeed, how to tell which ones they are. This is the general problem posed by lateral gene transfer — the movement of genes from one organism into another by a means other than by direct inheritance.[3] If genes circulate with the freedom of money in an economic union, it becomes virtually impossible to trace the descent of an organism — it may have inherited its genes vertically from its own ancestors, or laterally from an unrelated species. The further back we go in time, the more twisted and obscure this web becomes.

[3] The phenomenon of lateral gene transfer appears to be relatively common among bacteria over evolutionary time. As well as exchanging genes with close relatives by conjugation, bacteria in general are able to take up pieces of DNA from their environment, and occasionally these will become incorporated into their own DNA.

In the late 1960s, the web of genetic relatedness between organisms came to obsess a young researcher at the University of Illinois, a biophysicist turned evolutionary biologist by the name of Carl Woese. Woese recognized that if entire genomes could be sequenced, the 'average' relatedness of different species might still shine through the superimposed layers of lateral gene movement. At the time, however, sequencing such a massive number of genes was not feasible. What was needed instead was a single gene that could be relied upon to have stayed put — a gene that would not be transmitted sideways, but only vertically to the next generation. The fate of such a gene would be linked irrevocably with individual lineages, allowing, in principle, a grand reconstruction of all evolution.

This rare gene would also need to be highly resistant to change. The problem here is that the sequence of 'letters' in a gene gradually changes over evolutionary time, as a result of random mutations that change, insert or delete letters. Most genetic mutations that affect the protein or RNA product of the gene are harmful, but some are 'neutral', that is, they have no effect on the production or function of the gene's product, and a few are beneficial. As neutral or beneficial changes are not penalized by natural selection, they can accumulate over time. The outcome is that if you look at the 'same' genes from two species that have diverged from a common ancestor, their sequences will differ. In theory, the more closely related the species, the less the sequences will differ, as there has been little time for mutation to occur, whereas distantly related species will have more differences in sequence.

For example, the genes encoding the oxygen-carrying haemoglobins have diverged at a rate of about 1 per cent every 5 million years. This means that close relatives, which diverged only recently, have similar haemoglobin sequences, whereas distant relatives have quite different haemoglobins. Similar patterns apply to other essential and widely shared genes, such as that for the respiratory protein cytochrome *c*. Our gene for cytochrome *c* is approximately 1 per cent different from chimpanzees, 13 per cent different from kangaroos, 30 per cent different from tuna fish and 65 per cent different from that in the fungus *Neurospora*. Clearly, at this rate, genetic drift may result in the complete loss of any sequence similarity between genes over billions of years, even if they do share a common ancestor.

Some DNA sequences drift faster than others. The fastest changes take place in junk DNA, as these sequences do not code for anything and so are not subject to the restraining influences of natural selection. On the other hand, a few genes are so central to the life of the cell — as structurally important as a cantilever — that almost any tampering is detrimental. As any cell is likely to pay with its life for such changes, the 'cantilever genes' are the least likely to drift. Changes are almost never passed to the next generation because almost all the affected cells die. Even so, very rarely, a change will occur that is not penalized by natural selection. Changes in such genes in different species would accumulate very slowly over billions of years, and could be used to produce a branching tree of relationships that preserves a record of the earliest evolutionary patterns.

Do such genes exist? Woese reasoned that cells depend on a supply of building materials in the same way that a society depends on a supply of bricks and steel to build schools, factories and hospitals. Just as society would quickly grind to a halt if no building materials were available, Woese argued that life is unthinkable without proteins or the DNA code to ensure the subtlety and continuity of protein function. Protein synthesis must therefore be one of the most ancient and fundamental aspects of life, so it is no surprise to find that the pathways of protein synthesis are deeply embedded in the workings of a cell. As any changes in the genes controlling protein synthesis are highly likely to be fatal, these genes, more than any others, are likely to have been present in LUCA, to be very stable, accumulating relatively few genetic changes over time, and to be unlikely to move around the gene pool by lateral gene transfer.

We have seen that proteins are built on ribosomes. Ribosomes themselves are made from a mixture of proteins and yet another form of RNA, called ribosomal RNA. Both the proteins and the ribosomal RNA are encoded by DNA and so both are subject to the restraints of natural selection. Woese recognized that of all the components of a cell, ribosomes were the closest approximation to a cantilever — absolutely indispensable to all aspects of cellular function — and were therefore highly unlikely to undergo rapid mutation or wander around the gene pool. Furthermore, because the sequence of letters in ribosomal RNA is an exact replica of the gene, ribosomal RNA sequences could be compared directly, without recourse to the genes themselves. In the 1960s and 1970s this was invaluable, as ribosomal RNA was then much easier to isolate and sequence than the parent genes. Woese therefore settled on ribosomal RNA as a yardstick of evolution. He set about comparing ribosomal RNA sequences from his own lab and from the literature, to produce a map of the genetic relatedness of all life. This grand objective was taken up by many research groups, and the project quickly gathered momentum.

Along with everyone else working in the field, Woese expected to uncover an ancient ancestral genetic link between the prokaryotes and the eukaryotes — something analogous to the clear relationship between mitochondria and alpha-proteobacteria. Two great surprises were in store. First, the gap between the two domains continued to yawn. No microbial missing link could be found, nor indeed, any continuum between bacterial and eukaryotic ribosomal RNA sequence, as would be expected if the eukaryotes had simply evolved from bacteria. Instead, the RNA sequences clustered obstinately into two distinct groups, as if they had nothing in common. This could only mean that the split between bacteria and eukaryotes had taken place very early indeed, perhaps not long after the first stirrings of life itself. This in turn meant that the eukaryotes could *not* have evolved gradually from bacteria over 2 billion years, as everyone had expected. The split must have happened very quickly and very early.

Then came the second surprise, announced by Woese and Fox in 1977, and now seen as one of the great paradigm shifts in biology. A deep divide emerged within the prokaryotic domain itself. A little-known group of prokaryotes, most of which inhabited extreme environments such as hot springs and hypersaline lakes, confounded all expectations when their ribosomal RNA was analysed. The analyses showed that they shared little more with the bacteria than the absence of a nucleus. As more of their ribosomal RNA was sequenced and compared, it became clear that the divergence was not just a new kingdom within the prokaryotes, but something much more basic — an entirely new domain, which has become known as the Archaea (Figure 9). Today, instead of five kingdoms, we recognize three great domains of life: the Bacteria, the Archaea and the Eukaryotes. We ourselves, as animals, occupy no more than a small corner of the Eukaryotes (Figure 10).

The existence of the Archaea allows us to paint a far more convincing picture of LUCA. We can now compare the characteristics of three different domains of life. Archaea are obviously comparable to bacteria in that they lack a cell nucleus, and so are defined as prokaryotes. The organization of their genes is also similar to that of bacteria: they have a single circular chromosome, they cluster groups of related genes into operons, and they carry little junk DNA. Other aspects of their organization, such as the structure and function of proteins in the cell membranes, bear a more superficial resemblance to bacteria. Most archaea have a cell wall but, unlike bacteria, a few do not. Again unlike bacteria, the cell wall contains no peptidoglycans. The similarities quickly tail away.

In other respects, the archaea lie much closer to the eukaryotes. Although they do not have as many genes as eukaryotes, archaea have on average more than twice as many genes as bacteria. The DNA of archaea is not naked, but is wrapped in proteins similar to those used by eukaryotes. The detailed mechanism of DNA replication and protein synthesis is much closer to that of the eukaryotes. For example, they switch their genes on and off using mechanisms very similar to those in eukaryotes. The protein constituents of the ribosomes also resemble those of the eukaryotes in their structure. Other details of ribosomal function, includ-

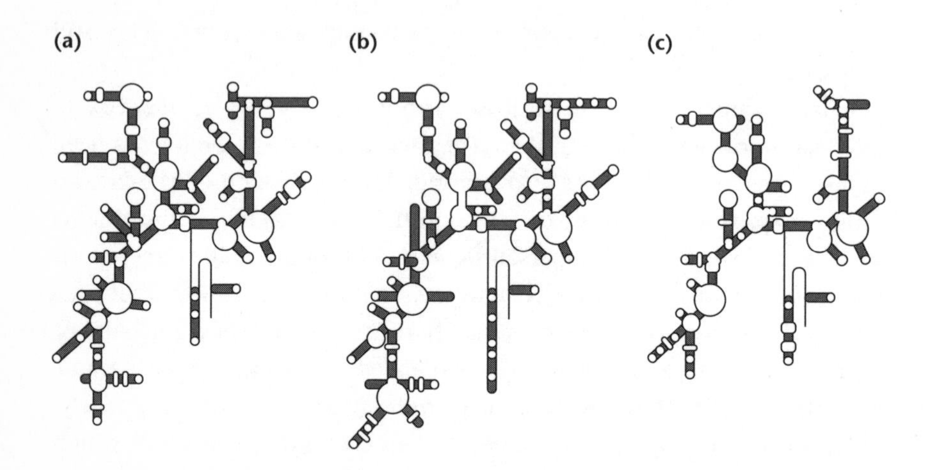

Figure 9: Predicted structures of ribosomal RNA from **(a)** the archaeon *Halobacterium volcanii*, **(b)** the eukaryote baker's yeast, and **(c)** cow mitochondria. RNA is single-stranded, but, as in DNA, the letters can pair up to form bridges between two chains. In the case of RNA, a single chain doubles back on itself to form loops and hairpins (whereas the famous DNA double helix is in fact two distinct chains entwined in a helix). The 'bubbles' in this illustration are single-stranded RNA, in which the letters have not paired up. A comparison of the three ribosomal RNAs shows that the overall shape and structure of ribosomal RNA (its secondary structure) has been maintained throughout evolution. However, the actual sequence of letters has drifted substantially, and the sequence similarities are very low. The mitochondrial RNA structure is reminiscent of bacterial RNA, from which the mitochondria originated. Reprinted with permission from Gutell *et al.*, and *Progress in Nucleic Acid Research and Molecular Biology*.

ing the initiation of protein synthesis, the elongation of protein chains, and the termination steps, parallel the eukaryotic process. Finally, and most convincingly of all, genetic analyses of so-called paralogous gene pairs — the products of gene duplications in a common ancestor, followed by divergent evolution in different groups of descendants — indicates that archaea are indeed relatives of eukaryotes. In essence, archaea are prokaryotes with many features of eukaryotes. They are as close to a missing link as we are ever likely to find.

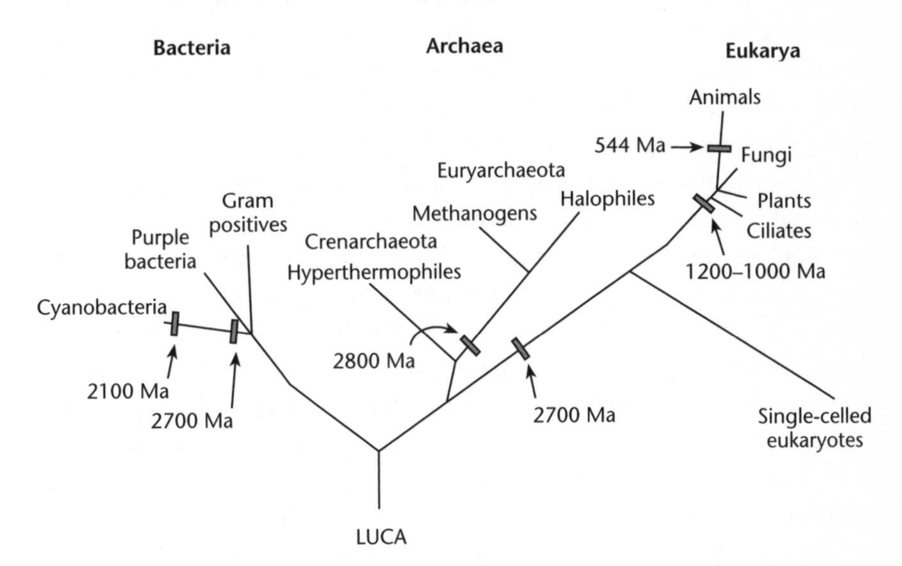

Figure 10: A 'rooted' universal tree of life, showing the three domains of life. The tree is based on sequence comparisons of ribosomal RNA, analysed by Carl Woese and his colleagues. The order and length of branches are proportional to the sequence similarities within and between the domains and the kingdoms of life — in other words, they are directly proportional to the genetic similarities between species. It is humbling to note that the animals, plants and fungi account for just a small corner of the Eucarya domain, and that there is less variation in ribosomal RNA sequences within the entire animal kingdom than there is between different groups of methanogen bacteria.

The common 'root' represents LUCA, the Last Universal Common Ancestor of bacteria, archaea and eukaryotes. Notice that the Archaea are intermediate between the Bacteria and Eucarya, as inferred from many of their detailed morphological and biochemical properties, as well as their ribosomal RNA sequences. The boxed dates indicate the minimum age of selected branches, based on fossil evidence and biochemical fingerprints, such as the characteristic membrane steroids found by Jochen Brocks and his colleagues in the shales underlying the Hammersley iron formation in Australia. Adapted with permission from Andrew Knoll and *Science*.

What does all this say about the identity of LUCA? It seems likely that the split between the Archaea and the Bacteria occurred very early in the history of life, perhaps 3.8 to 4 billion years ago. We assume that both archaea and bacteria retain some of the original features of LUCA herself. Calculations suggest that eukaryotes split from archaea later, perhaps around 2.5 to 3 billion years ago, as they share far more fundamental traits with archaea than with bacteria (see Figure 10). We know that eukaryotes acquired mitochondria and chloroplasts around 2 billion years ago by engulfing bacteria. We also know that some of these bacterial genes became part of the chromosomes of the eukaryotic cells. Here we return to the problem of lateral gene transfer. If the eukaryotes are essentially a fusion of archaea and bacteria, then it is plain that lateral gene transfer has taken place across domains. If we wish to draw a portrait of LUCA by comparing the properties of the different domains, can we be sure that they are not completely mixed up?

Luckily, there is some evidence that lateral gene transfer is not common across domains. The development of the eukaryotes seems to have been a singular event, possibly propelled by the unique environmental conditions around the time of the snowball Earth of 2.3 billion years ago (Chapter 3). In general, however, the archaea have kept themselves very much to themselves, and give every appearance of having changed little since the beginnings of time. No archaea are pathogenic, which means that they do not infect eukaryotes, and so do not have much opportunity to mix their genes with eukaryotes in the course of intimate warfare. Nor do they compete with bacteria in other settings. Their predilection for extreme conditions isolates them from most other organisms, even bacteria. Hyperthermophilic archaea, such as *Pyrolobus fumaris*, live at searing temperatures, well over 100°C, and high pressures in deep-sea hydrothermal vents. Other archaea, such as *Sulfolobus acidocaldarius*, add acidity to the heat and live in sulphur springs in places like Yellowstone National Park, at pH values as low as 1, the equivalent of dilute sulphuric acid. At the other end of the pH scale, some archaea thrive in the soda lakes of the Great Rift Valley in East Africa and elsewhere, at a pH of 13 and above — enough to dissolve rubber boots. The halophilic archaea are the only organisms that can live in hypersaline salt lakes, such as the Great Salt Lake in Utah and the Dead Sea. The psychrophiles prefer the cold and grow best at 4°C in Antarctica (their growth is actually retarded at higher temperatures).

Many of these favoured environments have barely changed for

billions of years. Without calamity or competition, the selection pressure for innovation and change must have been negligible. While it is true that some archaea do live in more normal environments — among plankton at the ocean surface, for example, and in swamps, sewage and the rumen of cattle — the genes of their extreme and reclusive cousins have surely had little traffic with the rest of life.

The extraordinary features of archaea quickly stimulated scientific and commercial interest, and the field blossomed as a distinct discipline during the 1990s. Enzymes that function normally at high temperatures and pressures are an answer begging for an application. Already enzymes extracted from archaea have been added to detergents and have been used for cleaning up contaminated sites such as oil spills. To enlist the skills of a microbe on an industrial scale, however, requires a working knowledge of its genes. Complete genome sequences have now been reported for representatives of all the known groups of archaea. These sequences at once confirm the great antiquity of archaea, and their splendid isolation over the aeons. But the greatest surprise is how many genes the archaea *do* have in common with bacteria.

Of the genes involved in energy production through respiration (aerobic or anaerobic), at least 16 have been found in both archaea and bacteria. From the close similarities in their sequences, it seems likely that these genes were present in LUCA, and were later inherited by both archaea and bacteria, as they diverged from each other to occupy their distinct evolutionary niches. This conclusion — that the 16 respiratory genes were present in LUCA herself — is supported by two independent lines of evidence, as argued by José Castresana and Matti Saraste of the European Molecular Biology Laboratory in Heidelberg.

The first line of evidence relates to evolutionary trees. The similarities between the DNA sequences of the 16 respiratory genes can be used to construct a tree of relatedness. This family tree is then superimposed over the family tree based on ribosomal RNA sequences. If the respiratory genes had been passed horizontally by lateral gene transfer, then closely related respiratory genes would be found in organisms that were otherwise only distantly related to each other. Put another way, the evolutionary histories of the respiratory genes would differ from the true evolutionary roots of their host organisms, just as the history of mitochondrial genes differs from that of the nuclear genes of eukaryotes. On the other hand, if the respiratory genes had stayed put in their respective organisms, then the evolutionary trees constructed from ribosomal RNA and the respiratory genes should correspond to each other. This is in fact the case: the evolutionary trees of the respiratory genes that have been analysed so far *do* broadly correspond to the reference tree constructed from ribosomal RNA, implying that lateral gene transfer did not occur between bacteria and archaea.

The second line of evidence relates to more recent metabolic innovations such as photosynthesis. LUCA, it seems, could not photosynthesize. No form of photosynthesis based on chlorophyll is found in any archaea. A completely different form of photosynthesis, based on a pigment called bacteriorhodopsin, similar to the photoreceptor pigments in our eyes, is practised by the so-called halobacteria, archaea that live in high-salt conditions. This mode of photosynthesis is not found in any bacteria. These disparate forms of photosynthesis presumably evolved independently in bacterial and archaeal lineages some time after the age of LUCA, and subsequently remained tied to their respective domains. If a metabolic innovation as important as photosynthesis did not cross from one domain to another, there is no reason to think that other forms of respiration would have done so. We should certainly be wary of postulating that respiratory genes crossed domains unless we have evidence that they did so; and the evidence from evolutionary trees suggests that they did not.

If we accept that lateral gene transfer between archaea and bacteria has been extremely rare, then the 16 respiratory genes must have been present in LUCA, and were later vertically inherited by different lines of bacteria and archaea. As these genes code for proteins involved in generating energy from a variety of compounds, including nitrate, nitrite, sulphate and sulphite, LUCA must have been a metabolically sophisticated organism. One gene in particular, however, shares a striking sequence similarity in archaea and bacteria, and it is this that Castresana and Saraste have used to paint an unexpected portrait of LUCA.

This gene codes for a metabolic enzyme called *cytochrome oxidase*, which transfers electrons to oxygen, to produce water, in the final step of aerobic respiration. If cytochrome oxidase was present in LUCA, then the logical, if ostensibly nonsensical, conclusion is that aerobic respiration evolved before photosynthesis. LUCA could breathe before there was any free oxygen. As Castresana and Saraste put it, no doubt relishing every word, "This evidence, that aerobic respiration may have evolved before oxygen was released to the atmosphere by photosynthetic organisms, is contrary to the textbook viewpoint."

The way in which cytochrome oxidase reduces oxygen is a marvel of nanoscale engineering. It receives electrons derived from the oxidation of glucose. It presses four electrons in turn, along with protons, onto an oxygen molecule to produce two molecules of water. This reaction is exactly the opposite of the water-splitting reaction of photosynthesis.

$$O_2 + 4e^- + 4H^+ \rightarrow 2H_2O$$

This is the single most important reaction in aerobic respiration — in effect, the combination of oxygen with hydrogen. Most people will remember the explosiveness of the reaction from school chemistry lessons. Like all reactions of oxygen, it must take place one electron at a time. The task of cytochrome oxidase is therefore a tightrope walk between harvesting the large amount of energy that can be released from this reaction, while preventing the escape of reactive free radicals. This is achieved with almost miraculous precision. In modern mitochondria, virtually no free radicals escape from cytochrome oxidase (almost all escape from other proteins in the mitochondrial chains of electron-transporting proteins). In fact, the ability of cytochrome oxidase to soak up any oxygen in the cell and transform it into water, without releasing toxic intermediates, makes it an antioxidant without equal. The fact that it also enables four times as much energy to be produced by a cell from a single molecule of glucose, compared with any other form of respiration, can almost be regarded as a bonus.

For many years, the antioxidant effect of cytochrome oxidase was cited as an explanation for its original evolution. It was logical to assume that the enzyme had first evolved in response to rising oxygen levels following the evolution of photosynthesis, and was only later pressed into service as a respiratory enzyme. This story was supported by the existence of a second (evolutionarily unrelated) form of cytochrome oxidase in some proteobacteria, including *Escherichia coli* and *Azotobacter vinelandii*. The alternative enzyme is 100 times less selective for oxygen (it does not discriminate between oxygen and similar molecules such as nitric oxide, NO) but works much faster, consuming excess oxygen very rapidly. Even more tellingly, the oxidase is only switched on when bacteria wander into high-oxygen environments, where it functions as a kind of vacuum cleaner, sucking up oxygen, or any similar molecules, that stray into its path.

We therefore have two different types of cytochrome oxidase, the activities of which vary according to the amount of oxygen in the vicinity. This does look very much like a defence against oxygen. The revisionist hypothesis of Castresana and Saraste thus poses a riddle. If a cytochrome oxidase was indeed present in LUCA, this protein could not have evolved in response to rising oxygen levels more than a billion years later; so why on earth did it evolve? In Chapter 7 we saw that lakes and sheltered ocean basins on the early Earth were probably under oxidative stress from ultraviolet radiation, which split water to produce oxygen free radicals and hydrogen peroxide. Antioxidant enzymes such as superoxide dismutase (SOD) are found in all three domains of life, and may well have been present in LUCA. Could cytochrome oxidase have evolved in the same way as an antioxidant defence against ultraviolet radiation, rather than rising oxygen levels?

The answer is not yet certain, but seems likely to be no. If it had evolved as an antioxidant to protect against ultraviolet radiation, then its respiratory function, where it harvests energy from the transfer of electrons to oxygen rather than simply mopping up oxygen gas, would have evolved later and independently in several different archaeal and bacterial lineages. If this had been the case, we would expect the detailed mechanism of energy-harvesting to vary in different lineages, whereas in fact it seems to be similar, suggesting descent from a common mechanism in a common ancestor.[4] But if we dismiss a primordial antioxidant role for cytochrome oxidase, we are left with the rather stark alternative: that the enzyme evolved for its present purpose — to generate energy by the transfer of electrons to oxygen. Is this any more credible? We have seen that oxidative stress is feasible without oxygen, but can we really have oxygen respiration when no oxygen was present? As President Clinton might have said, that depends on exactly what you mean by 'no oxygen'.

[4] This is a testable hypothesis, but it has not been fully answered at the time of writing. All true cytochrome oxidases generate energy from oxygen by pumping protons across a membrane. The proton gradient is then tapped to generate ATP, the energy currency of the cell. The generation of a proton gradient and its conversion into ATP is known as *chemiosmosis*, and the process unifies energy generation in most forms of respiration as well as photosynthesis — another example of the fundamental unity of life. If it turns out that the cytochrome oxidases in archaea and bacteria pump protons by the same mechanism, this will be good evidence that their respiratory function had already evolved in a common ancestor. On the other hand, if the detailed mechanism of energy generation is different in archaea and bacteria, this would imply that cytochrome oxidase had evolved in LUCA for a different purpose — as an antioxidant, for example, or for denitrification (conversion of nitrates into nitrogen) — and had later adapted to oxygen metabolism independently in different lineages. Evidence to date suggests that cytochrome oxidases do pump protons in the same way in different lineages, and so had probably evolved for that purpose in LUCA.

The term 'anoxia', meaning 'no oxygen', is surprisingly slippery and has different connotations for geologists, zoologists and microbiologists. Geologists call an environment 'aerobic' (oxygenated) if the oxygen content exceeds about 18 per cent of present atmospheric levels, and 'dysaerobic' if it falls below 18 per cent. Levels lower than about 1 per cent are considered to be 'azoic' or 'anoxic'. Zoologists talk about 'normoxic' and 'hypoxic' conditions, where hypoxia refers to oxygen levels that limit the rate of respiration — usually less than about 50 per cent of present atmospheric levels. The ground is even more shifting for microbiologists. The so-called 'Pasteur point', at which some microbes switch from aerobic respiration to fermentation, is usually less than about 1 per cent of present atmospheric levels of oxygen. Some microbes, though, are affected by very low levels of oxygen, often much less than 0.1 per cent of present atmospheric levels. Such low oxygen levels — essentially anoxic to geologists — could have been attained on the ancient Earth by the splitting of water, particularly in shallow seas.

Remarkably, some microbes living today are able to use oxygen even at such low levels. Some species of proteobacteria, for example, are symbiotic and live inside nodules in the roots of leguminous plants. Here, in exchange for shelter and protection, they convert nitrogen from the air into the nitrates required for plant growth. The enzyme that catalyses this reaction is called nitrogenase and, crucially, its function is blocked by oxygen, even at very low levels. Leguminous plants and nitrogen-fixing bacteria are specially adapted to keep oxygen levels in the root nodules to a minimum. For example, the bacteria surround themselves with a thick capsule of mucus to restrict the entry of oxygen. If this fails, they activate an enzyme that consumes oxygen rapidly without generating any energy. Leguminous plants also produce an oxygen-binding protein related to haemoglobin, called leghaemoglobin, which regulates the free oxygen concentration. Together, these adaptations maintain oxygen levels within the bacteria at less than 0.01 per cent of atmospheric oxygen, a concentration that does not inhibit the nitrogenase enzyme.

The remarkable fact is that, despite these adaptations to minimize oxygen levels, some of these nitrogen-fixing bacteria, such as *Bradyrhizobium japonicum*, are still aerobic. They possess a form of cytochrome oxidase known as FixN, which has an extremely high affinity for oxygen. This enzyme is distantly related to mitochondrial cytochrome oxidase and probably evolved from a common ancestor. There is some evidence that the FixN oxidase is coupled functionally with leghaemoglobin, which only releases its bound oxygen when intracellular oxygen levels are very low. Thus, the system as a whole works as follows. Low oxygen levels are maintained by a variety of mechanisms, and any oxygen that slips through this net is bound by leghaemoglobin. At *very* low oxygen levels (below 0.01 per cent), leghaemoglobin relinquishes its oxygen to the FixN oxidase, which uses it to produce energy in the form of ATP. Altogether, the system has balance and poise — it is concerned with the regulation of oxygen rather than its elimination.

The root nodule system is an extreme example of metabolism at low oxygen levels, but in conceptual terms it is very close to what actually happens inside ourselves. Our obvious dependency on oxygen conceals the fact that individual cells within internal organs are not at all adapted to bathe in an oxygen bath. The development of multicellular organisms can even be considered an antioxidant response, which has the effect of lowering oxygen levels inside individual cells. Our elegant circulatory system, which is usually presented as a means of distributing oxygen to individual cells, can be seen equally as a means of restricting, or at least regulating, oxygen delivery to the correct amount.

It's worth dwelling on this point for a moment. Consider the following. The atmospheric pressure of dry air at sea level is about 760 mm of mercury (mmHg). Of this overall pressure, 78 per cent is contributed by nitrogen and 21 per cent by oxygen. The oxygen in the air therefore exerts a barometric pressure of about 160 mmHg. Within the lungs, oxygen is taken up by haemoglobin, which is packed tightly in the red blood cells circulating through the capillaries. In these capillaries, haemoglobin is usually 95 per cent saturated with oxygen. The pressure exerted by oxygen is about 100 mmHg. As the blood is transported around the body, haemoglobin gives up its oxygen and so the oxygen pressure begins to fall. As blood leaves the heart, the oxygen pressure has already fallen to about 85 mmHg; in the arterioles it falls further to about 70 mmHg; and in the capillary networks in our organs to about 50 mmHg. Here the saturation of haemoglobin is about 60 to 70 per cent. Oxygen dissociates from haemoglobin and diffuses into the individual cells from the capillaries down a concentration gradient. This gradient is maintained by the continuous removal of oxygen by respiration. In most cells, the oxygen pressure is approximately 1–10 mmHg. In the final stage, oxygen is sucked into the mitochondria, where active respiration lowers levels even further. The oxygen pressure inside the mitochondria is typically less than 0.5 mmHg. In percentage terms, this is less than 0.3 per cent of atmospheric oxygen levels, or 0.07 per cent of total atmospheric pressure. This is a surprisingly low figure, and not so far above the supposedly 'anoxic' conditions of the early Earth. Might it be that mitochondria have succeeded in preserving a ghost of times past?

There is a second point of comparison between the root nodules of leguminous plants and animal respiration, and that is the function of haemoglobin and its cousins (including the muscle protein myoglobin). After the twists of this chapter, it may no longer come as a surprise to learn of the discovery of a haemoglobin-like protein — complete with tell-tale sequence similarities — in the archaeon *Halobacterium salinarum*, a finding reported in *Nature* in February 2000 by Shaobin Hou and his colleagues at the University of Honolulu in Hawaii. The great antiquity of haemoglobins and myoglobins is not in itself a surprise; a haemoglobin-like sequence has been found in bacteria. The significance of Hou's finding is that haemoglobin-like molecules may have been present in LUCA.

Why should haemoglobin, the oxygen-transporting protein in the blood stream of animals, be found in LUCA, or for that matter in any other single-celled organism? The answer is clear enough if we change our perspective: we should not think of haemoglobin as an oxygen transporter, but rather as a molecule that regulates oxygen storage and supply. This is essentially how leghaemoglobin works in the root-nodule system — it maintains intracellular oxygen at a low level, releasing its own oxygen on demand. This is also how myoglobin works, the protein responsible for the red colour of muscles. Myoglobin is similar in structure to one of the subunits of haemoglobin, and has a greater affinity for oxygen than haemoglobin. This means that myoglobin in muscles can draw in oxygen from the blood stream and store it in the muscles until needed. Whales and other diving mammals have very large muscular stores of myoglobin, which can pool large quantities of oxygen, allowing them to remain under water without breathing for an hour at a time. The level of free oxygen in their muscles remains constant and low.

The same system is at work in single cells. Their haemoglobin-like proteins store, and then release, oxygen to maintain a constant and low intracellular concentration suitable for respiration. This regulatory role is emphasized by the findings of Hou and his colleagues. The haemoglobin-like molecule that they discovered in *Halobacterium salinarum* functions as an oxygen sensor, which enables the archaeon to sense oxygen levels and to migrate towards zones with ideal oxygen concentrations. Some bacteria also employ oxygen sensors in this way. The common denominator in all these settings is the ability of haemoglobin-like molecules to maintain intracellular oxygen at an appropriate level.

From this point of view, the arguments of Castresana and Saraste, that LUCA could breathe oxygen, begin to seem reasonable. She wouldn't have needed much oxygen — a barely detectable amount — and she could probably store her own to use when required. If this is true, then many descendants of LUCA presumably lost the ability to use oxygen to generate energy as they adapted to specialized niches. Others lost the ability to metabolize sulphites or nitrites. The eukaryotes must have evolved from a lineage that had lost the genes for most respiratory proteins, including cytochrome oxidase, before regaining some of them from the purple bacteria that ultimately evolved into mitochondria. The unrecognizable relics of these defunct eukaryotic genes might still make up a part of our junk DNA.[5] But the most startling point is that LUCA herself could use oxygen to generate energy nearly 4 billion years ago. She clearly knew how to cope with oxygen, and probably made use of a haemoglobin-like protein and antioxidant enzymes such as SOD to protect herself. The notion that antioxidants appeared later as a desperate response to rising oxygen levels in some kind of oxygen holocaust, is falsified once again, this time from a genetic point of view.

The story we have prised from the genes is inevitably speculative, but is backed up by intriguing and coherent evidence. The conclusion that LUCA was metabolically versatile solves a number of paradoxes that cannot easily be explained by the old textbook version of events — in particular, the evolution of photosynthesis, and the antiquity of haemoglobin and aerobic respiration. If the outlines of the story are valid, then conventional wisdom has been turned on its head at almost every point.

To place this detail in perspective, here's a summary of the new evolutionary scheme. LUCA lived in a world scorched by radiation. This in itself suggests that wherever life might have originated, LUCA herself

[5] Genes lost more recently are still recognizable, as their sequences have had less time to drift. They are called pseudogenes. The completion of the human genome project has given us some idea of how many genes have been lost in this way. A good example is the sense of smell. We once had 900 genes dedicated to detecting smell, but 60 per cent of them are 'broken' in that no protein can be copied from them. As our primate ancestors took to the trees, vision became more important than smell in the struggle for survival, so many genes for smell could be lost without adverse consequences.

must have lived, at least part of the time, in the surface oceans. As the archaea evolved from LUCA (rather than the other way around) the sulphurous, heat-loving extremophiles cannot truly represent the earliest forms of life, as some have argued. On the contrary: if LUCA was metabolically versatile, then she must have lived in a world demanding versatility, and this must have included the surface layer of the oceans.

The effect of radiation on the ocean surface was far from the barren sterility often portrayed. By splitting water to generate free radicals and hydrogen peroxide, ultraviolet radiation provided an additional source of energy. Hydrogen peroxide was stable enough to build up in shallow seas and lakes before dissociating to produce water and oxygen. Oxygen generated in this way within cells could be captured and stored, using haemoglobin. This oxygen was later released for energy production using cytochrome oxidase. The ubiquitous haem chemical group, a component of cytochrome oxidase and haemoglobin, may even have been the basis for the evolution of the chemically closely related chlorophylls, which could capture light energy and convert it into sugars using variants of the existing respiratory pathways.[6]

The first photosynthesizers probably split hydrogen sulphide or iron salts, but as these resources were depleted, in sheltered environments, by oxidative stress (see Chapter 7), selective pressure drove the adaptation to alternatives: hydrogen peroxide and finally water. Once oxygenic photosynthesis had evolved, free oxygen began to build up in the atmosphere and oceans, transforming the world for ever. Yet the ideal oxygen levels for respiratory metabolism remained close to those in which the respiratory enzymes had evolved in the first place. Even today, our own cytochrome oxidase functions best at an oxygen concentration of less than 0.3 per cent of atmospheric oxygen pressure. Our bodies strive to maintain mitochondrial oxygen concentrations at this level.

While a striking departure in its own right, this new story leads to a conclusion that dovetails with a great deal of recent medical research, in particular the complexities associated with antioxidant treatments. The continuous strain between external and internal oxygen levels lies at the root of many human ailments — a strain that is ultimately worked out in the antioxidant balance of individual cells. The retention of a primordial antioxidant balance is analogous to the salt composition of fluids in the

[6] The similarity in the reaction mechanism between cytochrome oxidase and the oxygen-evolving complex has been noted by Curtis Hoganson and his colleagues at Michigan State University, among others; see Further Reading.

human body, which still parallels the sea water in which our single-celled ancestors once swam. (J. B. S. Haldane memorably referred to the collection of cells that make up the human body as a sea monster.)

The antioxidant balance is too integral to the innermost workings of the cell to behave predictably when subjected to the ruder assaults of medicine. Several of the molecules that we have discussed in this chapter are antioxidants. Catalase breaks down hydrogen peroxide to produce oxygen without generating hydroxyl radicals. Haemoglobins and myoglobins bind oxygen, only releasing it when dissolved oxygen levels have fallen to safe limits. Cytochrome oxidase hoovers up excess oxygen, again without generating free-radical intermediates. Each of these steps regulates intracellular oxygen levels and so attenuates the generation of free radicals. Yet it is quite conceivable that catalase evolved as much to generate oxygen from hydrogen peroxide as to lower toxic levels of oxygen. Haemoglobin may well have evolved to hoard oxygen at a time when it was a scarce and valuable resource. Cytochrome oxidase probably evolved as a metabolic enzyme rather than as an antioxidant. Their evolutionary past is built into their modern function, and we ignore it at our peril.

Such functional opacity highlights the surprising dilemma over the definition of the word 'antioxidant', as well as the misguided assumption that molecules evolve for a single purpose. In fact, many so-called antioxidants also serve other purposes, and are tightly woven into the regulatory system of cells. As such, antioxidants maintain oxygen levels within physiological limits, rather than simply mopping up free radicals. This is a crucial distinction. Antioxidant treatments often aim to eliminate free radicals, but by doing so may instead strain the regulatory balance. In thinking about our health, then, it is not enough to consider diseases on their own terms — we must also ask, from an evolutionary perspective, *how* things got to be that way and what might happen when we fiddle with them. We shall see in the next two chapters just how tightly antioxidants are integrated into the fabric of life. Then, we will examine what this means for our prospects of ameliorating human ageing and disease.

CHAPTER NINE

Portrait of a Paradox

Vitamin C and the Many Faces of an Antioxidant

An apple a day keeps the doctor away says the old adage; but is it true? And if it is true, why? The answer to the first question is formulated rather stiffly in our scientific age: a diet containing five 80-gram portions of fruit and vegetables each day reduces the risk of death from heart attacks, stroke and some cancers, especially those of the respiratory and digestive tracts. This is true regardless of our other habits or risks, such as smoking, weight, cholesterol levels and blood pressure. Most people currently eat about three portions a day. Several large epidemiological studies have indicated that increasing average consumption to five portions a day might reduce the risk of cancer by 20 per cent and the risk of heart attacks or strokes by about 15 per cent. Health-conscious individuals really do live longer — it is not just that it feels like it, as Clement Freud once remarked. A 17-year study that examined the mortality of 11 000 people recruited through health-food shops, vegetarian societies and magazines, found their mortality rate was half that of the general population (the study was carried out by doctors at the Radcliffe Infirmary in Oxford and reported in the *British Medical Journal*, not a partisan health-food magazine). Even allowing for the near-intractable methodological difficulties that blight such studies, and my own reluctance to eat fruit, there can be no question that a diet rich in fruit and vegetables is good for you. The problem is rather how to persuade children and adults, especially in northern Europe and the United States, to modify their diets to include five portions a day — a challenge encapsulated by the Europe Against Cancer slogan 'Take Five!'.

While the benefits of fruit and vegetables are indisputable, the epidemiology of diet, full as it is of bare associations and correlations, is two-dimensional science. For those of an inquisitive cast of mind, the question *why* is altogether more interesting and complex. Clearly, fruit and vegetables are filled with goodness; yet, perhaps surprisingly, this is about as much as we know for sure. The depth of our ignorance is conveyed pungently in an article by John Gutteridge and Barry Halliwell: "Twenty years of nutrition research have told us that for 'advanced' countries the way to a healthy lifestyle is to eat more plants, a concept familiar to Hippocrates. What it has not told us is exactly why."

If pressed for an explanation, I imagine that most people would say 'vitamin C', 'antioxidants' and suchlike. The reality is of course far more complicated. The health effects of hundreds, if not thousands, of biologically active compounds isolated from fruit and vegetables have been pored over without real consensus. Given the overwhelming detail, it is not surprising that we tend to fall back on a handful of vitamins that people have at least heard of, but which serve really as surrogates for the consumption of many other compounds. A good example is a study reported by Kay-Tee Khaw and her team at Cambridge in *The Lancet* in March 2001, which was reported widely and misleadingly by the press: the general projection being that vitamin C lengthens lifespan. In fact, the Cambridge team reported that the risk of death (from any cause) was higher in people with low plasma levels of vitamin C, and conversely, that people with high levels of vitamin C in their plasma were less likely to die within the period studied. The risk of death in people with the highest plasma levels of vitamin C was *half* that of people with the lowest plasma levels. In the article itself, Khaw and her colleagues were careful to point out that there was *no* association between vitamin C supplementation and mortality. The association was more generally with dietary intake. And the authors did not discriminate between the amount of vitamin C compared with other factors eaten in the same food at the same time (this is not laziness: asking a specific question often means ignoring superfluous details). No measurements were taken, for example, of plasma levels of vitamin E or beta-carotene. Had they been measured, a similar correlation would almost certainly have been found, for all kinds of antioxidants are abundant in fruit; but that does not mean that they were responsible for the lower number of deaths either. In the Cambridge study, then, plasma vitamin C levels were probably just a surrogate for overall fruit consumption. As to the role of vitamin C itself, we are none the wiser.

Because it is at once so familiar and so inscrutable, we will use the example of vitamin C in this chapter as a springboard to explore the wider function and behaviour of antioxidants. Although often defined simply as a water-soluble antioxidant, vitamin C illustrates many of the difficulties we face in trying to define an antioxidant. Here is Tom Kirkwood, eminent researcher on ageing at the University of Newcastle, and presenter of the 2001 BBC Reith lectures, giving a vivid depiction of the action of vitamin C:

> When a molecule of vitamin C encounters a free radical, it becomes oxidised and thereby renders the free radical innocuous. The oxidised vitamin C then gets restored to its non-oxidised state by an enzyme called vitamin C reductase. It is like a boxer who goes into the ring, takes a hit to his jaw, goes to his corner to recover, and then does it all over again.

Kirkwood's description is not wrong, but it is one-sided. His memorable simplicity conceals a can of worms. The molecular action of vitamin C is as simple and repetitive as flipping a coin, yet the effects are varied, unpredictable and utterly dependent on the milieu in which it operates. Just as flipping a coin leads to diametrically opposed outcomes, so too, vitamin C may on the one hand protect against illness and on the other kill tumours, or even people. The food chemist William Porter summed up the conundrum nicely, rising to an anguished eloquence rarely matched in scientific journals: "Of all the paradoxical compounds, vitamin C probably tops the list. It is truly a two-headed Janus, a Dr Jekyll-Mr Hyde, an oxymoron of antioxidants."

Few subjects have polarized medical opinion more violently or senselessly than vitamin C. If one man can be held responsible for this division, it is the great chemist, peace advocate and double Nobel laureate Linus Pauling. We will consider Pauling's life briefly, because his views should not be taken lightly. Neither, we shall see, should they be taken uncritically.

It is unfortunate, not least for the memory of Pauling himself, that his legacy should be tainted by his controversial views on vitamin C. No researcher had a more profound impact on the advances of chemistry in the twentieth century. One reviewer of his classic 1939 textbook, *The Nature of the Chemical Bond and the Structure of Molecules and Crystals*, went so far as to say that chemistry could now be understood rather than being memorized. Pauling was awarded his first Nobel Prize in 1954 for his "research on the nature of the chemical bond ... and its application to the elucidation of complex substances". In effect, this meant that he had been awarded the prize for the body of his work over the previous twenty years rather than a specific discovery, a move unprecedented in the history of the Nobel Institute. Yet a continuous thread did run through much of Pauling's early research — an application of the laws of quantum mechanics to the structure of chemical bonds. Pauling set about calculating the length and the angles of individual bonds, using x-ray diffraction, magnetism and measurements of the heat emitted or absorbed in chemical reactions. From the values he obtained, he went on to plot the three-dimensional structure of complex molecules. One of Pauling's earliest and greatest contributions to chemistry was the idea of *resonance*, in which electron delocalization stabilizes the molecular structure (the electron 'spreads out' in space to dilute the charge density). This feature is critical to the action of vitamin C and other antioxidants, as we shall see.

By the mid 1930s, Pauling had begun to apply his analytical methods to the structure of proteins. He demonstrated the importance of minute electrical charges (hydrogen bonds) in stabilizing the three-dimensional shape of proteins, and was the first to describe their grand architectural structures, familiar to all students of biochemistry, such as the alpha-helix and the beta-pleated sheet. By the early 1950s, Pauling was turning his attention to the unsolved structure of DNA. In his famous book *The Double Helix*, James Watson describes the foreboding that he and Francis Crick felt when they heard that the 'world's greatest chemist' was considering the problem of DNA. The pair raced to apply Pauling's own methods to pip him to the post, and were overjoyed when they realized that, on this occasion, their rival had committed an elementary blunder.

By now Pauling, spurred on by his indefatigable wife Ava Helen, was becoming increasingly committed to anti-war protests. From 1946, through the 1950s and 1960s, Pauling spoke out about the perils of atomic fallout, in particular the risk of birth defects and cancer. In 1957, he drafted a petition to end nuclear weapons testing, and eventually presented the signatures of 11 000 scientists to the White House, including those of Albert Einstein, Bertrand Russell and Albert Schweitzer. The petition was widely credited with precipitating the Nuclear Test Ban Treaty, in which the United States and the Soviet Union agreed to cease testing nuclear weapons. Pauling was awarded the Nobel Prize for Peace on 10 October 1963, the same day that the Nuclear Test Ban Treaty went into effect.

Pauling's anti-war activities inevitably raised the suspicions of the United States government in the early years of the Cold War, when the House Un-American Activities Committee and Senator McCarthy were embroiled in their notorious communist witch-hunt. During the early 1950s, Pauling had been investigated by the FBI and was refused renewal of his passport, receiving the explanation that "Your anti-communist statements haven't been strong enough". Only in 1954, when he won the Nobel Prize for Chemistry, and the *New York Times* brought the controversy to light, was he permitted to travel again. Similar struggles plagued his position at the California Institute of Technology. His funding from the National Institutes of Health (NIH) was cut, along with that of 40 other scientists, and he was eventually forced to resign from the faculty in 1963. After an interim of several years at the Center for the Study of Democratic Institutions in Santa Barbara, where he devoted himself to the problems of peace and war, he finally took up a chair in chemistry at Stanford University in 1969. There, he pursued his burgeoning interest in 'orthomolecular' substances, such as vitamin C, which he defined as substances normally present in the human body and required for life. He went on to establish the Linus Pauling Institute for Orthomolecular Medicine, to which he devoted his remaining years.

This brief biography must stand as a measure of the man who, in 1970, published the hugely popular book *Vitamin C and the Common Cold*, in which he claimed that large doses of vitamin C could prevent or cure the common cold. Pauling and his wife practised what they preached, consuming between 10 and 40 *grams* of vitamin C each day (several hundred times the recommended daily allowance), even adding spoonfuls to their orange juice. Over the next two decades, Pauling's claims became still more contentious: so-called 'mega-doses' of vitamin C could cure schizophrenia and cardiovascular disease, prevent heart attacks and ward off cancer, perhaps adding decades to our life expectancy. Most controversially of all, Pauling and the distinguished Scottish oncologist Ewan Cameron reported that mega-doses of vitamin C, given intravenously, could quadruple the survival time of patients with advanced cancer, even bringing about complete remission in some cases. The medical profession responded to these claims with suspicion, but the Mayo Clinic in Rochester, Minnesota, did at least conduct three small-scale clinical trials to test the effect of vitamin C in advanced cancer. All three trials failed to detect any benefit. Pauling and Cameron argued that the trials had been designed to fail: in particular, that vitamin C had been withdrawn too soon, and was given orally rather than intravenously. In 1989, the NIH agreed to review 25 case studies, to be selected by Cameron, for plausible evidence that mega-doses of vitamin C might have important effects in cancer. They concluded, in a letter to Pauling in 1991, that the case studies did not provide convincing evidence of a link.

Pauling, then, was a colossus of twentieth-century science, whose work laid the foundations of modern chemistry. He could be insufferably self-assured, but was far from infallible, as Watson and Crick were delighted to discover. His methods were unorthodox. In *The Double Helix*, written before Pauling's conversion to orthomolecular medicine, Watson described Pauling's approach to chemistry as intuitive rather than mathematical; and Pauling referred to his own application of intuitive guesses as the 'stochastic method'. Always the maverick, Pauling felt betrayed by the establishment and was quick to fight his corner, sometimes with barbed personal attacks. His experiences no doubt coloured his attitude to the pharmaceutical industry and the medical profession, which he termed the 'sickness industry', accusing them of misleading the public to bolster drug sales. For their part, doctors dismissed Pauling's claims for vitamin C as quackery and fraudulence. Journals became reluctant to publish his papers, and the dispute degenerated into a public slanging match. So began a stand-off that continues to this day. Pauling died in 1994, at the age of 93, an embittered man. If he was right, he had solved one of the world's greatest problems — how to age gracefully — and we would be fools to turn our backs on his simple solution. On the other hand, the woes of old age have hardly been vanquished, even among those who follow Pauling's example. One might be forgiven for assuming that he must have been mistaken. Was there any truth in his claims?

Vitamin C acquired its status as a vitamin — an essential trace constituent in the diet — for a curious reason. With the exception of higher primates, guinea-pigs and fruit bats, almost all other plants and animals manufacture their own vitamin C. In contrast, we must eat ours, because a common ancestor of the higher primates once lost the gene coding for an enzyme called gulonolactone oxidase, which catalyses the final step in the synthesis of vitamin C. As a result, the entire human race suffers from what amounts to an inborn error of metabolism. Pauling was fond of drawing attention to this deficit in his talks; he would hold up a test tube containing the amount of vitamin C produced by a goat in a single day and say, conciliatory to the end, "I would trust the biochemistry of a goat over the advice of a doctor".[1]

This apparently strong argument must be flawed. Loss of the gene for gulonolactone oxidase could not have been counterproductive for our primate ancestors, or they would have been eliminated by natural selection. Indeed, the fact that the individuals that lost the gene eventually prevailed among the primates suggests that there may even have been some benefit in its loss. In their authoritative text *Free Radicals in Biology and Medicine*, Halliwell and Gutteridge suggest one possibility. They note that gulonolactone oxidase produces hydrogen peroxide as a by-product of vitamin C synthesis. This means that high rates of vitamin C synthesis in animals such as the rat could, ironically, impose an oxidative stress. Given an adequate diet of fruit, which is rich in vitamin C, it might indeed be beneficial to consume, rather than synthesize, vitamin C. Of course, this is only true if we do eat enough. One of Pauling's arguments was based on the observation that gorillas consume nearly 5 grams of vitamin C daily in their normal diet. Our own Palaeolithic forebears are thought to have consumed about 400 milligrams daily.

Failure to eat enough vitamin C causes the once-dreaded deficiency disease *scurvy*. Scurvy is no longer a familiar sight, but once devastated the lives of sailors, who were deprived of fresh foods on long voyages. Scurvy was also endemic among soldiers on military campaigns from the Crusades to the First World War. The disease was a liability for global explorers, who were sometimes at sea for months or years at a time. One outbreak in 1536 afflicted all but 10 of the 110 men wintering aboard the ships of the French explorer Jacques Cartier, founder of Montreal, in the frozen St Lawrence river in Canada. Cartier wrote that "the victims' weakened limbs became swollen and discoloured whilst their putrid gums bled profusely." Other symptoms of scurvy include anaemia, spontaneous bruising, fatigue, heart failure and finally death. Thirty years after Cartier's

[1] This quantity (5–10 grams) was based on extrapolations made from measurements in homogenized liver samples, and may bear little resemblance to the amount actually produced by a goat each day.

winter of sorrows, the Dutch physician Ronsseus advised sailors to eat oranges to prevent scurvy, and in 1639 the English physician John Woodall recommended lemon juice. With characteristic phlegm, the British Admiralty disregarded this advice, standing firm even after the crew of Lord Anson's round-the-world expedition of 1740 was cut down by scurvy. Of the 1955 men who set sail, 320 died from fevers and dysentery and 997 from scurvy before Anson returned to England in 1744.

Protesting against the appalling conditions faced by sailors, James Lind, a Scottish naval surgeon, produced a *Treatise on Scurvy* in 1753, in which he, too, recommended citrus fruits. Unlike his predecessors, however, he had actually proved his theory in the world's first controlled clinical trial, on board HMS *Salisbury* in 1747. Lind tested a variety of reputed remedies on 12 members of the crew who had succumbed to scurvy. Two of them received a quart of cider each day; two had oil of vitriol; two received vinegar; two drank sea water; two had oranges and lemons; and the final pair took a medicament prepared from garlic, radish, Peru balsam and myrrh. The two seamen who received oranges and lemons made a speedy recovery and were put to nurse the others. Of the rest, only those receiving cider showed any signs of recovery. Curiously, despite the practicality of his conclusions, Lind did not regard scurvy as a deficiency disease, but rather as the result of a contagion in moist air. He thought of lemon juice as a detergent that could break down toxic particles.

Lind's recommendations were acted on by Captain Cook on his two circumnavigations of the globe between 1768 and 1775. Cook had exacting standards, and emphasized the importance of good diet, cleanliness, ventilation and high morale. He supplied his sailors with fresh lemons, limes, oranges, onions, cabbage, sauerkraut and malt. Only one seaman died from scurvy in a total of nearly six years at sea. Even so, the British Admiralty did not capitulate to Lind's demands until 1795, when it finally agreed to issue lemon juice on British ships. Thanks to the publicizing efforts of Sir Gilbert Blane, physician to the navy, the effect was dramatic. From an average of over 1000 patients with scurvy admitted to the Haslar naval hospital each year, the number dwindled to a mere two between 1806 and 1810. As the late social historian Roy Porter observed wryly, lemons might have done as much as Nelson to defeat Napoleon. The situation did not last long. In a cost-cutting measure typical of the British over the ages, the Admiralty replaced lemons with cheaper limes, which contain barely a quarter as much vitamin C. Scurvy soon reappeared. To add insult to injury, British sailors acquired the nickname 'limeys'.

The idea that scurvy might be a deficiency disease, rather than an infection, was advanced in the 1840s by George Budd, professor of medicine at Kings College in London, earning him the grand Victorian epithet 'the prophet Budd'. In a series of articles published in the *London Medical Gazette*, entitled "Disorders Resulting from Defective Nutriment", Budd prophesied that scurvy was due to the "lack of an essential element, which it is hardly too sanguine to state will be discovered by organic chemistry or the experiments of physiologists in a not too distant future".

In the event, fulfilment of Budd's prophesy had to wait another 93 years, in part because the concept of deficiency diseases was set back by Pasteur's germ theory of disease, which was then applied with indiscriminate enthusiasm to almost any condition. By the late 1920s, though, a number of researchers were racing to isolate the 'antiscorbutic factor' from oranges, lemons, cabbages and adrenal glands. Several workers, notably the Hungarian biochemist Albert Szent-Györgyi, succeeded in isolating white crystals of an acidic sugar, whose properties corresponded to vitamin C, but whose chemical identity remained a mystery. A bit of a joker, Szent-Györgyi proposed the name 'ignose', the *-ose* ending signifying its relationship to sugars, and the *ign-* prefix his ignorance of its nature. When this name was rejected, he proposed 'godnose'; and finally, in a single sentence in *Nature* in 1933, the term *ascorbic acid*, in reference to its antiscorbutic properties. Progress was fast. The same year, ascorbic acid was synthesized independently by the Polish émigré Tadeus Reichstein in Switzerland and Sir Walter Haworth in Birmingham, UK, making it not only the first vitamin to be assigned a chemical formula, but also the first to be synthesized by purely chemical means.

Ironically, the concept of vitamin C as the dietary element that prevents scurvy has hampered our understanding of its positive role in the body. The general approach to how much vitamin C we need to eat each day (the recommended daily allowance or RDA) is derived from this negative stance — the prevention of scurvy — rather than any positive criterion. The amount of vitamin C required to prevent *clinical* scurvy, in other words to hide any obvious signs of disease, is surprisingly small. A series of studies on the inmates of Iowa jails in the 1960s showed that only about 10 milligrams a day are required to abolish the signs and symptoms of scurvy. When the dose is raised to about 60 milligrams a day, we begin to excrete vitamin C in our urine, implying that the excess is superfluous to requirements. The notion that our body pool is saturated by about 60 milligrams a day is supported by the rate of breakdown of vitamin C: the Iowa studies suggested that breakdown products are excreted in the urine at a rate of about 60 milligrams each day. These three factors, then — prevention of scurvy with a margin for error, excretion of vitamin C, and excretion of breakdown products — form the basis of the long-standing RDA of 60 milligrams vitamin C daily.

Although this analysis sounds like a closed case, it is in reality misleadingly simplistic. The case is confounded by both practical and conceptual difficulties. These were scrutinized during the 1990s by Mark Levine, of the NIH. Levine had been a member of the panel convened by the NIH to review Cameron's 25 case studies in 1989, and has since done more than anyone to bridge the gap between mainstream medicine and the advocates of vitamin C.

Besides querying the accuracy of the early measurements of vitamin C and its breakdown products, which were carried out using insensitive and non-specific tests, Levine questioned the assumptions underlying each of the three factors used to estimate the RDA. First, he said, the amount of vitamin C required to prevent scurvy may be much less than the ideal intake for maintaining bodily functions. We do not know how much less. Second, the threshold of urinary secretion may or may not correspond to the saturation of body pools — it does for some substances and doesn't for others. For vitamin C, we do not know. Third, the rate of breakdown of vitamin C depends on a variety of factors, including the size of the dose consumed. High doses are broken down faster than low doses, perhaps because the body has less need to conserve a precious resource. This means that estimates of breakdown based on low doses (such as 30 or 60 milligrams) may be misleading. Thus, Levine stripped away the conceptual basis underpinning the current RDA of vitamin C.

Far from being merely critical, Levine worked up his own recommendations for a rational daily dose of vitamin C, based on the ideal amount required for known reactions, and the saturation of blood levels and other body pools. For the impatient, let me say immediately that he recommends 200 milligrams daily for healthy individuals; that doses above 400 milligrams have no evident value; and that doses of more than 1 gram may not be safe, as they can provoke diarrhoea and induce the growth of kidney stones. Five portions of fruit and vegetables each day corresponds to a daily intake of between 200 and 400 milligrams of vitamin C, so given a sensible diet there is no need for supplementary vitamin C. We shall see that there are other good reasons for not relying on supplemental vitamins. On the other hand, the RDA of 60 milligrams (raised to 90 milligrams in the United States in April 2000) is, according to Levine, too low. To understand his reasoning, and especially the wider ramifications in terms of antioxidant function, we need to look in more detail at what vitamin C actually does in the body.

Quite apart from its antioxidant effects, we need vitamin C for a wide variety of biochemical reactions that help to maintain our normal physiological function. The best known requirement for vitamin C is as a cofactor (a necessary accessory for enzyme function) in collagen synthesis.

Collagen fibres make up about 25 per cent of the total protein content of our bodies, and are familiar to all of us in a melted form as gelatine. In their normal bodily environment, collagen fibres are the most important structural and shock-absorbing component of connective tissues, including bone, teeth, cartilage, ligament, skin and blood vessels. In the absence of vitamin C, collagen fibres do not form properly. Many of the symptoms of scurvy can be attributed to defects in collagen production and maturation. As a result, blood vessels become fragile and wounds heal slowly, if at all. Such vascular degeneration probably accounts for the putrid bleeding gums, swollen joints, spontaneous bruising and, ultimately, as fluids seep out of leaky blood vessels and blood pressure falls, heart failure.

Other symptoms are characteristic of scurvy but not specific to it, including general malaise, fatigue and anaemia. Fatigue affects many millions of people. In some cases, fatigue might be a form of 'sub-clinical' scurvy; in other cases it is not. In his *Treatise on Scurvy*, Lind reported lassitude as an early and invariable symptom. While it is possible that errors in collagen synthesis may contribute to lassitude, such vague symptoms are more likely to relate to the synthesis of a small amino acid called carnitine — which requires vitamin C. We need carnitine to burn fats. When fats are broken down, the component fatty acids must be transported into the mitochondria where they are oxidized to produce energy. The problem is that fatty acids cannot get into the mitochondria by themselves, but must be ferried in attached to carnitine. Carnitine is also responsible for removing left-over organic acids from the mitochondria *en route* back to the cytoplasm. Without vitamin C, we cannot make enough carnitine to generate energy from fats, and the mitochondria eventually pollute themselves with toxic organic acids, reducing their ability even to generate energy from glucose. Fatigue might seem a small price to pay.

Vitamin C also has a variety of neuronal and endocrine functions, which are crucial to our physiological and psychological well-being. For example, we need vitamin C to make noradrenaline (norepinephrine), a cousin of adrenaline (epinephrine) which has an important role in modulating our response to stress. We also need it for the correct function of an enzyme called PAM (peptidyl alpha-amidating mono-oxygenase), which is found in many parts of the body, but especially in the pituitary gland. PAM bites the end off a large number of immature peptide hormones and neurotransmitters, and in so doing activates them. Without activation by PAM, the hormones remain inert. If only to downplay the perception of vitamin C as just a water-soluble antioxidant and cofactor in collagen synthesis let me list a few of the peptides that are activated by PAM. They include: corticotrophin-releasing hormone, which stimulates production of steroid hormones; growth hormone-releasing hormone, which promotes growth and influences energy metabolism; calcitonin, which promotes calcium phosphate absorption and distribution in the bones; gastrin, the most powerful stimulant of gastric acid secretion; oxytocin, which stimulates milk ejection and uterine contraction; vasopressin, which regulates water balance and stimulates intestinal contraction; secretin, which stimulates pancreatic and bile secretions; and substance P, a potent vasodilator and sensory neurotransmitter, which mediates our sense of pain, touch and temperature. Given this wide range of actions, the extent to which our normal physiology is fine-tuned by vitamin C is virtually anybody's guess.

Nor is this all. Vitamin C is also taken up by white blood cells. When we are infected with bacteria, white blood cells called neutrophils mount the first defence. In the course of this defence, neutrophils vacuum up vitamin C from their surroundings, using miniature protein pumps in their membranes. The level of vitamin C inside the neutrophils increases tenfold within minutes, and if the infection persists, may reach 30 times the level of resting neutrophils, or 100 times that in plasma, even in someone taking massive oral supplements.

Here, finally, we see an action of vitamin C that seems to be an antioxidant effect, along the lines of that described by Tom Kirkwood at the start of this chapter. Neutrophils need this extra protection to survive the wrath of their own assault, as they turn their immediate environment into a battlefield. The effect is a little like soldiers strapping on their gas masks before releasing chlorine gas onto the enemy. Instead of chlorine gas, neutrophils produce a burst of free radicals and other powerful oxidants (including hypochlorous acid, derived from chlorine), which are responsible for bacterial killing.[2] Vitamin C prevents or delays the demise of neutrophils in a chemical cesspool of their own making, and hastens the death of bacteria, which cannot take up vitamin C or continue to benefit from its presence in their locally denuded surroundings. Levine, among others, sees the uptake of vitamin C by neutrophils as a promising avenue for pharmaceutical development in the emerging era of antibiotic resistance.

Finally, what of anaemia? This, too, is a symptom of scurvy, but is not a physiological failure in the sense of those discussed above. In this case, vitamin C acts on the inorganic iron in food in the stomach and intestines, converting it from the insoluble form usually found in food (Fe^{3+}) to the soluble form (Fe^{2+}) that we can absorb in our intestines. (This is the reverse of the reaction that took place on a huge scale in the Precambrian oceans, leading to the precipitation of insoluble iron into banded iron formations, discussed in Chapter 3). Without adequate supplies of vitamin C, we cannot absorb enough iron to stock red blood cells with haemoglobin (which contains iron), and so we develop anaemia.

Such a diverse array of functions lends vitamin C an aura of magic. Yet in each of these cases, vitamin C behaves in exactly the same way at the molecular level, as repetitively as flipping a coin, even if the outcomes are opposed. To see how, let us take a single example in a little more detail. Collagen synthesis illustrates not only how vitamin C works, but also throws light on its antioxidant properties, as well as its more dangerous side.

Collagen can only be manufactured in the presence of molecular oxygen (as we noted in Chapter 4). Oxygen is needed, along with vitamin

[2] Chronic granulomatous syndrome is an unpleasant condition caused by a mutation in the gene coding for the enzyme NADH oxidase, which generates oxygen free radicals in activated neutrophils. People with the syndrome cannot kill bacteria efficiently and develop chronic suppurating granulomas (abscesses) in the skin, lymph nodes, lung, liver and bones. They also suffer frequently from opportunistic infections. The disease usually develops in childhood and can be treated to some extent with antibiotics.

C, to modify some of its amino acids after they are incorporated into the final protein. The amino acids are *hydroxylated*, which means that they have an additional hydroxyl (–OH) group attached to them. This hydroxylation enables cross-links to be formed between the individual collagen chains, first to form triple-chain collagen molecules, and then to link these molecules into thicker collagen fibrils. These cross-links are responsible for the tremendous tensile strength of collagen. Without vitamin C and oxygen, collagen cross-links cannot form and the connective tissues are weakened. Not only this, but non-hydroxylated collagen is retained inside the cells that make it, rather than being exported for building purposes. It is also less stable, more sensitive to heat, and more readily broken down by digestive enzymes. Jelly made from scorbutic collagen would be a sorry sight at a tea party.

The mechanism of collagen hydroxylation betrays the secret of vitamin C: it is an electron donor. The oxygen atom in the hydroxyl group comes from molecular oxygen. To attach this oxygen, a single electron must be added to each of the two atoms in molecular oxygen. Because electrons tend to double up as pairs, few compounds can give up a single electron without becoming unstable and reactive themselves. Metals such as iron and copper, which exist in several stable oxidation states, can do so, as can vitamin C. In biological reactions, vitamin C *always* donates electrons. This and no more. Having said that, it is not profligate with its electrons; under physiological conditions it is most likely to offer them to iron or copper.[3] In the process of collagen synthesis, this is exactly what happens. Vitamin C donates an electron to iron, which is embedded at the core of the enzymes that carry out the hydroxylation reaction. The iron thrusts this electron onto oxygen, which can now be attached to amino acids in collagen. In the process, iron is oxidized to the biologically inactive form (Fe^{3+}), until it receives an electron back from vitamin C.

The role of vitamin C in this case is therefore to regenerate the biologically active form of iron by passing an electron to the oxidized form. The hydroxylating enzyme acts as a kind of merry-go-round that uses iron to attach oxygen to amino acids. By providing iron with electrons, vitamin C provides the motive force that keeps the merry-go-round turning.

The triumvirate of iron (or copper), vitamin C and oxygen is at the

[3] In more technical terms, vitamin C has a relatively neutral reduction potential: it donates electrons to strong oxidants, rather than thrusting them onto weak oxidants, so does not interfere with normal workings of the cell.

heart of virtually all the physiological actions of vitamin C. A total of at least eight enzymes use vitamin C as a cofactor. All of these enzymes contain iron or copper at their core. All attach oxygen atoms to amino acids, using the iron or copper. All use vitamin C to regenerate active iron or copper. Essentially the same reaction also accounts for the ability of vitamin C to promote iron absorption in the intestine. In this case, vitamin C donates an electron to oxidized iron, converting it into the soluble form that can be absorbed.

Why is vitamin C used so extensively as an electron donor? There are two main reasons. First, vitamin C is very soluble in water, so it can be concentrated in confined spaces surrounded by membranes (which are made of lipids impermeable to vitamin C). The synthesis of noradrenaline from dopamine, for example, takes place in small membrane-bounded spaces, or vesicles, within cells of the cortex of the adrenal glands. The vitamin C concentration inside these vesicles reaches about 100 times that of blood plasma. As vitamin C is consumed by the enzyme dopamine mono-oxygenase, electrons are passed across the vesicle membrane (via an iron-containing protein, cytochrome b_{651}), to regenerate vitamin C within the vesicles. Thus, for periods of days or weeks, the intracellular vitamin C needed for physiological tasks can be 'insulated' from changes in plasma levels caused by variations in diet, and maintained at the ideal levels for a particular reaction.

The second reason that vitamin C is used as an electron donor is that the reaction product is fairly stable and unreactive. When vitamin C gives up an electron, it becomes a free radical called the *ascorbyl radical*. By free-radical standards, the ascorbyl radical is not very reactive. Its structure is stabilized by electron delocalization — the resonance effect first described by Linus Pauling in the late 1920s. This means that vitamin C can block free-radical chain reactions by donating an electron, while the reaction product, the ascorbyl radical, does not perpetuate the chain reaction itself.

Despite its slow reactivity, the ascorbyl radical usually gives up a second electron to produce dehydroascorbate. This molecule is unstable and needs to be caught quickly if it is not to break down spontaneously and irrevocably, and be lost from the body. The continual seeping loss of vitamin C in this way accounts for our need to replenish body pools by daily intake. Even so, we can minimize losses by recycling dehydroascorbate. Several different enzymes bind dehydroascorbate to regenerate vitamin C. These enzymes usually take two electrons from a small peptide called *glutathione*, and transfer them to dehydroascorbate. Because a pair of electrons are transferred, the regeneration of vitamin C does not produce free-radical intermediates.

The repetitive, coin-flip action of vitamin C is thus to donate single electrons (it donates two of them to form dehydroascorbate). It is regenerated from dehydroascorbate by receiving two electrons from glutathione. This cycle explains not only the enzyme-cofactor effects of vitamin C, but also its antioxidant activity. Even though vitamin C usually gives up electrons to iron or copper, other molecules in search of a single electron can extract one from vitamin C. These include many free radicals, which of course are defined as atoms or molecules with a single unpaired electron (see Chapter 6).

When a free radical reacts, it usually snatches an electron from the reactant, turning it into a free radical. This in turn will steal a single electron from another nearby molecule. A chain reaction ensues until two free radicals react together, effectively neutralizing each other, or alternatively, until an unreactive free-radical product is formed. Free radicals are said to be 'quenched' by vitamin C, because the free-radical product — the ascorbyl radical — is so unreactive. As a result, free-radical chain reactions are terminated. Lipid-soluble vitamin E (α-tocopherol) works in the same way, in membranes rather than in solution, often in cooperation with vitamin C at the interface between membranes and the cytosol (the watery ground substance of the cytoplasm that surrounds the intracellular organelles). When vitamin E reacts with a free radical, it too produces a poorly reactive (resonance-stabilized) free-radical product, called the α-tocopheryl radical. Tocopheryl radicals can be reconverted into vitamin E using electrons from vitamin C.

As I intimated at the beginning of this chapter, these simple, repetitive reactions conceal a hidden danger — the darker side of vitamin C. We have noted the link between iron, vitamin C and oxygen. When vitamin C interacts with iron and oxygen, it is acting as an electron donor, *but not as an antioxidant*. Quite the contrary. By regenerating the active form of iron inside an enzyme, vitamin C aids and abets the addition of oxygen — in other words, it helps to *oxidize* the substrate. Thus many of the beneficial actions of vitamin C are actually *pro-oxidant* actions, not antioxidant actions at all.

When iron is bound to the core of a protein this is safe enough — the iron is restrained in the manner of a horse wearing blinkers, and kept focused on a narrowly defined task. But it is a different matter when the iron is in solution. Now the iron is free to react in an uncontrolled way. We saw the danger coming in Chapter 6. Remember the Fenton reaction? Iron reacts with hydrogen peroxide to produce brutal hydroxyl radicals and the oxidized, inactive form of iron. Hydroxyl radicals react immediately with their nearest neighbours to initiate free-radical chain reactions. These damaging chain reactions can only be started when active iron is available, and so they eventually peter out after the active iron is used up. We saw that superoxide radicals are dangerous because they can regenerate active iron, perpetuating the Fenton reaction. They do this by donating electrons. Vitamin C can also donate electrons to regenerate active iron. In principle, then, there is no reason why vitamin C should not *exacerbate* free-radical damage rather than acting as an antioxidant. Theoretically, it can act as a pro-oxidant as well as an antioxidant.

The possibility that vitamin C could act as a pro-oxidant is not just theoretical. A common test for the potency of an antioxidant actually depends on vitamin C to promote injury. The method begins by stimulating free-radical chain reactions in a preparation of cell membranes, then measures how well antioxidants can stop the process in its tracks. The chain reactions are started using a concoction of iron and vitamin C — iron to catalyse the reaction, and vitamin C to regenerate the active iron. If this situation were ever to take place in the body, the effect could be catastrophic.

Two questions spring to mind. Does vitamin C ever act as a pro-oxidant in the body to cause damage? And if not, why not? What stops it from doing so? Few subjects have raised so many hackles among normally placid researchers, and the answers are by no means certain. Nonetheless, ambiguities aside, the issues raised by even the possibility of pro-oxidant effects give us a perspective on antioxidant 'networking' in cells — and also on why Linus Pauling and Ewan Cameron might just have been right about the anti-cancer activity of vitamin C after all.

I should say that there is very little evidence that vitamin C ever acts as a pro-oxidant in people. There are a few indications that the body is aware of the danger, however. In particular, we control our blood plasma levels of vitamin C carefully. Even if we take mega-doses of vitamin C, plasma levels hardly rise. There are two main controls on vitamin C levels in blood plasma — absorption and excretion. The absorption of vitamin C from the intestine falls off drastically the larger the dose. Mega-doses have a laxative effect and cause diarrhoea.[4] Some advocates of mega-dose vitamin C even recommend 'titrating to bowel tolerance' — eating nearly enough to cause diarrhoea — to ensure that the maximum possible amount is absorbed. It doesn't work. Less than 50 per cent of a 1-gram dose is absorbed from the intestine, and most of this is then excreted in the urine. Being freely soluble, vitamin C is filtered by the kidneys, and not all of it is reabsorbed unless an acute shortage leaves the body grasping at straws. Excretion in the urine starts at doses of about 60 to 100 milligrams daily and builds up from there. Nearly all of a 500 milligram dose is excreted in this way. In fact, blood and body pools are saturated by about 400 milligrams of vitamin C each day. It doesn't matter how much extra you take, you won't increase your body levels.

While this information is valuable in itself, the fact that the body controls vitamin C levels so tightly suggests that there might be a problem if it did not. Although nobody ever proved that vitamin C is toxic in this way, we should appreciate that neither has anybody proved that vitamin C works as an antioxidant in the body. We know that it *can* act as an antioxidant, and probably does, but we are far from the kind of rigorous proof that a rocket scientist might demand. Here, for example, is Baltz Frei, current head of the Linus Pauling Institute, writing in 1999: "the current evidence is insufficient to conclude that antioxidant vitamin supplementation materially reduces oxidative damage in humans." So what about the other side of the coin, actual bodily harm? If the potential toxicity of vitamin C relates to its interactions with iron, then the first place to look for danger is a disease in which iron metabolism is disrupted in some way.

One such condition is iron overload. Rather surprisingly, we have no specific mechanism for removing excess iron from the body, save menstrual bleeding or the shedding of gut-lining cells. The rate of *absorption* of iron therefore needs to be regulated tightly. The inherited form of iron overload, called haemochromatosis, stems from a failure of the normal mechanism for regulating iron absorption in the intestine. Too much iron is absorbed, and in time — 40 years or more, depending on diet — this exceeds the body's capacity to store it safely. Free iron appears in the blood stream. The effect is devastating. Without treatment, the victim can

[4] It is conventional to view the laxative property of vitamin C as a non-harmful side-effect — nothing in comparison with the side-effects of many pharmaceutical drugs — but it may be wiser to view the laxative effect as a deliberate physiological response on the part of the body to potentially toxic levels of vitamin C.

expect to suffer from liver failure (caused by cirrhosis or cancer), weight loss, skin pigmentation, joint inflammation, diabetes and heart failure. Nor is this a rare condition. In western populations, the prevalence is nearly 0.5 per cent, making it one of the most common genetic disorders.[5]

In theory, vitamin C can affect people with iron overload in two different ways. First, it might increase the rate of iron absorption from the intestine. Although there is no evidence that mega-doses of vitamin C cause iron overload in healthy subjects, it is not at all clear how they affect iron levels in people with haemochromatosis. Second, vitamin C might convert excess iron into the active form that can catalyse free-radical reactions. Whether this might tip the balance between vitamin C acting as an antioxidant or a pro-oxidant in people with haemochromatosis is unknown. Most evidence suggests that it does not, although a few case reports have claimed a detrimental effect. One unfortunate young man in Australia, for example, took high doses of vitamin C for a year before being admitted into hospital with severe heart failure. He died eight days later, diagnosed at autopsy with haemochromatosis. The doctors concluded that the progression of the disease might well have been accelerated by his excessive intake of vitamin C.

But let us consider the risk from another point of view. If there is a dark side of vitamin C, then it might offer an advantage in the treatment of cancer. We know that vitamin C can kill tumour cells grown in the test tube, and that the way in which it does so is dependent on supplies of iron and oxygen. Vitamin C can also kill malarial parasites at certain points in their life cycle, when they are actively accumulating iron from haemoglobin. Is this perhaps an explanation for the Pauling and Cameron findings? It is certainly plausible. The cores of large tumours are often composed of dead or dying cells, which release iron as they degenerate. The fine control of iron metabolism in normal cells may also be lost in cancer cells, which are chaotic and opportunistic in their behaviour. In addition, radiotherapy and chemotherapy cause transient plasma iron overload, which presumably derives partly from the tumour itself. For all

[5] Iron overload can also result from the medical treatment of other conditions, notably the thalassaemias, which are caused by an inability to produce haemoglobin at the normal rate. Unless treated with regular blood transfusions, patients with thalassaemia major die in childhood. On the other hand, given regular transfusions, many thalassaemia patients eventually develop iron overload.

these reasons, iron is likely to be present in tumours at a higher concentration than in normal tissues. In the presence of oxygen and vitamin C, then, tumours might conceivably be placed under enough oxidative stress to kill them.

If so, why were Cameron's results so hard to reproduce? In an incisive reappraisal of the anti-cancer effects of vitamin C in the *Canadian Medical Association Journal* in 2001, Mark Levine and Sebastian Padayatty (also at the NIH) argued that the *route* of administration holds the key. Pauling and Cameron infused vitamin C intravenously, while the Mayo clinic, in attempting to replicate their data, used high oral doses. When given orally, the low rate of absorption and the high rate of excretion maintain virtually constant plasma levels. Intravenous dosing, on the other hand, side-steps absorption altogether, while the kidneys take some time to remove vitamin C from the blood. For short periods, then, blood levels of vitamin C may reach 50 times their normal saturation point; and this might make all the difference. Padayatty and Levine called for rigorous new trials to test the possibility.

In the meantime, one study of an innovative new cancer therapy does suggest that vitamin C can help kill tumours by exacerbating the effect of free radicals. The treatment is called photodynamic therapy, and was touched on in Chapter 6. The procedure uses light to activate a drug. Once activated, the drug transfers chemical energy onto oxygen, generating singlet oxygen and various free radicals, which attack the tumour. Research teams at the University of Iowa and in China have shown that giving high-dose vitamin C in conjunction with photodynamic therapy enhances the destructive effects of treatment on the tumour. If this turns out to be an important clinical effect (it is still too early to say), the tarnished reputation of the prophet Pauling may yet be restored.

I have used the example of vitamin C to explore the wider function and behaviour of antioxidants. What have we learnt? The first conclusion is that vitamin C has a repetitive molecular action, constrained by its chemistry. It is not some kind of infinitely flexible superhero, capable of assuming any shape to protect us from evil. All molecular antioxidants are, of necessity, constrained within tight bounds by their chemistry. This does not prevent them from having a wide range of actions. Our second broad conclusion is that a single repetitive action can have many physiological roles besides an antioxidant effect. We have seen that vitamin C is a cofactor for at least eight enzymes, which affect completely different aspects of bodily function, from regular housekeeping tasks, such as collagen synthesis and fat metabolism, to survival measures, such as our response to stress (noradrenaline synthesis) and our perception of pain (activation of substance P). In fact, of all the actions of vitamin C, its antioxidant properties are probably the least well documented. This is equally true of many other reputed antioxidants.

The most compelling evidence that vitamin C *does* behave as an antioxidant in the body comes from its rapid uptake by neutrophils, where it protects them against their own noxious anti-bacterial effusions. In this respect, it is worth noting that neutrophils only accumulate vitamin C when they are activated by bacteria. Such a sporadic response may reflect no more than the energetic futility of vacuuming up vitamin C when it is not needed, but might equally be a precaution against vitamin C toxicity. This brings us to the third broad conclusion that we can draw from vitamin C. The precise behaviour of an antioxidant depends on its surroundings. Whether vitamin C acts as an antioxidant, or a pro-oxidant, or somewhere in between, depends primarily on its interactions with other molecules. We have seen that vitamin C interacts with some free radicals directly, but also with iron, copper, vitamin E and glutathione. For vitamin C to have a beneficial antioxidant effect, each of these needs to be present in the right amount at the right place, and so each requires its own network of support molecules. If we take a bird's-eye view, then each of these factors should really be considered antioxidants. Where do we draw the line? To appreciate the difficulty of defining an antioxidant, let us draw this chapter to a close with a quick look at the behaviour of activated neutrophils.

Although neutrophils accumulate vitamin C to 100 times plasma levels, they do not absorb vitamin C itself, but only dehydroascorbate, the oxidized form of vitamin C. Neutrophils have protein pumps in their cell membranes that recognize dehydroascorbate and pump it into the cell. Once inside, the dehydroascorbate is useless until it is regenerated into vitamin C. In neutrophils, this step needs an enzyme called glutaredoxin, which takes electrons from glutathione to regenerate vitamin C. If the whole system is not to grind to a halt, the depleted glutathione needs to be regenerated in turn. Glutathione regeneration is achieved by an enzyme called glutathione reductase, using electrons that would otherwise be used to convert oxygen into water in the course of cellular respiration. This amounts to a long-odds gamble on life itself. The physiological balance of the neutrophil is shifted away from normal respiration — from what amounts to breathing — into an emergency holding pattern, which is dedicated to regenerating glutathione and thus vitamin C. In other words, activated neutrophils trade taking a breath for protection, in the hope that they will survive long enough to kill the bacteria.[6]

Such heavy betting on the long odds begs the question *why vitamin C?* As a water-soluble vitamin, it accumulates within the cytosol of the cell. The 'thin red line' that holds bacteria at bay is not the inside of the neutrophil, but its surrounding cell membrane, made of fats immiscible with vitamin C. Even when bacteria have been ingested by neutrophils, they are still held in isolation inside the phagocytic vesicles formed from invaginations of the external cell membrane. The neutrophils pour their toxins into these vesicles (as well as the surrounding environment). If they are not to be killed by their own toxins, they must maintain the integrity of their external and internal membranes. If this membrane is damaged in the battle with bacteria, and loses its integrity, the neutrophil will die just as surely as we ourselves die when flayed of our skin. In fact, vitamin C is used to rally and, quite literally, revitalize the front-line forces.

Lipid-soluble vitamin E is the foremost defender of the cell membrane. Vitamin E donates electrons directly to free radicals that can damage the membrane's integrity and so neutralizes them, leaving behind its sacrificial corpse, the α-tocopheryl radical. Vitamin C breathes life back into this near-inert radical, resurrecting it as vitamin E. The reaction takes place without the aid of an enzyme, but its speed depends on the amount of vitamin C relative to vitamin E. The more vitamin C, the quicker the regeneration of vitamin E, hence the stockpiling of vitamin C by neutrophils. At the same time, however, high levels of vitamin C present a danger, especially in the presence of free radicals such as superoxide, which can release iron from proteins (Chapter 6). Vitamin C might change sides and start acting as a pro-oxidant. To stop this from happening, the main

[6] Advocates of mega-dose vitamin C argue that regeneration of vitamin C by glutathione is a liability: it is slow, drains cellular energy and ultimately deepens the crisis. One rationale for taking mega-dose vitamin C when ill is to side-step this dangerous regeneration step. Such an approach might work outside cells, but cannot work inside cells, where the protection is most urgently needed. The problem is that most cells only recognize the oxidized form of vitamin C, dehydroascorbate. For mega-dose vitamin C to protect cells from within, it must first be oxidized in the blood, then taken up by the cell as dehydroascorbate, and finally regenerated to vitamin C using glutathione. No short cuts here.

agents provocateurs, iron and copper, must be stowed away. Battening down the hatches demands molecular sensors that detect any free iron or copper within the cell, and the capacity to lock away the excess in protein cages (ferritin and caeruloplasmin, respectively). If the storage capacity is limited, new protein cages may need to be manufactured, which in turn requires the transcription and translation of various genes. In all, some 350 genes are expressed by human neutrophils within 2 hours of activation, including the genes for ferritin and caeruloplasmin.

Each link of these entwined chains is critical for the system as a whole to work. The fact that neutrophils protect themselves by accumulating vitamin C, but bacteria do not, hangs by a single thread: bacteria cannot detect dehydroascorbate in their surroundings or pump it into themselves. In neutrophils, the entire response is activated by the presence of dehydroascorbate in the surroundings. The more dehydroascorbate there is, the faster the pumps work. Indeed, neutrophils can be activated in the absence of bacteria by the simple expedient of adding a little dehydroascorbate. Bacteria, on the other hand, remain dormant, even when drowning in pools of dehydroascorbate. They have all the cellular machinery they need to regenerate vitamin C, vitamin E and glutathione, or to hide iron and copper, but are blind to the presence of dehydroascorbate. This single failure may cost them their life. If so, the neutrophils' gamble on the long odds will have paid off.

The most striking feature of this scenario is the way in which the whole metabolism of the neutrophil is re-routed in the presence of dehydroascorbate. All of these changes operating together contribute to the overall antioxidant response. We cannot simply define an antioxidant as a molecule with a particular type of action. 'Seeing' dehydroascorbate is an antioxidant response. Hiding iron is an antioxidant response. Regenerating glutathione is an antioxidant response. Even lowering the metabolic rate — holding the breath — is an antioxidant response. There is no hard and fast dividing line between factors normally described as antioxidants, such as vitamin C, and physiological adaptations not usually classed as antioxidant responses, such as a reduction in cellular respiration. To have any sense of how such large-scale webs of interactions work in organisms as a whole, we will need to step back from vitamin C. In the next chapter, then, we will take a wider look at how organisms deal with oxidative stress.

CHAPTER TEN

The Antioxidant Machine

A Hundred and One Ways of Living with Oxygen

GOVERNMENTS WORK WITH TIGHT DEFINITIONS of words like 'unemployed', 'literacy', or 'no more taxes!'. Opposition spokesmen and newspaper editors dispute the validity of their definitions. Words fly back and forth, all sound and fury, signifying little. Scientists are considered to be above this kind of thing. The terminology of science brooks no opposition: it is clearly defined and quantifiable, albeit unreadable. Scientists strive to render a definition into a mathematical symbol. Only when the term sits comfortably in an equation are we happy. Yet even within the perfect discipline of mathematics, precision is not always possible. The 'fiddle factor' is an ever-present bogey, which symbolizes the resistance of the world to petty classification.

In biology, the trouble with definitions is far worse than in mathematics. Few biologists ever use the word 'proof' — it is far too exact. Doctors rarely use the word 'cure'. Who knows, after all? Remission is a more comfortable word, if only because it means very little: "the illness seems to have gone away for now, for all I can see; I have no idea if it will come back". Nature deftly sidesteps our clumsy swings at a definition. How does one define 'life?' Reproduction and some form of metabolism seem to be musts, but then is a virus alive? It has no metabolism of its own and so falls outside most definitions of life. If you can define life to encompass a virus, then how about the infectious prion, which is simply a protein?

How does one define ageing? A relentless decline in faculties ending in death? Or was that just a description? If we can't define life, what on earth is death? If a prion is not alive, is it dead? If so, does that mean we can't kill it?

I've no intention of wallowing in this sea of semantics. Of course there are answers, albeit rarely simple. My concern now is to find a wider definition of the term 'antioxidant'. We saw the potential for confusion in Chapter 9. The problem is one of precision: just how precise can we be in defining such a slippery concept?

The original definition of 'antioxidant' came from chemistry. As befits a science full of symbols, it had a precise meaning. An antioxidant is an electron donor that prevents a substance from being oxidized (or stripped of electrons). The word came of age in food technology in the 1940s. Fatty foods, such as butter, go rancid if left in the air. Technically, they 'peroxidize'. Peroxidation is a chain reaction started by oxygen free radicals, such as hydroxyl radicals, which attack lipids to get an electron. They may seize an electron and run, or become mired in the lipid like a rugby player who got the ball but couldn't escape the scrum. Either way, the lipid loses an electron: it has become a free radical and so attacks its neighbours to get an electron back. Such chain reactions spread like wildfire through the densely packed lipids in butter. An antioxidant stops all this by 'scavenging' free radicals. It gives up electrons to stop the chain reactions from spreading. Antioxidants, such as BHA (butylated hydroxyanisole), are added routinely to manufactured foods.

Such a precise definition is fine for chemistry or food technology but is not helpful in biology. In the presence of iron, we could just as easily define an electron donor as a pro-oxidant. Everything depends on context. In this chapter, then, I propose to explore the context and drop the precision — to put aside the reductionist approach and see how a broad synthesis works. We'll therefore look at how whole organisms, whether single cells or multicellular creatures, avoid being oxidized. Rather than thinking only in terms of reactions, I'll include traits such as morphology and behaviour.

Antioxidant defences can be divided into five categories: *avoidance* (sheltering), *antioxidant enzymes* (prevention), *free-radical scavengers* (containment), *repair mechanisms* (first aid) and *stress responses* (entrenchment). Some organisms, particularly those that hide from oxygen, rely on just one or two of these mechanisms, while others, including ourselves, draw on the entire armoury. We are, truly, antioxidant machines. To see how these defences work, we will examine a few instances of each mechanism in turn. Rather than an exhaustive analysis, I will highlight some of the insights that these mechanisms offer into our own physical and physiological makeup.

The simplest form of protection against oxygen toxicity is to hide. Tiny bacteria have a wealth of possible hiding places. Some strictly anaerobic bacteria, which are killed by trace amounts of oxygen, even shelter within larger cells to escape the long reach of oxygen. An extreme example is the methane-producing bacteria that live in the rumen of cattle and sheep. Like a set of Russian dolls, they hide inside the symbiotic microbes that break down cellulose from grass, which in turn hide inside the rumen.

The hind-guts of animals, from wood-eating termites to elephants, are comfortable hiding places for all manner of anaerobic microbes. Our own guts are inhabited by heaving colonies of so-called commensal bacteria, which are usually harmless or beneficial, but sometimes as noxious as their smell. The total population of gut bacteria is said to have a metabolic capacity equal to that of the liver. Together, the undigested organic matter and bacteria sponge up oxygen, rendering the large intestine virtually anoxic, with an oxygen concentration typically less than 0.1 per cent of atmospheric levels. Under these conditions, anaerobic bacteria such as *Bacteroides* outnumber their aerobic cousins a hundred-fold.

On a much larger scale, free-living anaerobic bacteria can hide from oxygen by buffering the outside world. A good example is the sulphate-reducing bacteria, which produce hydrogen sulphide gas as a waste product (see Chapters 3 and 4). Hydrogen sulphide reacts with oxygen to regenerate sulphate, thus simultaneously replenishing the raw material of the sulphate-reducing bacteria, and depleting dissolved oxygen. The effect is to keep the outside world out and sustain an unchanging stagnant environment. Following their evolution, perhaps 2.7 billion years ago, sulphate-reducing bacteria dominated the deep ocean ecosystem for more than 2 billion years. They are still to be found today in the Black Sea, and indeed in any stagnant, stinking sludge the world over, including our own guts. Their success in turning the world to their own advantage is comparable only to the cyanobacteria, which for 3.5 billion years have used sunlight to fill the world with oxygen. The antithesis is almost a biblical pitting of the forces of darkness against the forces of light. Sulphate-reducing bacteria shun the grace of light and air for darkness and slime, the sulphurous, foul-smelling pits of the underworld. Yet their noxious waste helps to maintain ecological diversity. Just as the opposition of light and darkness is central to many religions, with most of us inhabiting the moral middle ground, so too the opposite poles of the living world create a spectrum of conditions that fosters the variety of life.

Many single-celled organisms do not hide from oxygen, but are quick to move away from places where the oxygen levels are too high, exploiting the spectrum of conditions maintained by microbes like sulphate-reducing bacteria. We saw in Chapter 3 that free-living ciliates actively swim away from oxygen. This is actually quite a complex response. To do this, they need sensors that can detect the oxygen level in their surroundings. The information gleaned must then be coupled to the beating of their cilia, to drive them away from oxygen. The sensors themselves are proteins containing haem, similar to haemoglobin. Haem proteins are well suited to this purpose, because their physical properties change in the presence of oxygen, as in the colour change of haemoglobin from blue to red. Representatives of all three of the great domains of life use haem proteins as oxygen sensors, implying that LUCA herself (the Last Universal Common Ancestor, see Chapter 8) might have used them for this purpose.

When used in this way, haem proteins are acting as antioxidants: they play an integral part in maintaining external oxygen levels within congenial limits. Even when no movement is involved, as in the root nodules of leguminous plants (Chapter 8), haem proteins act as antioxidants by binding to excess oxygen and releasing it only slowly, maintaining a constant and low concentration of oxygen in the immediate environment.

Some microbes tolerate oxygen in their surroundings by physically screening themselves from the worst atrocities. The simplest screen is a pile of dead cells. In the same way, the riddled body of a dead comrade may protect a soldier from the bullets of an execution squad. Anaerobic cells living in stromatolites (Chapter 3) have relied heavily on this device, sheltering behind successive layers of dead cells. They have been doing this for 3.5 billion years.

A more sophisticated approach to the same problem is to secrete a layer of mucus and lurk within it. All free-living aerobic microbes live inside a mucus capsule as habitually as a crab lives in its shell. Rather surprisingly, mucus has some big advantages over calcite shells. James Lovelock was confounded by mucus while trying to sterilize hospital wards with high-intensity ultraviolet rays in the early 1950s. He succeeded in killing bacteria that had been denuded of mucus, but had no impact at all on their intact cousins, even at hundreds of times the normal atmospheric levels of ultraviolet. We saw in Chapter 6 that the damage wrought by radiation on living cells is mediated by the formation of free radicals from water. By protecting against free-radical damage, mucus goes some way towards explaining the remarkable ability of bacteria to survive in outer space and in other highly irradiated environments. Survival in outer space! This is a startling property for a substance we tend to associate with a blocked nose. The horror-movie concept of repulsive slimy aliens may be more soundly based in biology than their creators imagined. Perhaps we should not be too surprised to learn that the way in which mucus deals with free radicals is far more clever than just a high slime index.

Bacterial mucus is a mixture of long-chain polymers, analogous to plastics, which have one thing in common: they all carry a negative electrical charge. This means that mucus binds tightly to positively charged atoms such as iron and manganese, removing them from the surroundings. The affinity is so strong that some bacteria are exploited industrially to trap heavy-metal contaminants in waste water.

What advantage does a full metal jacket offer to bacteria? The answer is somewhat counterintuitive. We saw in Chapter 6 that hydrogen peroxide and superoxide radicals are not very reactive, and so are likely to diffuse some distance before reacting. They are only dangerous in the presence of metals such as iron, which can catalyse the formation of ferociously reactive hydroxyl radicals. Given that metals catalyse dangerous free-radical reactions, a metal jacket might seem to be a liability — it must be fizzing with menace all the time. Yet by stockpiling iron in their jackets, bacteria make sure that any dangerous free-radical reactions take place at arms length, rather than inside the cell itself. The mucus acts as a sacrificial target and the iron is converted into biologically inactive rust. The effect is equivalent to disposing of a bomb by exploding it at a safe distance. Not only this, but the controlled explosions detonate invaders such as bacteriophages (viruses that infect bacteria) and possibly even immune cells that try to engulf them. There is, certainly, a direct correlation between the thickness of the jacket and both the survival and the infectiousness of many bacteria.

As the bacteria become encrusted with metals, their jackets grow ever more cumbersome. In the end, they may be obliterated by their accumulated wealth. The presence of hollow microscopic pits in the structure of some banded iron formations suggests that these rocks might have been formed from the buried corpses of innumerable iron-encrusted bacteria. As the bacteria themselves dissolved away, only the metal jackets remained to testify to the existence of a mass graveyard.

We should not think of these rudimentary tricks as fit only for microbes. Each of them has an equivalent in ourselves. We too hide behind a layer of dead cells, otherwise known as the skin. Like the ciliates, we too use haem proteins as sensors to maintain internal oxygen at a constant low level. Like the sulphate-reducing bacteria, we too use sulphur as a buffer against oxygen (a trick we shall look into later in this chapter). We too produce mucus to protect against oxygen or bacterial infection in our nasal passages, airways and lungs. In the same way that anaerobic bacteria shelter within the intestine to avoid oxygen, so our own cells might be said to shelter within the body itself, where the oxygen level is kept at a fraction of that in the hostile outside world.

In this regard, we can reasonably view the gigantism discussed in Chapter 5 as an antioxidant response. The increase in body size compensates for the higher external oxygen levels, especially in diffusion-limited creatures such as the Bolsover dragonfly. An increase in size presumably lowers the oxygen levels within the end-users, the mitochondria. As we saw in Chapter 8, mitochondria operate best at oxygen levels not far above those tolerated by sulphate-reducing bacteria. If external oxygen levels rise, an increase in size will counteract the internal rise, and maintain oxygen at the right level.

The mitochondria themselves contribute to this balance. Mitochondria were once free-living bacteria that found shelter and protection within larger cells. The bargain was not one-sided: the internalized bacteria gained protection, but also lowered oxygen levels for the host cell by actively respiring. The relationship is now far more complex, but we should not forget that mitochondria do still lower the oxygen level inside cells. If mitochondria are defective, but the blood stream continues to deliver oxygen at the same rate, then our cells would presumably become oxidized. As we grow older, our mitochondria do begin to fail and our cells do become oxidized. Such oxidation is often attributed to free radicals escaping from defective mitochondria, but it might also be the result of higher oxygen levels in the rest of the cell, due to lower mitochondrial uptake of oxygen.

Oxygen-dependent organisms cannot rely on shielding as the only solution to oxygen toxicity. In all these organisms, oxygen must be consumed continuously to stay alive, because it is an essential ingredient in their only, or main, means of producing energy. To hide is not just impossible, but also metabolically detrimental. A different balance must be found, in which the dangerous effects of free radicals are prevented or contained by antioxidant enzymes and free-radical scavengers — the second and third defence mechanisms on our list. I will deal with the enzymes first.

Two of the most important antioxidant enzymes are superoxide dismutase and catalase. Virtually all organisms that spend any time in the presence of oxygen possess copies of these two enzymes. Their near ubiquity in aerobic organisms sharpens the paradox of cyanobacteria. These bacteria were the first photosynthetic organisms to split water and produce oxygen. If they had evolved in a world without oxygen, they should have been poisoned by their own waste. The fallacy of this textbook argument is discussed in Chapter 7, and we saw that the cyanobacteria were, in all likelihood, already protected against their poisonous product by superoxide dismutase and catalase. These enzymes, and probably others, evolved in response to the formation of reactive oxygen intermediates by ultraviolet radiation very early in the history of life. Indeed, we saw that there is good evidence that antioxidant enzymes were present in LUCA.

Superoxide dismutase, with the memorable acronym SOD, holds a special place in the affections of free-radical biochemists. It is, symbolically, the proof of the pudding. Ever since the early 1950s, when researchers first started to speculate about the role of oxygen free radicals in ageing and disease, proof of their importance has been hard to come by. Free radicals are fleeting. For years, their existence was supported only by tell-tale traces of the damage caused — evidence as flawed and controversial as deducing the existence of a yeti from outsized footprints in the snow. Then, in 1968, Joe McCord and Irwin Fridovich, at Duke University in North Carolina, showed that an abundant blue-green protein called haemocuprein, long thought of as an inert depository for copper, did actually have catalytic activity. It converted superoxide radicals ($O_2^{\bullet-}$) into hydrogen peroxide (H_2O_2) and oxygen. Despite an intensive search, no other target for this enzyme could be found. Moreover, the rate at which it disposed of superoxide almost beggared belief. Superoxide radicals are unstable and react with each other within seconds to produce hydrogen peroxide, but haemocuprein sped up this natural reaction by a factor of a billion. Surely this could not be accidental![1] McCord and Fridovich renamed the protein superoxide dismutase (SOD) in their seminal 1969 paper in the *Journal of Biological Chemistry* — in the opinion of many, the most important discovery in modern biology never to win the Nobel Prize.

The implications transformed the field. If an enzyme as efficient as SOD evolved specifically to eliminate superoxide radicals, then superoxide radicals must be biologically important. Seen the other way round, free radicals are a normal feature of biology and life has evolved a remarkably effective mechanism for dealing with them. Given the tendency of mutations to corrupt the superfluous, this mechanism can only be that good because it needs to be that good. So what would happen if SOD failed in some way, swamping the cell with free radicals? Ageing? Death? The myriad possibilities had a galvanizing effect that has persisted to this day.

It was not long before other forms of SOD were discovered. The first, reported in 1970 by McCord and Fridovich again, was isolated from the bacterium *Escherichia coli*. This was a startlingly different kettle of fish: a pink, manganese-containing enzyme, but with a similar capacity for eliminating superoxide radicals. Even more intriguingly, it turned out that many eukaryotic cells actually possess both forms of SOD. As I write, 30 years on, it has become a commonplace that many eukaryotes produce several different types of SOD — usually one in the mitochondria, another in the cytosol, and a third that is secreted from the cell. Although their detailed structures differ, all contain metal atoms at their catalytic centres: copper and zinc, manganese, and iron or nickel.

[1] Fridovich, now venerated as the godfather of the free-radical field, points out that the speed of this reaction is close to the maximum rate of diffusion and on the face of it seems impossibly fast. The copper site at the centre of the enzyme takes up less than 1 per cent of the surface of the enzyme. On the basis of random diffusion, one would expect more than 99 per cent of collisions between enzyme and substrate to be fruitless. In fact, superoxide radicals are guided into the active site down an electrostatic gradient, like an aeroplane being guided into an airport by runway lights. Thus the reason that the enzyme can match the speed of diffusion is because it streamlines the chaotic natural process. Substituting amino acids in the enzyme alters the electrostatic fields and affects the speed of reaction, without necessarily having any impact on the active site itself. It is actually possible to produce even faster versions of superoxide dismutase.

The importance of these enzymes is illustrated by the fate of so-called 'knock-out' mice, which lack part of a gene for one of the SOD enzymes but are in other respects genetically 'normal'. In a 1996 study, Russell Lebovitz and his colleagues at the Baylor College of Medicine in Houston reported that mice born with a defective mitochondrial form of SOD died within three weeks of birth. The mice were pathetic runts, severely anaemic and with a form of motor neuron disease that led to weakness, rapid fatigue, and what Lebovitz described as 'circling behaviour' — presumably a disoriented tendency to stumble sideways in search of their tails. At autopsy, they were also found to have cardiac abnormalities and fatty deposits in the liver. The mitochondria of mice that had survived beyond seven days were shot to pieces, especially in tissues with a high metabolic rate, such as heart muscle and brain. Another knock-out strain of mice, with a different mutation in the same enzyme, failed to survive beyond five days. In people, smaller defects in mitochondrial SOD have been linked with ovarian cancer and insulin-dependent diabetes, among other conditions. In contrast, loss of the cytosolic form of SOD is less devastating, although this, too, causes problems in later life, notably infertility, neurological damage and cancer.

There could hardly be more convincing evidence of the importance of SOD. Yet the enzyme demonstrates once again the need for networking between antioxidants. Far from disposing of a toxic chemical, SOD merely defers the problem. The product of SOD is hydrogen peroxide, a potentially dangerous chemical in its own right. We may well wonder if it is necessarily beneficial to generate hydrogen peroxide at a billion times the natural rate. There are, in fact, circumstances in which too much SOD may be detrimental. For example, people with Down syndrome have an additional copy of chromosome 23. Why this should be so damaging is unknown, but researchers have noted that the gene for SOD sits on chromosome 23. People with Down syndrome therefore make too much SOD. The syndrome itself is characterized by oxidative stress, leading to neurological degeneration. It may be that people with Down syndrome suffer oxidative stress because they have too much SOD.

In normal physiological conditions, however, hydrogen peroxide is removed equally quickly by catalase, which converts it into oxygen and water. We saw in Chapter 7 that there are other enzymes that can remove hydrogen peroxide and organic peroxides safely, without producing oxygen, using a variety of electron donors such as glutathione and vitamin C.

Enzymes that dispose of peroxides are being reported all the time. In 1988, for example, a new family of antioxidant enzymes, now known as peroxiredoxins, were discovered by Sue Goo Rhee and her colleagues at the National Heart, Lung and Blood Institute in the United States. These enzymes do not have metals at their centre, but instead have two adjacent sulphur atoms that accept electrons from a small sulphur-rich protein called thioredoxin. By the mid 1990s, similar peroxiredoxins had been isolated from representatives of all three domains of life, suggesting that they, too, dated all the way back to LUCA. At least 200 related peroxiredoxin genes have now been found, five of which are found in people, and their DNA sequences determined,

I mention the peroxiredoxins, in part, because they solve a long-standing puzzle relating to human parasites, including *Plasmodium falciparum*, the protozoan responsible for the most severe form of malaria, and helminth worms such as *Fasciola hepatica*. When invading the body, these parasites are subjected to a barrage of oxygen free radicals from neutrophils and other immune cells. The ferocity of this attack can trigger inflammation and fever serious enough to kill the host, let alone the parasite. Most parasites defend themselves using antioxidant enzymes such as SOD, but, curiously, few possess catalase to remove hydrogen peroxide. To researchers in the 1980s, this seemed a contradiction: the action of SOD in the absence of catalase should have flooded the parasites in hydrogen peroxide and so exacerbated the effects of the immune attack. It did not. The parasites held their own. Unless the theory was full of holes, the parasites must have been using an unknown enzyme that could detoxify hydrogen peroxide. The search for this 'missing link' was finally rewarded by the discovery of the peroxiredoxins, which have since been found in virtually all parasites lacking catalase. In fact, representatives of all three domains of life produce related peroxiredoxins, implying that they too may have been present in LUCA.

Our understanding of the peroxiredoxins is beginning to open up new targets in the struggle against parasitic diseases. One possibility, for instance, is to vaccinate against the parts of the parasite enzymes that are appreciably different from the human versions, thus gearing up the immune system to attack one of the critical bastions of the parasite antioxidant defences.

The combination of SOD with some sort of enzyme to remove hydrogen peroxide is almost mandatory if cells are to defuse free radicals before too much damage is done. Because there are so many enzymes that can dispose of hydrogen peroxide, catalase deficiency is less injurious than SOD deficiency. Moreover, as we have seen, hydrogen peroxide is only really toxic in the presence of iron or copper, either of which can catalyse the formation of hydroxyl radicals. Under normal circumstances, these metals are kept locked away in their respective protein cages, ferritin and caeruloplasmin. Tomasz Bilinski, a free-thinking evolutionary microbiologist at the University of Lublin in Poland, has argued that stowing away metals might be the single most successful strategy to prevent formation of hydroxyl radicals. Even so, despite the many precautions, some hydroxyl radicals are undoubtedly still formed. We noted in Chapter 6 that the rate at which oxidized fragments of DNA are excreted in the urine suggests that each cell in the body sustains thousands of 'hits' to its DNA every day. Even allowing for experimental error, we must conclude that the enzyme networks are not fool-proof. This conclusion is supported by our need for dietary antioxidants such as vitamin C and vitamin E. They are technically referred to as 'chain-breaking' antioxidants, because they quench free-radical chain reactions already started by hydroxyl radicals. Chain-breaking antioxidants are the third defence mechanism on our list.

Most chain-breaking antioxidants work in essentially the same way as vitamin C, by donating electrons. Many of the best-known antioxidants, including carotenoids, flavenoids, phenols and tannins, must be obtained in the diet from plants. Their relative importance in the body's antioxidant balance is tricky to determine, but their overall contribution is generally held to be responsible for the benefits of fruit and vegetables. On the other hand, we should not think of chain-breaking antioxidants as merely dietary components. A few are products of our own metabolism, such as uric acid, bilirubin (a bile pigment and breakdown product of haem) and lipoic acid. These are powerful chain-breaking antioxidants in their own right, at least as potent as vitamins C and E. Some conditions that we tend to think of as pathological, such as jaundice in newborn babies, may really be an evolved physiological adaptation. In the case of jaundice, evidence suggests that bilirubin is stockpiled in the skin to *protect* against the oxidative stress of birth. The baby emerges from the cloistered security of the womb into the high-oxygen world outside, but has not yet built up protection from dietary antioxidants, hence the need for bilirubin. Similarly, the disfiguring colours of a bruise — courtesy, again, of bilirubin — help to protect against the oxidative stress of traumatic injury, when blood-borne antioxidants may not be able to get through.

In many cases, the balance of risks and benefits with chain-breaking antioxidants is far from clear. Take uric acid, for example — a potent antioxidant, certainly, but one that at high levels causes the painful inflammation of gout as it crystallizes in the joints. Uric acid is sometimes considered to be a risk factor for cardiovascular disease as well, because the people with the highest levels of uric acid in their blood stream are also at the greatest risk of a heart attack. In reality, this simple association may be quite misleading. Those at greatest risk of cardiovascular disease tend to have the lowest intake of dietary antioxidants. What could be more sensible for the body than to step up levels of endogenous antioxidants? The worse the disease, the more uric acid is needed to combat it; hence the association between the severity of disease and plasma uric acid levels. This line of argument is speculative, but highlights the danger of bare associations. In this case, treatments aimed at lowering plasma uric acid might easily prove detrimental, unless coupled with a better diet; and of course, if we change diet, we can hardly come to any sensible conclusions about the true role of uric acid. There are few straight relationships between antioxidants and health.

Two small sulphur-containing compounds, glutathione and thioredoxin, have now cropped up several times in this chapter and the last. Both donate electrons, either to regenerate antioxidants such as vitamin C, or to detoxify hydrogen peroxide and organic peroxides directly. Far from being bit players, these sulphur compounds are the gatekeepers between genes and diet, between health and disease. It is time they took centre stage, for they are responsible for orchestrating the last two of the five antioxidant mechanisms that I outlined at the start of this chapter — repair mechanisms and stress responses. Sulphur is now seen as the most important counterbalance to oxygen inside individual cells, as well as in the wider ecosystem.

Here I must beg indulgence for a single paragraph of unmitigated biochemistry, a diffident request after the geneticist Steve Jones's dismissal of biochemistry as a subject unfit for popularization. However, the concept is not difficult, and is critical to understanding how sulphur can act as a molecular switch that tells the cell when it is ill. I'm picking out a single example of how such processes work; a baffling array of other mechanisms are superimposed over this one, eroding or strengthening the signal. Nonetheless, the behaviour of sulphur is as exemplary as can be, and is without doubt one of the most important areas in the hot science of biological signalling.

Sulphur, bound to hydrogen (–SH), is a constituent of just one of the 20 amino acids — a nondescript little molecule with 14 atoms called cysteine. The unique sulphur group in cysteine, the –SH group, is called a *thiol* group. Thiols are delicate structures, easily susceptible to oxidation. I imagine them waving their yellow sulphur heads gently, delicate as seeding dandelions. When thiol groups are oxidized, one of two things can happen. First, if the hydrogen atom (proton plus electron) is extracted, the sulphur stubs of adjacent thiols may bind together to form sulphur-to-sulphur bonds, known as disulphide bridges. In the presence of oxygen, disulphide bridges are more robust than unoxidized thiols and have long been recognized as important structural features of extracellular proteins, helping to stabilize their three-dimensional shapes. The second possible fate for delicate thiols, formidably named *S-nitrosylation*, began to emerge in the late 1990s under the searching intellect of Jonathan Stamler, latest in a distinguished line of free-radical biochemists at Duke University. Stamler and his colleagues argue that oxidative stress increases the production of another free radical, nitric oxide ($NO^{\bullet}$). Nitric oxide itself is sluggishly reactive with most substances, but it acts synergistically with other free radicals to oxidize thiols. In this case, the product is not a disulphide bridge, but an *S*-nitrosothiol (–SNO), which is almost as stable. The formation of either a disulphide bridge or an *S*-nitrosothiol modifies the structure of the protein in a reversible way. It is possible to regenerate the delicate thiol group using hydrogen atoms from glutathione or thioredoxin.

For proteins, structure corresponds to activity. If thiols influence the structure of proteins, then they also affect their activity. In other words, the oxidation state of thiols can act as a molecular switch for thiol-containing proteins, turning their activity on and off. The list of proteins with sensitive thiol groups is growing all the time, and is now known to include some of the most important transcription factors (proteins that bind to DNA and stimulate the transcription of genes to produce new proteins). Whether these factors bind to DNA or not, even whether they journey into the nucleus at all, depends on the status of their thiol groups.

The inside of a healthy cell is full of delicately waving thiols. They are kept in the unoxidized state by the continuous policing of glutathione and thioredoxin. Any thiols oxidized by 'mistake' are reconverted back into the pristine state. Glutathione and thioredoxin are regenerated by tapping the energy of cellular respiration, as we saw in Chapter 9 while discussing vitamin C. Under normal conditions, this is not much of a drain on the resources of the cell. The situation is reversed, however, when the cell is under oxidative stress.

Picture a cell under oxidative stress. The problem might be too much oxygen, or perhaps an infection or a disease. The end result is the same: there are free radicals everywhere. Chain-breaking antioxidants such as vitamin C are used up quickly. They are regenerated by glutathione, but not without some loss. As antioxidants are depleted, the same number of free radicals causes more damage. Thiol groups in proteins are now being oxidized. Some are patched up by glutathione and thioredoxin, but the balance is shifting. This is a war zone. The defenders cannot rebuild bombed bridges every night for ever. Now half the proteins in the cell have oxidized thiols. Some of the proteins are switched off by the oxidized thiols, shut down to await the end of the war. Others are switched on. A guerrilla militia is taking charge of the nucleus, the last stronghold. The guerrillas are mostly transcription factors. Inside the nucleus they bind to DNA and stimulate the production of new proteins. But which new proteins? This is no random selection. The cell has a serious decision to make: entrench for battle or commit suicide (apoptosis) for the good of the body as a whole. The decision depends on the odds of success, in particular on the number and state of the transcription factors that made it safely to the nucleus. At the molecular level, Nietzsche's aphorism holds sway — what doesn't kill you makes you stronger.

If a cell opts to slug it out rather than take an honourable exit, the kinds of defences it draws on are qualitatively similar in all living things, from *E. coli* to humans. At present, the system in *E. coli* is better understood, in part because bacterial genes are organized into operating units called operons. This makes it easier to identify exactly which genes are involved. In *E. coli*, two main transcription factors are activated by the oxidation of sulphur groups. One is a thiol-containing protein known, in the opaque shorthand of molecular biology, as OxyR, and the other is the protein SoxRS, which contains sulphur in the form of an iron–sulphur cluster, (this acts in much the same way for our purposes here). When oxidized, each of these transcription factors controls the transcription of a dozen or more genes, the products of which strengthen the antioxidant defences of the cell.

In people, the list of transcription factors whose activity is controlled by the oxidation of thiol groups is growing all the time, and includes NFκB (pronounced NF-kappa B), Nrf-2, AP-1 and P53. For our purposes, the two most important are NFκB and Nrf-2. NFκB marshals the 'stress' response by activating a mixture of offensive (inflammatory) and defensive (antioxidant) genes. Nrf-2 has a purely defensive (antioxidant) role, actually switching off the offensive inflammatory genes. Both NFκB and Nrf-2 therefore strengthen the cell, but otherwise their effects are opposing. They act like two generals in a council of war, one of whom advocates all-out war, the other appeasement. The outcome depends on how successful they are at persuading the rest of the council to accept their view. For transcription factors, this comes down to numbers on the ground. If 1000 activated NFκB proteins reach the nucleus but only 100 activated Nrf-2 proteins, the cell will go to war, launching an inflammatory attack on the invader and bolstering its own defences. If Nrf-2 holds sway in the nucleus, then the cell effectively barricades itself in bunker and waits out the war. In either case, the extra defences provide immediate benefits, but they also confer resistance to future attacks, regardless of their exact nature. Forewarned is forearmed, whatever the threat may turn out to be.

What are the products of these defensive genes? Some have not yet been identified, others are familiar to us already. As we might expect by now, the levels of SOD, catalase and other antioxidant enzymes are stepped up; new metabolic proteins divert cellular respiration towards the regeneration of glutathione and thioredoxin; and extra iron-sensors and iron-binding proteins are made that detect and store free iron. A number of so-called 'stress' proteins are also produced, which busy themselves with clearing up the mess and dusting down the salvageable, like the emergency services picking over rubble in the aftermath of a bombing raid. Irredeemably broken proteins are earmarked for destruction and recycling. Less seriously damaged proteins, bruised but unbroken, are refolded and repackaged with the aid of proteins known as molecular chaperones. Other proteins set to work on DNA, cutting out oxidized fragments, inserting replacements and re-joining breaks in the DNA strands.

Taken together, these actions restore the cell to health, but (with the exception of antioxidant enzymes) have little effect on its longer-term capacity to withstand a similar assault. There *are* some proteins that can strengthen a cell's resistance to future assault, however. Two of the most powerful are metallothionein and one of the versions of haem oxygenase — the stress protein HO-1. These proteins bolster the resistance of cells to a whole range of insults, from heavy-metal contamination to radiation poisoning and infection — conditions linked by the common thread of oxidative stress. They are more powerful protectors than any dietary antioxidant, although they achieve this by imposing what amounts to a curfew, in which the normal daily life of the cell is held in check.

The firmness of their rearguard action unveils a final irony: antioxidant supplements, which behave more like a scattered vanguard on a looting trip, have the potential to exacerbate some diseases. The reason is as follows. The signal to produce metallothionein and haem oxygenase is the oxidation of thiols. Thiols become oxidized when the supply of antioxidants is depleted. Raising our intake of antioxidants can therefore suppress the signal by sparing thiols, thereby robbing the cell of its most powerful allies. This is not just an idle fancy. Roberto Motterlini and Roberta Foresti, at the Northwick Park Institute for Medical Research, London, have shown that adding antioxidants to cells under oxidative stress (especially thiol-regenerating antioxidants, like *N*-acetylcysteine) does indeed prevent the cell from making new haem oxygenase. This renders the cells more vulnerable to injury. In the body, suppression or loss of haem oxygenase activity can have some dire consequences.

Just how important such a loss might be is illustrated by an exceptional case study reported in 1999 by a team at Kanazawa University in Japan: the first known case of haem oxygenase deficiency in humans. An unfortunate six-year-old boy suffered from severe growth retardation, abnormal blood coagulation, haemolytic anaemia and serious renal injury. All this misery stemmed from the loss of a single enzyme, a stress protein supposedly produced only sporadically in times of oxidative stress. Similar problems have been reported in animals. Kenneth Poss and Susumu Tonegawa, at the Massachusetts Institute of Technology, have shown that 'knock-out' mice deficient in haem oxygenase succumb to a serious inflammatory condition, which resembles haemochromatosis in people (the iron-overload syndrome discussed in Chapter 9). The mice lacking haem oxygenase suffer from serious iron overloading in tissues and organs, which leads to pathological enlargement of the spleen, liver injury and fibrosis, various immune disorders, weight loss, decreased mobility and premature mortality. Mice without haem oxygenase also have shrunken testes, 25 per cent smaller than similarly sized littermates, and a lack of libido, both of which are characteristic of hereditary haemochromatosis in people.

For the effects of haem oxygenase deficiency to be so profound we may well wonder whether some degree of oxidative stress is in fact the norm. If so, then haem oxygenase levels presumably change continuously as part of a flexible and self-regulating feedback mechanism. We shall see in Chapter 15 that a continuously high haem oxygenase activity probably does have beneficial effects on our health in old age.

We have come a long way from vitamin C. If nothing else, I hope that this chapter has exploded the notion of antioxidants as dietary elements that will help us live forever. Our whole body is an antioxidant machine, from the physical structure of individual cells to the physique of a human being. This perspective is obscured by our deeply seated, almost childlike, yearning to allocate purpose, which is seldom echoed in biology. We want to say a car is 'for' driving, a business is 'for' making money, life 'for' living. Similarly, mitochondria are 'for' producing energy, haemoglobin is 'for' delivering oxygen and vitamin C 'for' protecting against free radicals. I compared the skin to a layer of dead cells in a stromatolite. Certainly it is. We cannot breathe through our skin like amphibians. But the skin is equally a barrier that prevents evaporation and infection, and an important criterion of beauty. The point is that few things serve a single purpose, least of all in biology, and we must fight our instinct to impose one. Apart from anything else, this means we must resist the scientific urge to apply tight — and therefore often singular — definitions.

In Chapter 9, we saw that vitamin C is used in diverse ways, united only by the same molecular action. This also applies to SOD, catalase or haem oxygenase. At the molecular level, their actions are always identical. The effects, however, are diverse and may serve quite different purposes. Take SOD. Its action is simple: to remove superoxide radicals. But is this purely and simply an antioxidant action, or is it also a signal? If the formation of superoxide radicals outstrips their removal by SOD, some of the extra free radicals will oxidize the thiol groups in proteins, sending transcription factors scurrying to the nucleus. In the nucleus, these transcription factors bind to DNA and stimulate the production of new proteins, which help restore the cell to health. In other words, the cell adapts to a small change in circumstances, such as a slight increase in oxidative stress, by a subtle change in its repertoire of proteins. Are we really helping if we block this pathway by swamping it in an excess of SOD or other antioxidants? Who knows. We have noted one reason why dabbling in this way might be detrimental: haem oxygenase is produced in response to thiol oxidation. If it is true that the levels of haem oxygenase flux continuously in response to small changes in oxidative stress, then smothering this subtle interplay may dull our response to sudden challenges such as infection. Perhaps this is why it has been so hard to confirm any protective effects of vitamin C against the common cold, as claimed by Pauling. The benefits of vitamin C may be offset by the disadvantages of suppressing the synthesis of stress proteins like haem oxygenase.

Despite these uncertainties, we cannot dispute that eating fruit and vegetables is good for us. If this has anything to do with antioxidants, we might conclude that our sophisticated antioxidant machine falls short of perfection. Presumably, the extra antioxidants mop up free radicals that would otherwise damage DNA and proteins, undermining the vitality of cells and ultimately the whole body. On the other hand, it is quite possible that the benefits of fruit also relate to other factors, such as mild doses of toxins that stimulate the production of stress proteins like haem oxygenase. These toxins are designed to prevent fruit from being eaten before it is properly ripe, or by the wrong sort of vegetarian. Perhaps it is the balance of antioxidants and mild toxins that confers the benefits of fruit. Certainly the benefits of fruit and vegetables have never been reproduced simply by taking antioxidant supplements.

Half a century ago, the pioneer of free-radical biology Rebeca Gerschman questioned whether a continuous small 'slipping' in antioxidant defences might contribute to aging and death. This idea is still the basis of the free-radical theory of ageing. I hope this chapter has shown that our antioxidant defences are far more complex than Gerschman, or for that matter Pauling, could ever have imagined. These are the fruits of molecular biology in the past few years. But we are left on the horns of a dilemma. Can ageing be slowed down by blocking free-radical slippage, or is some degree of slippage *necessary* for marshalling our resistance to stress? If chain-breaking antioxidants are relatively unimportant compared with the stress response, might we be able to slow down ageing by manipulating the stress response directly? Armed with our new understanding of antioxidants, we will look at the balance of evidence in the next four chapters.

CHAPTER ELEVEN

Sex and the Art of Bodily Maintenance

Trade-offs in the Evolution of Ageing

GILGAMESH, KING OF URUK, sets out on a quest for immortality, the pain of loss in his heart. Passing through twelve leagues of darkness, where there is no light from the rising of the sun to the setting of the sun, he seeks Utnapishtim the Faraway, who knows the secret of everlasting life. Utnapishtim tells Gilgamesh that his quest is like a search for the wind, but then reveals the mystery of the gods: a plant that grows under the water, which restores lost youth to a man. By its virtue, a man may win back all his former strength. Its name is "The Old Men Are Young Again". Gilgamesh finds the plant, only to have it snatched away by a serpent before he can eat of it. The serpent sloughs away its skin and disappears into a spring of water. Gilgamesh weeps and returns to Uruk empty-handed. He has his story engraved on clay tablets. His achievements are celebrated by his people, and when he dies, like a hooked fish stretched on the bed, the people of the city, great and small, are not silent; they lift up the lament. All men of flesh and blood lift up the lament.

The *Epic of Gilgamesh*, the oldest surviving masterpiece of the Sumerian dynasty of ancient Mesopotamia, is at least 1500 years older than Homer. From the dawn of recorded history, its tale of friendship and heroism resonates with the eternal concerns of mankind: bereavement, ageing, death, the dream of immortality. These themes recur throughout history, and not just in a vague sense — we cling to the idea that everlasting youth is attainable through the possession of some kind of magical artefact, be it a plant, the nectar of the gods, the Holy Grail, the grated horn of a unicorn, the Elixir of Life, the philosopher's stone, or growth hormone.

Biologists fall victim to the same yearning for an antidote to ageing. The history of biology affords a succession of bizarre claims. In 1904, for example, the Russian immunologist and Nobel laureate Elie Metchnikoff claimed that ageing was caused by toxins released from bacteria in the intestinal tract. He regarded the large intestine as a necessary evil, a reservoir of waste material that relieved the need for constant defecation stops while on the run from predators (or after them). He was fascinated by Bulgarian fables of centenarians, and ascribed their longevity to yoghurt, which was unknown in western Europe at the time. Metchnikoff championed the idea that we would all live to 200 if only we ate more yoghurt, full of "the most useful of microbes, which can be acclimatised in the digestive tube for the purpose of arresting putrefactions and pernicious fermentations." He did have a point — gut bacteria do influence health, if not maximum human lifespan.[1]

Other theories linked ageing in men with diminishing testicular secretions. In 1889, Charles Edouard Brown-Sequard, a prominent French physiologist, announced to the Société de Biologie in Paris that he had rejuvenated his mind and body by injecting himself with a liquid extracted from the testicles of dogs and guinea-pigs. Apparently the injections not only increased his physical strength and intellectual energy, but also relieved his constipation and lengthened the arc of his urine. Later, a number of surgeons tried implanting whole or sliced testicles into the scrotums of recipients. Leo L. Stanley was resident physician in San Quentin prison in California. He began transplanting testicles (removed from recently executed prisoners) into inmates in 1918. Some of the recipients reported full recovery of their sexual potency. By 1920, the scarcity of human gonads induced Stanley to substitute ram, goat, deer and boar testes, which he said worked equally well. He went on to perform hundreds of

[1] We must distinguish between *average* life expectancy and maximum lifespan. In the West, average life expectancy has risen dramatically over the last century: fewer people die of infectious disease, in childhood or in childbirth, and far more people live into old age. Despite these massive demographic changes, maximum lifespan has changed little, if at all. A few people have always survived past 100; today many more do, but we are hardly more likely now than we were in biblical times to live beyond 115–120. Short of a major breakthrough, this can be considered the maximum human lifespan.

operations, treating patients with ailments as diverse as senility, asthma, epilepsy, tuberculosis, diabetes and gangrene.

The high demand for gonadal implants made the fortunes of at least two surgeons during the 1920s and 1930s. In France, the Russian émigré Serge Voronoff transplanted monkey glands to extend the life of his wealthy and famous clients. A respected biologist, Voronoff experimented on eunuchs in the courts of Egypt, and even tried grafting monkey ovaries into women, with dire consequences. In America, the notorious quack, 'Doctor' John R. Brinkley, transplanted hundreds of sliced goat testicles into his ageing customers in Milford, Kansas, where he became so popular that he was nearly voted governor in 1930. Each patient had the privilege of selecting his own goat from the doctor's herd. The financial success of this venture enabled him to build and operate the first radio station in Kansas — KFKB, or Kansas' First, Kansas' Best — through which he brazenly promoted his own secret remedies, including goat gland transplants. After a series of court cases, and opposition from both the American Medical Association and the Federal Radio Commission, Brinkley fled to the Mexican borderlands, where he set up a new, even more powerful, radio station, and continued his shady medical operations, amassing an estimated $12-million fortune. He is said to have kept penguins and giant Galapagos turtles on his estates in Texas. It was not to last. Endless lawsuits and punitive taxes eventually obliged him to file for bankruptcy in 1941. His health collapsed, and after suffering a heart attack, kidney failure and the amputation of a leg, he died penniless later that year, at the age of 57, the most famous charlatan in American history.

The craving for renewed youth is not just a historical curiosity. In more recent times, vitamin C, oestrogen, melatonin, telomerase and growth hormone have all been touted as miracle cures. Each retains a faithful band of adherents but, whatever their merits, their failure to extend maximum lifespan is plain to most. Medicine has distanced itself from this kind of wishful thinking, but the effect has not been altogether positive. Rather, the traditional attitude of mainstream medicine, that ageing is in some sense necessary or inevitable, and thus beyond the domain of medical science, has perpetuated the mystique. Even today, ageing is rarely considered a proper field of study within the discipline of medicine — it is still contaminated by the legacy of quacks like Brinkley. In most countries, medical school curricula devote no classes to ageing. Yet while turning a blind eye to the study of ageing, medicine has accumulated a tremendous wealth of information on the infirmities of old age. This mass of information is all part of gerontology, even though most specialists do not think of themselves as gerontologists: they are cardiologists, neurologists, oncologists or endocrinologists. Few refer to each other's journals, so there is little sense of overall perspective.

Examining diseases in isolation, medical researchers have historically paid little attention to evolutionary theories, and tend to consider illness from a mechanistic point of view. In the case of heart disease, for example, we know in minute detail how oxidized cholesterol builds up in coronary arteries, producing atherosclerotic plaques, how these plaques rupture, and how thrombosis causes myocardial infarction. As a corollary, we know how to deflect the impending calamity, for a period at least: how to lower blood cholesterol or dilate the coronary arteries using drugs, how to salvage heart muscle after an infarction. What we are far less certain about is how heart disease relates to other diseases of old age, such as cancer, and whether it is possible to prevent both diseases by targeting a common underlying cause. As we saw in the last two chapters, the closest we have come, in the face of all the advances of modern medicine, is to say 'Eat your greens!'; and even then we are not quite sure why. This bleak situation is gradually changing. With so much at stake in a greying world, many more researchers are applying themselves squarely to the problem of ageing. The field of gerontology is now one of the most fertile in biology, and generates more interest than at any time since the alchemists. At last the dismal mountains of biological and medical evidence are being remodelled into broader, testable theories.

Half a century ago, in a celebrated inaugural lecture as professor of zoology at University College London, Peter Medawar described ageing as a great unsolved problem in biology. For many people outside the field, his assertion still holds true today. This is not the case. The two main theories of ageing — which we might loosely call the programmed and the stochastic theories — are daily growing closer together. Theories of programmed ageing hold that ageing is programmed in the genes, and is equivalent to other developmental processes such as the growth of the embryo, puberty or the menopause. Stochastic theories hold, in contrast, that ageing is essentially an accumulation of wear and tear over a lifetime and is not programmed in the genes. As is so often the case in science, the reality lies somewhere in between, drawing on elements from each theory. We do not know all the answers, and many details are perplexing, but in broad terms I think it is true to say that we do now understand *why* and *how* we age. As the eminent British gerontologist Tom Kirkwood has argued, ageing is not biologically inevitable, and does not follow a fixed genetic programme — although it is most certainly written in the genes. We shall see in future chapters that oxygen is central not just to ageing and death, but also, through the deepest of connections, to sex and the emergence of gender. There remain a number of outstanding questions. To what extent is ageing distinct from age-related diseases? Most important of all, how tractable is the ageing process, given the evolved architecture of the human body? We will explore these issues in later chapters. First, though, we must consider a few basic principles of biology.

Ageing, or rather senescence — the loss of function over the years — is not inevitable. We saw in Chapter 8 that we are all connected to LUCA, the Last Universal Common Ancestor, through an unbroken chain of ancestors. We know this because we, in common with our most distant cousins, the archaea and the bacteria, have inherited a few of LUCA's genes almost intact. At the most basic level, life shares a unity that would be baffling if we were not all related. In the sense that life has not remained static — that we have *evolved* — life itself has aged; but in no sense are we the *senescent* products of primordial DNA. Wines, cheeses and some human beings improve with age; nothing improves with senescence. We need only look around us to see that life is flourishing at the grand old age of 4 billion. If we accept that all life is descended from a common ancestor, then clearly senescence is avoidable.

The mechanism that has prevented the senescence of life — indeed, propelled life's evolution — is natural selection. Darwin's idea is most commonly expressed in the phrase coined by the English philosopher Herbert Spencer: the 'survival of the fittest'. This phrase is often criticized by evolutionary biologists as misleading, as natural selection is concerned not with survival as such, but with reproduction. The individuals that *reproduce* themselves most successfully are most likely to pass on their genes to the next generation. Those that fail to reproduce perish (unless, of course, they live for ever). But if we step back for a moment, we can see that there is something to be said for Spencer's misrepresentation. Why on earth do individuals want to reproduce themselves at all? Where does the reproductive imperative come from? The way in which even simple viruses seem compelled to replicate themselves is uncanny. It is hard not to succumb to the idea of a mysterious life force, an urge to reproduce. If we shun the idea of a life force, we need an explanation: why do living things want so desperately to reproduce?

The answer is that only reproduction can ensure survival. All complex matter is eventually destroyed by something. Even mountains are eroded over the aeons. The more complex the structure, the more likely it is to be broken down. Organic matter is fragile and will be shredded by ultraviolet rays or chemical attack sooner or later. Its atoms will be recycled in simpler combinations. Carbon dioxide, being a simple molecule, is more stable than DNA. On the other hand, if a piece of matter happens to have a propensity to replicate itself, for a little while its chances of persisting intact are doubled. It is still only a matter of time before the daughters are destroyed; but if one of the daughter molecules succeeds in replicating itself in the meantime, then the process can continue indefinitely.

The ability to replicate is not a magical property. As the Glasgow chemist Graham Cairns-Smith argues in his thought-provoking book *Seven Clues to the Origin of Life*, clay crystals replicate themselves in stream beds through a purely physical process. It is hard (unless you are Cairns-Smith) to see a life force here. Even so, the reason that life appears to have such a powerful urge to reproduce is simply because it would not exist if it did not. Only replicators can survive, so all survivors must replicate.

Given the tendency to destruction, the *rate* of replication is profoundly important. If we assume a steady rate of destruction, then to ensure survival, the rate of replication must surpass the rate of destruction. The importance of this relationship was addressed by the chemist Leslie Orgel, at the Salk Institute in San Diego, in 1973. Orgel theorized about the likely behaviour of populations of 'immortal' cells in culture, if subjected to different levels of irradiation. In this sense, immortal refers to a *population* of cells that has the potential to continue dividing indefinitely without becoming senescent; it does not mean that *individual* cells cannot die by accident or old age. Two cell types that behave in this way are bacteria and cancer cells. Both can be grown in cell culture without any appearance of senescence, but at any one time a proportion (perhaps 10 to 30 per cent) are unable to divide. They are doomed to die. Their place is taken by the progeny of cells that do continue to divide. Orgel made the point that if these 'immortal' cell populations are irradiated, so that the probability of any daughter cells surviving is less than 50 per cent (in other words, the rate of replication fails to surpass the rate of destruction), then the overall population will gradually decline and die out, at least in theory. At a lower radiation dose, the population might continue to grow, albeit at a lower rate than the unirradiated one.

We have already seen that the effects of radiation on biological molecules are mediated largely by the splitting of water (by the radiation) to produce oxygen free radicals such as hydroxyl radicals. The hydroxyl radical is not discriminating in its targets: almost all organic molecules are damaged by its attack. Such attacks take place continuously, in a more-or-less random manner. Unless the damage inflicted undermines the integrity of the cell at one fell swoop, the destruction of proteins and lipids is not necessarily a calamity. Given a suitable supply of energy, they can be replaced, new for old, following instructions in the DNA.[2] The problem comes when the code itself, the DNA, is damaged. If the damage to DNA results in the production of a faulty protein, which is incapable of carrying out a critical function such as the manufacture of other proteins, then the cell will almost certainly die. The central question in biology is therefore how to maintain the integrity of DNA from generation to generation.

Let us think again about the 'immortal' cells in culture. Imagine that they are being irradiated at an intensity calculated to kill more than 50 per cent of them. For a while, the population dwindles, as predicted by Orgel; but then it shows signs of recovery, even though it is still being irradiated at the same intensity. After a little longer, we may have a thriving population once more, which seems to be immune to radiation. This is not at all what the theory projects. What is going on?

This is natural selection at work. Several changes take place in the cells. First of all, some of the cells divide faster. These faster replicators are disproportionately represented among the survivors, because they are more likely to have replicated themselves before their DNA was destroyed. For the population as a whole, each population doubling now takes place in a shorter period. The survivors produce a new set of genes in a shorter time than that required for radiation to dismember a single set. The progeny now have a greater than 50 per cent probability of surviving intact to the next generation.

As long as the cells have sufficient space and nutrients, this adaptation alone might suffice. However, many cells have probably made a second, closely related, adaptation. As the population growth steadies, we

[2] DNA of course codes for RNA and proteins, which in turn provide the cellular infrastructure that enables the replication of more DNA. The origins of this system are beyond the scope of this book; if you are interested I recommend the writings of Leslie Orgel himself, in Further Reading.

see that these cells have extra copies of their own DNA. They now have multiple identical chromosomes. The effect is similar to increasing the speed of replication, but is much more profound. The reason is as follows. If each cell has only one copy of each gene, then a single unfortunate hit to any gene has the power to knock out a critical protein and kill the entire cell. On the other hand, if the cell has multiple copies of all its genes, it can take an equivalent number of hits with a good chance that the *same* gene will not be destroyed on all the chromosomes. As a back-up plan, this is much less costly than producing a whole extra cell, with its proteins, mitochondria, vesicles and membranes, which then faces exactly the same problem as its parent.

We are also likely to see two other adaptations: the first is an increased rate of bacterial conjugation, in which two bacterial cells become temporarily connected and one passes additional copies of its genes to the other; the second is a stress response. Bacterial conjugation is, in principle, similar to sex. Accumulating genes that have come from different places, with different histories, reduces the likelihood of having two copies of a gene with the same error in the same place, which would be the case if you simply replicated the chromosome containing the aberrant gene. If you own a suit with torn trousers, and are given an identical suit by an equally careless friend, there's a fair chance that the rip in your friend's suit would be in the jacket. Wear your jacket with your friend's trousers, and you have a fully functional suit. Such mixing and matching is the basis of both bacterial conjugation and sex in higher organisms.

The stress response, too, is characteristic of virtually all organisms from bacteria to humans. We met a few of the proteins involved in Chapter 10. These proteins are equivalent to the emergency services, helping to repair damaged DNA, degrade damaged proteins and prevent free-radical chain reactions getting out of hand. Cells that successfully respond in this way to the altered conditions have a selective advantage. They will survive and reproduce, while their less well endowed cousins, even those with extra chromosomes, are more likely to accumulate damage and finally die.

By subjecting a bacterial population to radiation, therefore, we will select, over many generations, for a population able to withstand the effects of radiation. Now imagine that we switch off the radiation source. We have selected for a battle-scarred population of bacteria, laden with as much armour as a mediaeval knight. We drop them back into peace time. Suddenly, the extra protection is an unnecessary burden, costly to reproduce. Each time a stress-resistant cell reproduces itself, it must replicate multiple copies of its genes, and it is also funnelling a substantial proportion of its energy into the production of more stress proteins. Any bacterium that loses a few chromosomes and switches off its stress response is more likely to replicate quickly. Our armour-clad knight has to get rid of his armour before he can father children. In just a few generations, the stress-resistant bacteria may be but a distant memory. From an anthropocentric point of view, seeking purpose in nature, this endless cycling seems as blind as it is futile, but such is evolution. That is why bacteria are still bacteria.

What does all this have to do with ageing? Bacteria, for the most part, do not age. There is no reason for them to do so. They maintain the integrity of their genes by rapid reproduction. They can produce a new generation every 30 minutes. They protect themselves by hoarding multiple chromosomes, by exchanging genes through conjugation and lateral gene transfer (see Chapter 8, page 154), and by fixing damaged DNA wherever possible. Disabling errors in DNA are quickly eliminated by natural selection. Any selective advantages are embraced just as quickly. Bacteria have behaved in this way for nearly 4 billion years. Certainly they have evolved, and in this sense they have aged, but in every other respect they are as youthful now as they were all those countless generations ago.

The critical point is that the survival of bacterial life involves death on a massive scale. In 24 hours, a single bacterium can produce 2^{48}, or 10^{16}, cells, with a total biomass of about 30 kilograms. Clearly, such exponential growth cannot be sustained. In most natural environments, the size of bacterial populations remains roughly constant. Bacteria die from prolonged starvation or dehydration, or are food for other organisms, such as nematode worms — or fail to divide because of cellular damage. When death outweighs life on such a massive scale, natural selection cleanses the population of genetic damage. Only the fittest survive. Perverse as it may sound, the main criterion for immortality is death.

Shifting the perspective, we are left with the following conclusions. If senescence is not necessary to life as such, and has not always been with us, then presumably it has evolved. If senescence evolved, then it must be determined at least partly by genes, as only traits that are genetically determined can evolve and be passed on to the next generation. And if it has persisted, it must confer some kind of selective advantage.

Ageing evolved at the same time as sex. By sex (I should come clean) I mean the production of sex cells such as sperm and egg and their fusion to form a new organism. The terms 'sex' and 'reproduction' are often used interchangeably, but technically they have completely different meanings. As John Maynard Smith and Eörs Szathmáry put it, "the sexual process is in fact the opposite of reproduction. In reproduction, one cell divides into two: in sex, two cells fuse to form one." This poses a puzzle, which we shall see applies as much to ageing as to sex: what is the benefit to the individual?

Sex is thought to benefit *populations* in numerous ways. Perhaps the most important of these benefits is the swift dispersal of a new version of a gene through an entire population, thereby promoting genetic variability. Most genes exist in several different forms, and sexual reproduction brings about new and continuously shifting combinations of these versions. This variety is clear in the human population, in which out of 6 billion people, you'd be hard put to find anyone genetically identical to you, apart from an identical twin. This is important because genetically variable populations are more adaptable to changing environmental conditions or selection pressures.

But these benefits can only be enjoyed *after* the evolution of sex, and as we saw in Chapter 7, we cannot use hindsight to explain evolution. For a trait to become established in a population, it must first be beneficial to *individuals*, who then thrive at the expense of other individuals lacking the trait. The advantage to individuals of recombining genes is not immediately obvious. In sexual reproduction, two robust individuals, who have succeeded against the odds in surviving until sexual maturity, and then mating, have their robust genetic constitutions shuffled and reconstituted into new combinations in the offspring that, in statistical terms, are likely to be less robust. The reasons for the evolution of sex are, in fact, still much debated, and we shall look into them in a moment. First, we should just note that the same considerations apply to senescence. If senescence evolved, then it must have been beneficial to individuals or it could never have become such an integral feature of a large part of the living world. Senescence is, in fact, even more widespread than sex, affecting essentially all plants and animals, implying that the advantage must be very pervasive.

Because sex and senescence are closely linked, we will examine the problem of sex first, from the point of view of individual cells. The fundamental problem with sex is the *rate* of reproduction. If an asexual microorganism, such as a bacterium, reproduces by simply dividing into two (*binary fission*), then one cell produces two, two cells produce four and so on. The population as a whole expands at an exponential rate. The rate of sexual propagation is necessarily much slower: two cells conjoin to produce one, and this must, at the very least, divide in two before it can produce daughter cells that can fuse with other cells to produce more offspring. Not only is the rate of reproduction slower, but the sex cells also have to find each other, and determine that they are right for each other, before they can fuse and then reproduce. The process is fraught with danger and is energetically costly. An asexual population should outnumber a sexual population in a handful of generations. Alternatively, if one individual in a sexual population were to revert to asexual reproduction, the asexual progeny should swiftly outnumber the sexual progeny. On the basis of simple arithmetic, sex should never have got started. Having evolved against all the odds, it ought to have been weeded out long ago. Why did this not happen?

To resolve the dilemma, we must return to the central problem in biology: how to maintain the integrity of the genetic instructions from generation to generation. We have seen that bacteria succeed by combining a high rate of reproduction with heavy selection. The cost of genetic cleansing in asexual reproduction is thus a high death rate, which translates into a lot of pain for little gain. This might not worry the blind watchmaker (the evolutionary biologist Richard Dawkins's famous coinage for natural selection), but it is wasteful of resources and so might cede advantage to more efficient ways of cleansing, or at least masking, genetic damage. Sex is certainly more efficient at masking damage. This is the basis of that mysterious property known as *hybrid vigour* in plant and animal breeding, in which the offspring of unrelated parents have qualities superior to either parent. Conversely, too much inbreeding has the opposite effect. To understand why, we need to consider the mechanics of sex, in particular the form of cell division known as *meiosis*, in a little more detail.

In most sexual species, including humans, the sex cells, or gametes, each contain half the genetic material of their progenitor cells. They are said to be *haploid*, meaning that they have a single set of chromosomes, selected more or less at random from the two parental sets. When the gametes fuse together, the two haploid cells each contribute a single set of chromosomes to the fertilized egg, thereby restoring the full complement of chromosomes. The fertilized egg thus has two equivalent sets of chromosomes, and so is *diploid*. Now if this cell were simply to fuse with another, the result would be *tetraploid*, with four sets of chromosomes. If we continued in this way, we would not have to wait long before the situation became untenable. The usual solution in sexual species is to regenerate more sex cells through meiosis. This type of cell division halves the number of chromosomes, to regenerate more haploid sex cells for the next generation.

As with all types of cell division, even where the ultimate aim is to halve the number of chromosomes, each chromosome is first replicated, with the two new daughter chromosomes remaining joined together. Then, in the first step of meiosis, the twin sets of duplicated chromosomes pair up and are shuffled like a pack of cards. During this process, corresponding pieces of paired duplicated chromosomes are exchanged with each other. It is as if the top half of a red queen was grafted on to the bottom half of a black queen, to produce a mixed-queen card (or the trousers and jacket of two suits swapped around). This process is called *recombination*, and it means that the different versions of genes are rearranged in new combinations on a chromosome in the next generation. This is why it is possible for a child to resemble his great-grandfather even though nobody else has done in the intervening generations. In the next step, the pairs of chromosomes separate to produce two nuclei, which are still technically diploid. In the final step of meiosis, the daughter chromosomes of each duplicated chromosome are separated, thus finally producing four new haploid cells, all with different gene combinations, from the original diploid cell. (This is a very much simplified picture but conveys the essential concepts.)

There are, then, two cardinal features of organisms that undergo meiosis and sexual reproduction. First, genes and chromosomes from the two parents can be inherited in various combinations by the offspring. Second, sexual reproduction results in organisms that alternate between haploid and diploid states, even if one or other of these states is much reduced in the life history.

The advantage of bringing chromosomes together from two different parents is fairly easy to see. When the two haploid cells fuse to produce a new organism, the two copies of every gene are derived from organisms with differing genetic inheritance and life histories. This means that any genetic damage or mutations are unlikely to overlap. Recombination randomizes this allocation process in the same way that shuffling a pack of cards equalizes the odds of a good hand. Any mutations in a gene from one partner will almost certainly be counterbalanced in the offspring by an unaltered copy of the same gene from the other partner. In the unlikely event that two bad copies of the same gene are inherited, the offspring is eliminated by natural selection, ridding the population of detrimental mutations in the same way as in bacteria. At an individual level, sex is thus a more efficient mechanism than binary fission for maintaining the integrity of the genetic instructions, without the need for quite such heavy mortality.

The advantage of cycling between haploid and diploid states is more enigmatic. Diploidy clearly makes sense. Diploid cells, with their two equivalent sets of chromosomes, are analogous to the stress-resistant bacteria that stockpile multiple sets of identical chromosomes. The diploid state is a trade-off between the cost of producing multiple chromosomes and the danger of only having one copy of a gene. With two equivalent chromosomes, DNA breaks or deletions affecting one chromosome can be repaired using the other as a template. But the haploid state is perplexing. It is, in principle, both avoidable and dangerous, as cells could easily cycle between the diploid and tetraploid states (four chromosomes) with much less risk. More perplexing still is the fact that haploid cells are not uncommon in nature, and are by no means restricted to sex cells. Indeed, the males of many species of Hymenoptera, including some wasps, bees and ants, are *entirely* haploid — they develop from an unfertilized egg. In contrast, the females develop from a fertilized egg, and are diploid. This is no accident. In some circumstances, male haploidy is even maintained by behavioural adaptations. In the honey bee, for example, 8 per cent of newborn males are diploid, the rest being haploid. Within six hours of emergence, the worker bees actually find and eat all of the diploid males.

Why should this be? The answer is far from certain, but Wirt Atmar, an eminent computer scientist with a deep personal interest in ecology and animal behaviour, put forward a strong argument in the journal *Animal Behaviour* in 1991. His argument seems to have been inexplicably ignored by many molecular biologists. Perhaps too few read *Animal Behaviour*. Atmar argued that haploid males act as 'auxiliary defect sieves', to purge the population of genetic error by exposing latent gene defects to selection. In other words, because genetic errors cannot be concealed in haploid animals, any haploid males with demonstrable vigour *must* possess a near-perfect complement of genes. In this sense, haploid males are an extreme version of haploid sperm. But why go the whole hog and produce an entirely haploid male? Sperm, after all, are far from costly — a single millilitre of semen contains 100 million of them.

According to Atmar, the answer lies in the distinction between 'housekeeping' and 'luxury' genes. Housekeeping genes are concerned with the basic metabolism of cells, and so are active in virtually all cells, including sperm. Luxury genes, on the other hand, code for specialized proteins that are only produced in particular cells, such as haemoglobin in red blood cells in mammals. Defects in haemoglobin would certainly be detected in a haploid human — diseases like sickle-cell anaemia would be eliminated in next to no time — but do not affect haploid sperm, which do not need haemoglobin. Incidentally, Atmar also notes that men are haploid for the X and Y chromosomes, whereas women are diploid for the X chromosome. To the extent that only men suffer from genetic disorders caused by mutations on the Y chromosome, such as haemophilia and colour blindness, it may be that we too use a mild form of haploidy to clean up our *germ line* (the inheritable DNA in sex cells). Atmar goes on to argue that this mild form of haploidy is supplemented in most species by the evolution of aggressive male traits and high mortality rates — only the vigorous, dominant males survive to fertilize the females.

Let us accept, for the sake of argument, that haploid males serve as defect filters. Seen in this light, their existence explains some unexpected experimental data that put paid to a once popular theory of ageing — the so called 'somatic mutation' theory (from the 'soma', meaning the body). According to this theory, ageing is caused by the accumulation of spontaneous mutations in somatic DNA during the lifetime of the animal, in much the same way that cancer is caused by spontaneous mutations. The theory is easily testable. If ageing is indeed caused by spontaneous mutations, and having two copies of the same gene masks any errors in one copy, then haploid animals should age more quickly than diploid animals, because they have only one copy of each gene. Further, irradiation should speed up ageing in haploid males more than in diploid females, because a single mutation would destroy function in a haploid animal, but not in a diploid animal. The idea was put to the test in wasps and found to be wanting. In fact, haploid males and diploid females live for similar periods. Although very high doses of radiation kill male wasps faster (as might be expected for an extreme situation not at all representative of normal ageing), low doses of irradiation do not affect their rate of ageing.

These results do not fulfil the predictions of the somatic mutation theory of ageing. Clearly, spontaneous mutations *alone* are not responsible for ageing in wasps. From an evolutionary point of view, perhaps this result should have been expected. If the function of haploid males is to eliminate defects in the germ line, then the vigorous males *must* be robust enough to survive long enough to pass on essentially faultless DNA to the next generation. This implies that the rate of spontaneous mutations cannot be cripplingly high, otherwise the haploid males would be unable to pass on anything approaching 'faultless' DNA. Most subsequent studies have confirmed that the rate of spontaneous mutation is not sufficiently high to be the sole cause of ageing in most individuals (although mutations almost certainly contribute).

Sex cleans up the germ line by recombining DNA from different sources, and by directing the full force of natural selection at a section of the population — the haploid sex cells — rather than the whole population. In humans, an average ejaculation contains several hundred million sperm, which are subject to intense competition to fertilize the egg. Perhaps a couple of thousand sperm eventually reach the vicinity of the egg, so 99.9999 per cent perish en route — this is natural selection on a par with bacteria. Inherent in this idea of sex is the concept of *redundancy*. To preserve an uncorrupted germ line, a redundant part of the population is held out for selection, and only the best bits are ploughed back into the germ line. It seems that the nagging feeling of redundancy experienced by many men runs deep. This is not all. Redundancy is also behind the differentiation of cells into germ-line cells and somatic cells.

At the most basic level, the function of sex cells is to pass on undamaged DNA to the next generation, while the function of the soma, the body, is to be selected for vigour, not to be perpetuated itself. The origins of this differentiation between sex cells and somatic cells stretch back to the earliest days of sex. How the differentiation came about is unknown, but its relationship to sex is discussed by William Clark, an immunologist at the University of California, Los Angeles, in his book *Sex and the Origins of Death*. Clark cites the tiny animalcule *Paramecium*, a single-celled eukaryote living in freshwater ponds, as an example of what might have been the first evolutionary links between differentiation (within a single cell in this case), sex, senescence and death.

Paramecium reproduces both sexually and asexually. Asexual reproduction is achieved by budding from the mother cell. However, budding cannot continue indefinitely. After about 30 cell divisions, a culture of *Paramecium* shows signs of senescence, even given perfect conditions for growth. Their growth rate slows down, they cease to divide, and the population dies unless reinvigorated by sex. This is quite unlike bacterial populations, which are theoretically immortal, even if many individual cells die. In the case of *Paramecium*, whole populations are evidently mortal. This peculiar state of affairs is explained by the complex life cycle of *Paramecium*. A single *Paramecium* has two nuclei, a large one called the macronucleus and a small one called the micronucleus. The macronucleus is in charge of the daily running of the cell, while the micronucleus wraps its DNA tightly in proteins and shuts itself off. When a *Paramecium* divides asexually, the micronucleus flickers to life briefly, replicates its DNA and divides to provide a micronucleus for each daughter cell, then shuts down again. At the same time, the macronucleus divides to provide a working macronucleus for each daughter cell.

It seems that it's the macronucleus that finally gives up the ghost and becomes senescent. Exactly what causes this senescence is uncertain. Despite leaning towards a theory of programmed ageing (in which I think he is wrong), Clark ascribes the demise of the macronucleus to wear and tear — the accumulation of random genetic mutations over 30 generations. Whatever the reason for the senescence, the *Paramecium* must now indulge in sex to reset its biological clock to zero. When two suitable cells conjugate, the micronucleus in each one springs to life, dividing in two by meiosis to produce two haploid micronuclei for each cell. One nucleus from each cell is exchanged, and the newly shuffled pairs of haploid nuclei fuse to produce one diploid micronucleus for each cell. These then divide by mitosis to form a new diploid macronucleus for each cell. The new micronucleus shuts down and the new macronucleus takes over running the reinvigorated cells. The old macronucleus disintegrates in an orderly, programmed fashion and its constituents are recycled for use by the rejuvenated cells. *Paramecium* therefore combines the speed of reproduction by asexual budding with the periodic genetic cleansing of sex, gaining benefits from each.

The disintegration of the old macronucleus echoes what must have been a pivotal moment in evolution. For the first time, we encounter DNA that is deliberately *not* transmitted to the next generation. Is this the origin of the soma and of senescence? It certainly is, according to Clark: "it is in the programmed death of the macronuclei of early eukaryotes like *Paramecium* that our own corporeal deaths are foreshadowed." Whether or not this statement is strictly true (Clark associates the programmed demise of the macronucleus with what he sees as our own programmed demise), it is certainly true in a general sense. The body is a useful but ultimately redundant subsidiary to the germ line — not only mortal, but designed from the very beginning to be thrown away. The advantages are obvious: a body enables the specialization of individual cells — a specialized team has every advantage over the individual amateur — and physical protection for the germ cells. But there is no need for the body to outlast its usefulness. As the old saying goes, a chicken is just an egg's way of making another egg; and a man is just an egg cell's way of making sure that the next egg cell is not riddled with errors.

The throwaway body is central to the evolution of ageing — *why* we age — though it says little about the actual mechanism of ageing. The idea is known as the *disposable soma* theory of ageing, and was formulated by Tom Kirkwood in the late 1970s, and later developed by Kirkwood and the eminent geneticist Robin Holliday. Today, the theory is seen by most researchers as the best framework for understanding ageing.

The theory draws on the distinction between the immortal germ line and the mortal soma, or body, first postulated by the great German biologist August Weismann in the 1880s. Kirkwood and Holliday considered the dichotomy between the germ line and the soma as the outcome of a trade-off between survival and reproduction. In essence, to be of any use, the body must survive at least to reproductive age. Survival costs. A substantial portion of an organism's energy is needed to maintain life — to keep body and soul together for long enough for the germ line to be propagated. Most of the food we eat is burnt up to keep the body working — the heart beating, the brain thinking, the kidneys filtering, the lungs breathing. The same is true at a cellular level. The high rate of DNA damage and mutations that we noted in Chapters 6 and 10 must be corrected through the synthesis and incorporation of new building blocks. Molecular proofreading mechanisms are needed to ensure that repaired DNA reads correctly. Damaged proteins and lipids must be broken down and replaced. The importance of protein turnover is illustrated by our dietary requirement for nitrogen, in the form of amino acids, and by the continual excretion of nitrogen in the urine as urea. Urea excretion reflects the destruction and elimination of damaged proteins. The disposable soma theory infers that the energy required to keep these various metabolic systems going detracts from the energy available for reproduction.

The validity of the disposable soma theory hangs on the veracity of its predictions. If survival and reproduction both require energy or resources that are in limited supply, then there should be an optimal balance, in which bodily maintenance is set against reproductive success. This optimal balance would be expected to vary between species, according to their environment, competitors, reproductive capacity, and so on. If so, there ought to be a general relationship between the lifespan of species and their fecundity (the potential number of offspring produced during the reproductive period). Moreover, factors that tend to increase lifespan should decrease fecundity, and vice versa. Are such relationships detectable in nature?

Notwithstanding difficulties in specifying the maximum lifespan and reproductive potential of animals in the wild, or even in zoos, the answer is an unequivocal yes. With a few exceptions, usually explicable by particular circumstances, there is indeed a strong inverse relationship between fecundity and maximum lifespan. Mice, for example, start breeding at about six weeks old, produce many litters a year, and live for about three years. Domestic cats start breeding at about one year, produce two or three litters annually, and live for about 15 to 20 years. Herbivores usually have one offspring a year and live for 30 or 40 years. The implication is that high fecundity has a cost in terms of survival, and conversely, that investing in long-term survival reduces fecundity.

Do factors that increase lifespan decrease fecundity? There are a number of indications that they do. *Calorie restriction*, for example, in which animals are fed a balanced low-calorie diet, usually increases maximum lifespan by 30 to 50 per cent, and lowers fecundity during the period of dietary restriction. We shall see in Chapter 13 that the molecular basis of this relationship is only now being worked out, even though the original discovery was made in the 1930s. Nonetheless, the rationale in the wild seems clear enough: if food is scarce, unrestrained breeding would threaten the lives of parents as well as offspring. Calorie restriction simulates mild starvation and increases stress-resistance in general. Animals that survive the famine are restored to normal fecundity in times of plenty. But then, if the evolved response to famine is to put life on hold until times of plenty, we would expect to find an inverse relationship between fecundity and survival. Are there any other, less stressful, examples?

In some circumstances, longevity can be selected for in the wild. One study, carried out in the early 1990s by Steven Austad, a zoologist then working at Harvard, examined mortality, senescence and fecundity in the Virginia opossum, the only North American marsupial. Opossums are claimed to be one of the most stupid animals in the world, with a ratio of brain size to body size well below that of most mammals. They fall easy victim to predators. Their favourite ploy is to 'play possum', or feign death, with sadly predictable consequences. In the mountains of Virginia, few opossums survive beyond 18 months (more than half are eaten by predators) and those that are not eaten age rapidly. The main reason opossums survive at all, and indeed proliferate, is their tremendous fecundity. An average female produces two litters, of eight to ten offspring each, in a single reproductive season.

Austad wondered what would happen to the lifespan and fecundity of opossums living in an environment free from predators. One place that fitted the bill was Sapelo Island, off the coast of Georgia, where opossums have probably lived without much predation for 4000 to 5000 years. The conditions provide a test case for the disposable soma theory. According to evolutionary theory, we would expect individuals freed from predation to age more slowly. This is because long-lived animals can have more offspring, and look after their offspring for longer, and so should be selected at the expense of short-lived animals. However, there are two possible outcomes, which give us an indication of the mechanism of ageing. If ageing results from a simple accumulation of damage, then slowing down the rate of damage should increase lifespan but have no effect on fertility earlier in life. Conversely, if the cost of repair is decreased fecundity, then we should see a different picture: longer lives should be gained at the expense of reproductive vigour earlier in life.

In the case of the Virginia opossum, this latter was exactly the case. Austad tagged about 70 opossums in the Virginia mountains and on Sapelo Island, and monitored their progress through life. On the mainland, he confirmed that the surviving opossums aged rapidly beyond 18 months, which effectively limited their reproduction to a single season. Only 8 per cent survived into a second reproductive season, and none into a third. Litters averaged about eight offspring. In contrast, ageing was much slower on the island. Here, half the female opossums survived into a second reproductive season, and 9 per cent into a third. Biochemical estimates of the rate of ageing (measurements of collagen cross-linking in the tail — the same changes that bring about the wrinkling of our own skin) suggested that island opossums age at virtually half the rate of their mainland cousins. Critically, litter size was reduced from eight to five or six. This litter size was sustained through the second reproductive season, implying that the total number of offspring was not reduced, rather the *distribution* of fertility was spread over a longer lifetime.

Similar patterns have been reported for other animals living in less insular circumstances. For example, the lifespan and fecundity of guppies (a South American freshwater fish, named after the Trinidad clergyman R. J. L. Guppy, who sent the first specimen to the Natural History Museum in London) is influenced by the rate of predation in different streams. Heavy predation and high mortality is linked with rapid ageing and the compression of breeding into a short lifespan. Fecundity is reduced, or rather spread out, in longer-lived fish. The same is true of some birds. Lars Gustafsson has reported an inverse relationship between clutch size and lifespan in collared flycatchers in Gotland, Sweden. Again, there is a cost to early reproductive effort — females with larger brood sizes early in life laid smaller clutches later in life, compared with those that had smaller early broods.

But wouldn't we expect animals to breed faster if faced with a threat to their survival? These findings may be no more than an ecological equivalent of the faster rate of bacterial replication in response to irradiation, or our own urge to find a sexual partner in the last few minutes after a nuclear warning. Certainly the evidence is supportive of the disposable soma theory, but it is far from proof. Can the relationship be emulated in the laboratory, with minimal exposure to predators or other life-shortening factors?

One classic experiment suggests that it can. In the fruit fly *Drosophila*, the cost of extended life is lower fecundity, even in the laboratory. This conclusion comes from the selective breeding experiments carried out by Michael Rose, an evolutionary biologist at the University of California, Irvine, and others. Rose postulated that the flies reaching sexual maturity fastest would be the most fertile and therefore the shortest lived. Conversely, the slowest to mature would be the least fertile but the longest lived. To test this idea, he maintained two populations of flies over a number of generations. In the first population, the first-laid eggs were collected and used for breeding the next generation. The procedure was repeated for each successive generation. In the second population, only the very last eggs laid were selected for breeding the next generation. Rose found that the average lifespan of flies propagated from the last-laid eggs more than doubled over ten generations. The total number of eggs laid in a lifetime was similar in both lines, but in comparison with the short-lived population, the long-lived flies reproduced more slowly while young, and faster later in life. Thus, even in the absence of predation or other life-shortening factors, there is a trade-off between longevity and fecundity, in which the price of longevity is the suppression of fecundity earlier in life.

A trade-off between sex and death sounds, on the face of it, the worst possibility in the worst of all possible worlds. Is chastity the price of a long life? Does this confirm Aristotle's gloomy dictum, that each act of sex has a direct life-shortening effect? Quite the contrary, in fact. The trade-off frees us from a tyrant: the belief that senescence is inevitable. Sex is not tied directly to lifespan, unless we have a heart condition. The two are linked because, as a species, the resources that our genes directed at reproduction, over evolutionary time, were subtracted from our investment in maintenance. After a million years of human evolution, we have found a stable balance; but this balance can, in principle, be shifted. The working assumption has always been that we must make the best of limited resources, invested over a probable lifespan in the wild. Neither of these conditions — limited resources or probable mortality in the wild from predation, starvation, infection and accident — is the same now as it was for our first human ancestors.

If the disposable soma theory is correct, we can make two predictions, both of which are supported by the findings outlined in this chapter. First, if lifespan is set to an optimum, then that optimum can be shifted by changing the parameters. The changes in lifespan that we have noted took place over generations. If we wish to extend our own lives within a single generation, we must find out the terms of the contract, in other words the genes or biochemistry in which it is written. Second, if we want to enjoy a life extension, we are not obliged to remain childless. The trick applied by nature is to *defer* sexual maturation.

In the next chapters, we will examine the terms of the contract, and whether there is anything we can do to rewrite them. First, however, it is worth thinking about ourselves for a moment. Is there any evidence in people that lifespan is an optimal trade-off between reproductive prowess and longevity? There are exceptions to every rule, especially in biology.

Are we, perhaps, the exception to this one? The question is difficult to answer directly, as we live so long anyway. Any direct measurements would require many decades. Even so, there are two indications that the rule does indeed apply to us.

First, our lifespan is much longer than other primates. The fecundity of the great apes has barely changed over evolution. The chimpanzee, gorilla, orangutan and human female all have an inter-birth interval of two to three years, and a similar number of offspring per female. Despite this, humans live twice as long as a gorilla or a chimpanzee. The discrepancy is simply explained: in primates, longevity has been purchased by deferring sexual maturity, by slowing down the rate of growth to adulthood. Humans live twice as long as gorillas, but take a third longer to reach sexual maturity.

In Western societies, we may be assisting this trend. Women give birth to their first child later and later. According to the Population Reference Bureau, just 10 per cent of women in Europe now give birth to their first child before the age of 20. In comparison, in the developing world, some 33 per cent of mothers have their first child before the age of 20; in West Africa the figure is around 55 per cent. Of the 15 million young women aged 15 to 19 who give birth every year, 13 million live in less developed countries. It is too early to say whether we are actively selecting for longevity in Europe, but it is hard to see why we should not. I suspect that this trend will prove more powerful than any foreseeable medical breakthrough.

The second indication that our own lifespan is an optimal trade-off comes from a study of genealogical records by Tom Kirkwood and Rudi Westendorp, an epidemiologist at the University of Leiden in The Netherlands. Kirkwood and Westendorp reasoned that the detailed records of births, deaths and marriages from the British aristocracy might provide buried evidence of a trade-off between human fertility and longevity. After making allowances for historical trends towards smaller families and longer lives, they still found that "the longest-lived aristocrats tended, on the average, to have had the greatest trouble with fertility". Taking the pattern as a whole, they concluded that a predisposition to above-average longevity is indeed linked to below-average fertility.

I am satisfied that the disposable soma theory applies to us. This is good news, in the sense that an obsolete optimum, which has served its purpose, can in principle be modified to a more congenial optimum with a new purpose — to abolish the misery of old age. The disposable soma theory argues that the rate of ageing is dictated by the level of resources that are committed to bodily maintenance. The question we must address now is *why* do these resources become less efficient as we grow older? If we can maintain ourselves perfectly well in youth, when our sexuality is at its height, why do we then decline?

The disposable soma theory does not discriminate between competing mechanistic theories of ageing, such as the programmed and stochastic theories. For example, are we programmed to commit maximum resources to somatic maintenance until reaching sexual maturity, and then decline after we have switched our resources to sex? Such a process could be envisaged easily enough in terms of hormonal changes, which in any case control development, puberty and the menopause. Why not ageing? Or is ageing not an example of programmed development at all, but a gradual accumulation of damage? If so, why are we not troubled incrementally from childhood? Why do we not 'feel' as if we are ageing until we reach middle age? We will explore these issues in the next chapters.

CHAPTER TWELVE

Eat! Or You'll Live Forever

The Triangle of Food, Sex and Longevity

August Weismann, the nineteenth-century German biologist who first made the distinction between the immortal germ line and the mortal body, was also the first to think about ageing from a Darwinian perspective. Since resources are limited, said Weismann, it is imperative that parents do not compete with their offspring. Weismann argued that ageing was a means of ridding the population of worn-out individuals, thereby clearing space for offspring, but not so fast as to lose the social benefits of experience. There are also genetic advantages to having a flux of individuals in the population. The problem is that a genetically static population is a sitting target for pathogens and predators. In the same way, it is much easier to rob a bank if you have memorized the unchanging patrols of security guards. In each generation, old genes are remixed in new combinations and so present a shifting and elusive target for predators and pathogens. According to Weismann and his followers, ageing is an adaptation to reap the benefits of social wisdom, while clearing space for new individuals and thus maintaining the species in a state of genetic flux.

Weismann's argument is nowadays dismissed by most evolutionary biologists as unworkable, as it places the emphasis on so-called group selection rather than individual selection. If ageing is programmed in the same way as, say, our embryonic development, the benefit is to the group, not the individual, who gains nothing from being displaced. As we saw with sex, any theory that seeks to explain the origins of a trait must do so in terms of individual selection. On the other hand, even if group selection does not explain the origins of ageing, it remains possible that group selection could maintain ageing once it had evolved. The idea persistently refuses to go away, and is, in fact, the conceptual bedrock of most theories of programmed ageing. Is there any evidence that group selection maintains a programme for ageing?

In the sense that animals and plants have fixed lifespans, longevity is obviously written in the genes. This does not mean there is a formal genetic programme, any more than a car is programmed to become obsolete over 20 years. In the case of a car, the parts are designed from the beginning to last for only so long, and the fact that they wear out simultaneously is no evidence of the workings of a hidden programme. An apocryphal story tells of Henry Ford looking at a junkyard filled with Model Ts. "Is there *any*thing that never goes wrong with any of these cars?" he asked. Yes, he was told, the steering column never fails. "Then go and redesign it", he said to his chief engineer. "If it never breaks we must be spending too much on it."

Natural selection works in the same way. If an organ works well enough for its deficiencies *not* to constitute an adverse selective pressure, then natural selection has no way to improve on it. Conversely, if an organ works *better* than required (in new circumstances), the random accumulation of negative mutations over generations will gradually degrade its performance to that required, at which point selection pressure will maintain the standard. For this reason, animals that have adapted recently (in evolutionary terms) to permanent darkness in a cave or at the bottom of the ocean often have vestigial eyes that are no longer functional. Degradation to a common denominator is alone sufficient to explain the apparently synchronous wearing out of organ systems as we age. As John Maynard Smith put it, "synchronous collapse does not imply a single mechanism of senescence."

The impression that ageing is programmed is strongest in animals that undergo 'catastrophic' senescence. The most famous example is the Pacific salmon, though there are several others, including mayflies, marsupial mice (*Antechinus*) and the octopus *Octopus hummelincki*. Pacific salmon hatch in small freshwater streams and migrate to the sea. When mature, the adults cover long distances to return to the stream of their birth, where they spawn, producing huge numbers of eggs and sperm. Within a few weeks, the adults degenerate and die, their rotting flesh augmenting local food chains, ultimately to the benefit of their own young. These dramatic events are brought about by hormonal changes, and benefit the group — the offspring — rather than the parents. If this is not an example of group selection leading to programmed ageing, as postulated by Weismann, then it is hard to imagine what might be. The process has even been termed 'phenoptosis', or the programmed demise of the phenotype, in sonorous contrast to *apoptosis*, or the programmed death of a cell.

The first point we should appreciate about the Pacific salmon is that it is an exception to the rule, even among salmon. The Atlantic salmon, for example, migrates shorter distances and is able to breed over several seasons; it does not undergo catastrophic senescence. We would be quite mistaken to consider the Pacific salmon a paradigm for human ageing. Nonetheless, we cannot dismiss the idea of programmed ageing in people unless we can explain why the Pacific salmon is different. In fact, we do not have to look far. The critical point, once again, is sex. The Pacific salmon, in common with mayflies, marsupial mice and *O. hummelincki*, are *semelparous*: they breed only once. In *iteroparous* organisms, which breed repeatedly, the decline is typically more gradual.

Think again about the disposable soma theory of ageing (Chapter 11, page 228) and the trade-off between sex and survival. If an individual breeds only once, and does not look after its offspring, then its survival afterwards has no effect whatsoever on the genetic make-up of the next generation. Looking at it the other way round, all the selective pressure is squeezed into a short time window early in life, during the reproductive period. To understand why this is important, imagine that one individual puts more effort into breeding, perhaps because it naturally produces a little extra testosterone or oestrogen. The heightened reproductive effort results in more offspring, but has a cost in terms of the survival of the parent. If that parent is engaged in rearing young, natural selection might notice the difference and select against it; but in the case of the semelparous salmon, which has no contact with its offspring, the blind watchmaker cares not a whit. The most fertile, shortest-lived salmon will come to prevail in the population. Almost incidentally, an intensified rush of reproductive hormones will be selected for before spawning, which ensures maximum breeding success. If we measure the hormonal changes in the Pacific salmon, we certainly see evidence of a pattern that looks like programmed senescence; but in fact, the hormonal changes are almost certainly secondary to the evolutionary imperative to breed. They are not the causal mechanism behind ageing. Thus, the catastrophic demise of the Pacific salmon is explained by the disposable soma theory as the total dedication of resources to sex, coupled with the complete decommissioning of all genes involved in longevity.

A similar argument, derivable from the disposable soma theory, applies to iteroparous species, which breed repeatedly. Rather than a single breeding opportunity, the important parameter in this case is the breeding window, bounded by the likelihood of death. Recall that long-lived animals breed more slowly. If they die by accident or from predation, they will leave fewer offspring behind. Short-lived, fertile animals will therefore predominate, and longer-lived variants will be eliminated by natural selection. This clearly happens in species subject to predation, such as opossums. On the other hand, if the threat of predation is lifted, longer lives will be selected for, if only because the risk of mortality in childbirth is lower with smaller litter sizes. Species that have a naturally low mortality in the wild would be expected to live longer because they are not penalized for their slow reproduction. This is true of opossums living on islands, and seems to be true of birds, bats, tortoises, social insects and humans. All live long lives because all are sheltered from predation, by virtue of aerial flight, hard shells, social organization or intelligence.

Longer lives can also be selected for if the parents affect the survival of their offspring. If, by bringing up our children, we increase the likelihood of their survival, then genes for longevity will be selected for. This is not a gift from the fairy godmother, a reward for altruistic spirit; it is simply the opposite of the Pacific salmon's catastrophic demise. Given a population of individuals with a normal scatter of longevity genes, then the parents that live the longest will, on balance, offer more support to their children. These children will have a better chance of surviving childhood. They will, of course, inherit the same longevity genes, so their own children will reap the same benefits. Eventually, a long-lived population will be selected for (as long as 'accidental' death is held at bay for long enough).

As Tom Kirkwood and Steven Austad have argued, the power of this effect is exemplified by the existence of the menopause. For older women, the balance between sex and survival adjusts itself towards survival. Older women have more to offer, in biological terms, by bringing up children than they do by having more children. The risk of childbirth is dangerously high for older women, whereas their prolonged survival benefits existing children or grandchildren; hence the menopause. The same cannot be said of men, who do not give birth or go through the menopause, and generally die earlier. Over evolutionary time, the prolonged survival of fathers has been less important, and they have less to lose from fathering more children.

In all the cases we have examined, the advantages of a long or a short life always accrue to the individual. We have come full circle. This view is exactly the opposite of Weismann's theory, with which we opened this chapter. Weismann argued that ageing is somehow programmed into individuals, an enforced act of altruism for the good of the species. In fact, even catastrophic ageing is explained far more believably by the disposable soma theory, in terms of selfish individual selection. If faced with a necessarily short lifespan, individuals pass on more of their genes if they breed quickly. The effort to breed undermines the capacity for survival, as both properties draw on resources from the same pot. The lifespan and the rate of breeding therefore find an optimal trade-off that fits the time-window available. If the time-window is less pressing, then longer lives are selected for, especially in cases where the parents help to rear their offspring.

In all of these cases, the genetic balance resets itself through selection. There is no need for a programme for ageing and no evidence that one exists. Even without a programme, however, these changes are clearly encoded in the genes. Thankfully, people do not senescence catastrophically after sex; a small nap is enough. If ageing is in the genes, but is not programmed, what does happen?

Ironically, the difficulty with Weismann's Darwinian argument is implicit in Darwin's own theory of natural selection. Survival of the fittest presupposes death of the weakest. When set against a backdrop of high mortality, selection pressure falls quickly with time. If our average life expectancy was 20 years, and our reproductive cycle was completed within this time, then there would be little selective pressure to extend life beyond 20. This argument was first put forward by J. B. S. Haldane and Peter Medawar in the 1940s and 1950s, and later developed by the American evolutionary biologist George C. Williams as the *antagonistic pleiotropy* theory of ageing (pleiotropy is from the Greek for 'many effects', of which some are opposing, or antagonistic).

The example of low selection pressure originally cited by Haldane in 1942 is still perhaps the most eloquent: Huntington's disease. The hallmarks of this cruel genetic disorder are a relentlessly progressive chorea (loss of motor control causing repetitive, jerking movements) combined with dementia. Typically starting with mild twitches and stumbling in early middle age, the disease eventually strips away the ability to walk, talk, think and reason. Historically, the lurching madness was mistaken for possession by witchcraft, and many victims were burnt at the stake, including some of the notorious Salem witches in 1693. Despite its severity, Huntington's disease remains among the most common genetic disorders, afflicting one in 15 000 people worldwide. In some areas, such as the villages lining the shores of Lake Maracaibo in Venezuela, the prevalence is as high as 40 per cent. In these villages, all the cases are thought to be descended from Maria Concepción, who had 20 children early in the nineteenth century. She is said to have had 16 000 descendants (so far).

Huntington's disease is caused by a single, dominant gene. Dominance means that only one copy of the gene is needed to cause the disease, unlike most genetic diseases, which are 'recessive'; that means they only occur when the person carries two copies of the 'bad' gene. As we saw in Chapter 11, in diploids, the negative effects of a 'bad' gene from one parent are often suppressed by a functional copy of the gene from the other parent. A number of such recessive traits are maintained in the population by hidden benefits. For example, the defective haemoglobin gene responsible for sickle-cell anaemia also protects against malaria, and has been maintained at a high frequency in regions where malaria is endemic, such as West Africa. Each year, hundreds of thousands of children die from sickle-cell anaemia, but the carriers, who have a single bad copy of the gene, rarely suffer from serious anaemia. They benefit by having an almost complete protection against malaria and its consequences. The frequency of the gene for sickle-cell anaemia is therefore determined by the balance of risks and benefits. (J. B. S. Haldane, incidentally, was the first to suggest that there might be a link between sickle-cell anaemia and malaria.)

The same cannot be true of Huntington's disease, where every carrier succumbs to the disease. The difference here, as Haldane pointed out, is the average age of onset — 35 to 40 years. For most of the history of mankind, most people simply did not live to that age. The selection pressure to remove the Huntington mutation from the population has therefore been weak. Imagine, in contrast, the fate of any variant of the gene that caused this disease at the age of 10 — it would be eliminated from the population, as those possessing it would not have children.

In this light, ageing could be seen as the result of the accumulation of deleterious late-acting mutations over many generations, not within an individual lifetime. Each individual inherits the late-acting mutational baggage of previous generations. Ageing is thus a kind of 'rubbish bin' of bad genes. The idea of antagonistic pleiotropy is a development of this concept. The problem with the rubbish-bin theory is that there is no selective force causing negative late-acting mutations to accumulate: there is no selective force that favours degeneration, other than the general tendency towards wear and tear. George C. Williams put forward one positive reason why genes with a detrimental effect might be selected for in evolution. He pointed out that many genes have more than one effect: they are pleiotropic. In the same way, we saw that vitamin C is involved in multiple cellular processes. Similarly, a gene might have some beneficial effects, but we can easily envisage that these benefits might be opposed, or antagonized, by other, detrimental, effects. In the case of vitamin C, we saw that its beneficial antioxidant properties are counterbalanced, in some circumstances, by potentially dangerous pro-oxidant properties. The theory of antagonistic pleiotropy posits that when genes have both 'good' and 'bad' effects, the outcome is an optimal trade-off between the good and the bad.

The theory of antagonistic pleiotropy assumes that rather than being merely a rubbish bin of late-acting mutations, individual genes involved in ageing would have beneficial actions early in life and detrimental actions later. If the benefits outweigh the disadvantages, then the gene will be selected for by evolution. As Medawar put it, "Even a relatively small advantage conferred early in the life of an individual may outweigh a catastrophic disadvantage withheld until later." Let us stay with Huntington's disease. A number of studies have suggested, tantalizingly, that the mutations in the Huntington's gene do in fact confer a competitive advantage in youth, although the mechanism is unknown. People with the gene for Huntington's disease, who go on to develop the disease in middle age, tend to have more interest in sex than the rest of us. Studies in Wales, Canada and Australia concur that fertility is enhanced in people who go on to develop Huntington's disease, compared with either their unaffected siblings or the general population. The slightness of this effect

— barely 1 per cent — emphasizes the stark conclusion that tiny benefits in youth can outweigh dreadful afflictions later in life, but only if the net result is to leave behind more children.

For many years, the theory of antagonistic pleiotropy dominated the field of ageing, and it is still one of the most prominent theories. There is certainly some truth in it. The theory is not at odds with the disposable soma theory — both are trade-offs, in which an individual's genetic resources are concentrated on reproductive prowess in youth at the expense of health later in life. However, the similarities between the two theories have often led one to be seen as a special case of the other. This is far from the truth.

The disposable soma theory argues that there is a trade-off between reproductive success and bodily maintenance. To live for longer, we must invest more in maintenance and less in fertility. This is essentially a life choice, a resetting of our resource allocation, over which the individual can, in principle, have an influence. In contrast, the theory of antagonistic pleiotropy argues that the trade-off is between the effects of early and late-acting genes, which, on balance, favour early vigour against later decline. The trade-off probably involves hundreds, possibly even thousands of genes. This is a critical difference. If senescence is the rubbish bin of hundreds or thousands of deleterious late effects, then there is very little we can do about it. To change our maximum lifespan would require altering our entire genetic make-up, at an unknown cost to our health in youth. For this reason, the theory of antagonistic pleiotropy has had a baleful effect on biology. Essentially, it argues that everything that can go wrong will go wrong. Bad genes cause disease, so we will inevitably become mired in disease in old age.

Is this really true? Is it not possible to die of old age, free from disease? Most people would think so, even if it only happens rarely. The 'oldest old', the centenarians, often die of muscle wastage rather than any particular disease. The implication is that there is indeed a distinction between ageing and age-related disease caused by late-acting genes. Perhaps the disposable soma theory can account for ageing in general, whereas the theory of antagonistic pleiotropy explains our susceptibility to age-related disease with a genetic basis? Perhaps. We will return to this possibility in Chapter 14.

The idea that ageing is more tractable than implied by the theory of antagonistic pleiotropy is supported by the flexibility of longevity in the wild. If a change in longevity requires the coordinated mutation of hundreds or even thousands of genes with late-acting effects, then any change should take place over prolonged periods. We have seen that opossums can double their lifespan in less than 5000 years — a blink of an eye in evolutionary time. Humans have doubled the lifespan of higher primates in a few million years, while the primates themselves quickly evolved long lifespans by the standards of other mammals. In the laboratory, we can double the lifespan of *Drosophila* in ten generations. The rapidity of such changes suggests that lifespan can be modulated by selecting only a handful of genes.

This hope has been confirmed by recent research. A number of genes, so-called *gerontogenes*, have now been discovered, whose effects can double or even triple the lifespan of simple animals like nematode worms. At first sight, these genes have bemusingly diverse effects, but as we have learned more we have come to see that they are linked by a common factor — oxygen.

The first life-extending mutation was reported in 1988 by David Friedman and Tom Johnson, then at the University of California, Irvine. The mutated gene, known as *age-1*, doubled the maximum lifespan of the tiny (1 mm) nematode worm *Caenorhabditis elegans*, from 22 to 46 days. The mutant nematodes seemed normal in every other respect, except that their fertility was reduced by 75 per cent. In 1993, a mutation in a related nematode gene called *daf-2*, was discovered by Cynthia Kenyon and her team at the University of California, San Francisco, which nearly tripled the maximum lifespan of *C. elegans*, to 60 days — the equivalent of humans living for nearly 300 years. It transpired that both genes had the power to arrest the development of *C. elegans*, diverting it into a long-lived, stress-resistant form known as a 'dauer' larva (from the German *dauern*, meaning enduring).

In all, more than 30 genes are known to influence dauer formation.[1]

[1] Other genes influence longevity in *C. elegans*, such as the *clock* genes, but do not affect dauer formation. They tend to have relatively small effects, in the order of 30 to 60 per cent extensions of lifespan. The *clock* gene products are thought to lower the metabolic rate, suppressing mitochondrial function, and may contribute to the effects of restriction of the intake of calories in nematodes.

Dauer larvae normally form in response to extreme environmental conditions: in particular, food shortage and overcrowding. The larvae wait out the hard times in a state of dormancy. They store nutrients, freeing them from the need to eat, and develop a thick cuticle, which helps to protect them against environmental insults. When conditions improve, the worms emerge from the dauer state and resume life where they left off. The time spent as a dauer larva has no effect on their subsequent lifespan as an adult. If a worm had ten days left to live before its dauer interlude, it will survive for ten days afterwards. In this sense, dauer larvae are non-ageing, though in fact they rarely revive after about 70 days of slumber. The larvae have two characteristics that might account for their longevity: low metabolism and increased stress-resistance. In particular, dauer larvae are resistant to oxidative stress induced by hydrogen peroxide or high oxygen levels.

Mutations in the genes that control the formation of dauer larvae sometimes cause the larvae to form inappropriately, despite perfect environmental conditions. In other cases, the worms are unable to enter the dauer state, even in extreme conditions. But the most exciting and significant finding is that the effect on longevity can be dissociated from dauer formation. Given appropriate conditions, mutations in *age-1* and *daf-2* can double the lifespan of normal adult worms, without any requirement to enter the dauer state. Curiously, one of the conditions required is the correct function of a third gene, called *daf-16*. If *daf-16* is mutated so that it does not work properly, lifespan cannot be increased by mutations in *age-1* and *daf-2*. The implication is that *age-1* and *daf-2* normally reduce longevity by inhibiting *daf-16*.

Regardless of the exact mechanism, one point is clear: all these genes interact in a regulated manner, designed to be modulated according to circumstances. As Cynthia Kenyon put it in *Nature*, "longevity of the dauer results from a regulated lifespan extension mechanism that can be uncoupled from other aspects of dauer formation. *daf*-2 and *daf-16* provide entry points into understanding how lifespan can be extended."

What do these genes actually do? The answer to this question begins to make sense of many of the results discussed in this chapter and the last. In the late 1990s, Heidi Tissenbaum, Gary Ruvkun and their team at Harvard cloned the genes for *age-1, daf-2* and *daf-16* successively, in an impressive blaze of productivity. The genes code for proteins that control the response of cells to hormones. Each of the genes encodes a link in a chain of signals. The sequence is as follows. A hormone binds to its receptor on the cell membrane, which is coded by *daf-2*. This receptor activates an adjoining enzyme, coded by *age-1*. When activated by the receptor, the enzyme amplifies the message by catalysing the production of a large number of 'second' messengers, as if it were setting free a host of molecular gossips. The second messengers migrate to the nucleus, where their whispering can either activate or deactivate transcription factors (proteins that bind to DNA, controlling the activity of genes). One of the most important of these transcription factors is coded by *daf-16*. By binding to DNA, the *daf-16* transcription factor coordinates the cell's response to the hormonal message, selecting a particular set of genes for transcription.

Such a relay is known as signal transduction. The details of these relays are learnt, somewhat resentfully, by all students of biochemistry and cell biology. Signal transduction pathways are the standard cellular communication system, allowing amplification of the original message and elimination of 'noise'. Describing one of these relays is a little like explaining how a telegraph network operates. The really interesting question, in both cases, is not *how* the message is transmitted, but rather what the content is.

The answer lies hidden in the detailed sequences of the genes themselves. Even though these genes are from the lowly nematode worm, they share sequence similarities with the equivalent genes in other species. As we saw in Chapter 8, sequence similarities normally imply not only inheritance from a common ancestor, but also a conservation of purpose. In the case of the *age* and *daf* genes, the gene sequences betray a deep evolutionary kinship that links nematode worms to flies, mice and men. All these species have genes with strikingly similar sequences to those in the nematode worm. In each case, the genes code for the components of a signalling pathway. The signal comes from a small group of hormones — the insulin family.

Insulin belongs to a group of related hormones, all of which have profound effects on cellular metabolism. The exact function of each hormone varies from species to species, but in broad terms insulin and its cousins control the triangle of nutrition, reproduction and longevity. Insulin induces a shift in metabolism towards growth. When insulin is present, glucose is taken up rapidly by all cells in the body and stored as the carbohydrate glycogen. Protein and fat synthesis is stimulated, leading to a gain in weight. Breakdown of glycogen and proteins for energy is inhibited. As glucose is used up, blood glucose levels fall. The actions of insulin are countered by the hormone glucagon, which restores blood glucose levels to normal. In a developmental sense, insulin is a signal of plenty. Glucose means that food is abundant. Insulin passes on the message: *now* is a good moment to grow, complete development, reproduce! Seize the day!

If this message is repeated insistently enough, because glucose is plentiful in the diet, for example, the clarion call is taken up by other hormones of the insulin family, which act over longer periods. High blood glucose stimulates the production of growth hormone, which in turn elicits the production of *insulin-like growth factors* (IGFs). These are structurally and functionally similar to insulin, but with even more potent effects. The IGFs stimulate the synthesis of new proteins, promoting cell growth, multiplication and differentiation. Critically, the IGFs also modulate the actions of sex hormones, influencing puberty, menstrual cycles, ovulation, implantation and foetal growth. Mutation of the gene for IGF-I leads to retarded development of the primary sex organs.

Here, if anywhere, is the switch between reproduction and longevity, which underpins the disposable soma theory of ageing. In the presence of plentiful food, insulin and the IGFs are produced. The organism gears up for sexual maturation and reproduction, throwing longevity to the wind. There is a moment of choice, controlled by a genetic switch: sex or long life. In nematodes, this switch looks very much like the transcription factor encoded by *daf-16*.

If the switch is indeed the daf-16 protein, then long life works like this. Persistently low blood glucose keeps insulin and IGF levels low. The receptors in the cell membrane that would normally pass on the message stand idle. The gossiping second messengers fall silent. These messengers would normally block the action of daf-16, but in their absence daf-16 springs to life and coordinates the transcription of a number of specific genes. The products of these genes confer longevity on nematode worms, enabling them to wait out the lean times. Daf-16 is also activated when the *daf-2* gene — coding for the insulin receptor in the membrane — is mutated. In this case, the insulin signals are not passed on. Daf-16 again springs to life and the organism behaves as if there was no insulin: it becomes *resistant* to the presence of insulin.

Mutation of *daf-2* thus confers insulin-resistance on worms. Interestingly, the same effect can be achieved by sensory deprivation.[2] If a nematode thinks there is no food available, it produces less insulin and survives for longer, even if food is in reality abundant — and even if the nematode actually eats it. In the worm, then, longevity can be decoupled from metabolism through the power of thought (or at least the power of delusion).

These effects of insulin and IGFs on longevity are consistent in both *Drosophila* and mice, so it seems that similar signals control ageing in worms, insects and mammals. In 2001, David Clancy, David Gems, Linda Partridge and their team at University College, London, described in the journal *Nature* a mutant strain of *Drosophila* which had the same defects in the insulin signalling system as the *daf-2* mutant nematodes. The result was a 50 per cent increase in maximum lifespan and, again, enhanced resistance to stress. Curiously, the long-lived mutant flies were dwarfs. Gems drew comparison with dwarf mice, which are also long-lived and stress resistant, and almost certainly deficient in IGF-I. There is some evidence that stature influences longevity in people too. Population studies show that small, wiry men — the human equivalent of dwarf mice — live on average five to ten years longer than taller, heavier men. The Napoleon complex, it seems, goes beyond abrasiveness into hardiness and longevity. All the more reason not to pick a fight.

Insulin-resistance confers longevity! This is an irony that typifies the swings and roundabouts of science. In people, resistance to insulin and the IGFs is not at all beneficial. The outcome is type 2 diabetes and metabolic disarray. In the Western world, this form of diabetes is approaching epidemic status, and is probably the biggest health problem associated with the Western lifestyle. Far from living longer, people with type 2 diabetes are at high risk of heart attacks, stroke, blindness, renal failure, gangrene and limb amputation. Average lifespan is at least ten years shorter than the general population.

Such a disappointing reversal has led many researchers to dismiss the relevance of nematode research to human ageing. I think they are wrong.

[2] *C. elegans* senses its surroundings through cilia located in sensory organs in the head and tail. Work from Cynthia Kenyon's lab shows that mutants with defective cilia have impaired sensory perception — and live longer.

A question mark must inevitably hang over the relevance of animal data to human conditions. Of course people are more complicated than nematodes: we should expect layers of complexity to be superimposed over the relative simplicity of tiny worms. Yet there are good grounds for thinking that similar processes are at work, even though the effects are very different.

To see the parallels between worms and men, we must step back from the small print and look at the terms of the contract in a general sense. Insulin-resistance is clearly important in humans and affects both lifespan and fertility. Susceptibility to type 2 diabetes has a genetic basis. The sheer number of people who are susceptible to the disease implies that the susceptibility genes were positively selected for in our recent evolutionary past. This idea is sustained by the startlingly high incidence of diabetes among certain races, notably the Micronesian islanders of Nauru in the Pacific and the native American Pima. The case of Nauru is vivid and well known. A remote Pacific atoll, with a population of about 5000 Micronesians, its rich phosphate reserves attracted American mining companies during the 1940s. As the islanders grew wealthy, their diet and lifestyle was Coca Colonized: nearly all their food was imported and they now live on a typically high-energy, Westernized diet. The frequency of obesity and type 2 diabetes, which had been virtually nonexistent, started to reach epidemic proportions in the 1950s. By the late 1980s, half the adult population had diabetes. The problem was not simply one of overeating: the incidence of diabetes is far higher among Micronesians, Polynesians, native Americans and Australian aboriginals than it is among Caucasians, given a similar diet and lifestyle. The Indians are said to have a 'thrifty' genotype. They are genetically geared to hoard energy during times of plenty, and they use these big energy reserves to help them survive extended bouts of starvation or hardship (this is true for all of us to some extent, but is far less marked in agricultural societies, where food has been relatively plentiful for thousands of years). In the case of the Micronesians and Polynesians, the thrifty genotype might have helped them survive their long ocean voyages. Unfortunately, the thrifty genetic make-up is utterly counterproductive when the times of plenty are sustained continuously.

Resistance to insulin is one of the central features of the thrifty genotype. Insulin normally stimulates the uptake of glucose from the bloodstream, and its conversion into glycogen, proteins and fats in readiness for mighty reproductive endeavours. In times of hardship, however, the body strives to maintain blood glucose at normal levels, lest the brain — which relies on glucose for all its energy — shut down and we lose consciousness. If hardship is the norm, with only occasional punctuations of plenty, then insulin-resistance helps to maintain blood glucose at normal levels by blocking its uptake in organs that can subsist on other fuels. As glucose availability inside individual cells falls, the metabolic rate is suppressed, preventing unnecessary energy expenditure. Insulin-resistance is not total. There are some aspects of insulin function that are unaffected or even strengthened. In particular, fats are still stored away. The process is not pathological: it is a carefully orchestrated response to likely circumstances. Taken together, the changes prepare the body for times of scarcity and are similar to those that take place in nematodes before they enter the dauer larva stage of their life cycle.

Insulin-resistance in people almost certainly has other effects that are similar to those noted in nematodes, especially increases in stress-resistance and long-term survival. In countries where poor nutrition is widespread, babies are often born with a low birth weight. As in all species, runts are more likely to die than bigger, stronger babies. When almost all babies are born with low birth weight, however, those most likely to survive are insulin-resistant. As in nematodes, insulin-resistance confers general stress-resistance. Children who are genetically insulin-resistant are thus more likely to survive into adulthood, and to pass on their genes for insulin-resistance. This only becomes a problem when a high-carbohydrate diet is imposed on a thrifty genotype.

Resistance to insulin persuades the body that food is scarce, that we are starving, even when we are obviously not — the equivalent of sensory deprivation in nematodes. The switch is the same, if grievously misguided, in both cases: towards longevity and away from reproduction. When plentiful food is superimposed over a thrifty genotype, blood glucose levels can only be controlled by producing more and more insulin. In the end, pushed past its limit for many years, the pancreas begins to fail. Less insulin is produced. Low insulin secretion, combined with insulin-resistance, means that blood glucose levels can no longer be controlled. Loss of glucose control marks the onset of type 2 diabetes. All the terrible afflictions of the disease are secondary to this failure to control blood levels of glucose and lipids.

Type 2 diabetes is less common in populations of European ancestry than in peoples whose immediate ancestors were hunter-gatherers. Presumably the ancestors of Europeans somehow escaped the most severe selection pressure for the thrifty genotype. This seems to be so. The reasons are not entirely clear, but may be linked to the origins of farming, and especially of drinking milk. Milk is rich in lactose, a valuable source of glucose. Lactose is broken down to release glucose by an enzyme called lactase. Lactase is present in all breast-feeding babies, but is often lost later in life. Loss of the enzyme accounts for lactose intolerance — the reason many people cannot digest cheese or other milk products. Most European and Asian peoples adapted long ago to drinking milk throughout their lives — cattle, for example, were common throughout Europe and Asia — but others, notably the native Americans and the Pacific Islanders, never herded milk cattle, and remained predominantly lactose intolerant. They were therefore denied the most plentiful source of sugar in a farming community. Whether lactose tolerance signalled the demise of the thrifty genotype in Europeans is unknown, but the fact is that all populations with lactose tolerance have a low susceptibility to diabetes. Conversely, all populations that are lactose intolerant are highly susceptible to diabetes.

There is nothing special about lactose. It is simply that when high blood sugar levels become the norm, natural selection penalizes the thrifty genotype. This is happening today on Nauru. The diet and lifestyle of the islanders have not changed since the 1980s when the incidence of diabetes was at a high point, yet even so the incidence of diabetes is falling. Underlying this fall, genetic insulin-resistance — the thrifty genotype — is today present in only 9 per cent of young adults, a fall of two thirds since the late 1980s. The decline in diabetes illustrates natural selection in operation. It happened because mortality exceeded fertility in people with diabetes.

In conclusion then, diabetes, rather than longevity, is the outcome of insulin-resistance in present-day conditions because we are not starving. The underlying mechanism offsetting reproduction against longevity is conserved, however, in nematodes and humans. Insulin presses a genetic switch, which makes organisms gear up for reproduction. In this light, the fact that many people with diabetes have problems with fertility makes more sense — diabetes is a forlorn attempt to survive by postponing reproduction. Insulin-resistance causes the organism to store away food and prepare to survive the coming period of abstinence. The adaptations for survival include stress-resistance and the suppression of metabolism at the level of individual cells — in other words, a redistribution of energy towards dormancy and weight gain, as in nematode worms. Unlike nematode worms, which pass out in a dauer state, we just keep on eating.

Nonetheless, both the studies of human diabetes and the genetic findings in nematodes point us to the central importance of metabolism and stress-resistance in ageing. Here we tie in with 70 years of empirical research from a completely different direction: oxygen and the rate of living. In the next chapter, we shall see why a fast metabolism, coupled with a low stress-resistance, shortens life, while a slow metabolism, coupled with a high stress-resistance, has the reverse effect. We will also look into the possibilities of combining a fast metabolism with a high stress-resistance.

CHAPTER THIRTEEN

Gender Bender

The Rate of Living and the Need for Sexes

THE DISTINGUISHED JOHNS HOPKINS UNIVERSITY BIOLOGIST Raymond Pearl cast a long shadow over biology from the early years of the twentieth century. An unusually tall, overbearingly arrogant and brilliant man, Pearl dominated his peers, publishing over 700 academic papers and 17 books, to say nothing of newspaper articles. Although one of the founding fathers of biometry (the application of statistics to biology and medicine), he is now chiefly remembered for the theory of ageing that he named the 'rate-of-living' theory in a book published in 1928. The observations underpinning his ideas related to the effect of temperature on the fruit fly *Drosophila*. The higher the temperature, the shorter the flies' lifetimes. The relationship between temperature and lifespan was similar to that between temperature and the rate of a chemical reaction, implying that the biochemical reactions sustaining life speed up at higher temperatures. Raising the temperature from 18°C to 23°C halved the lifespan of *Drosophila*, but doubled their metabolic rate, so that they consumed twice as much oxygen in an hour. Pearl also noted that the short lifespan of a mutant strain of *Drosophila*, the shaker mutant (which moves relentlessly), was related to its high metabolic rate. His conclusion was intuitively appealing: live fast, die young. He encapsulated his ideas in an article in 1927 for the *Baltimore Sun*, headlined "Why Lazy People Live the Longest".

Acceptance of Pearl's theory gained momentum from the empirical notion that animals have a fixed quota of heartbeats. If we measure the heart rate of a mouse and multiply it by the lifespan, we get a number that is similar for most other mammals, whether horse, cow, cat, dog or guinea-pig. The same is true if we measure the total volume of blood pumped, the quantity of glucose burnt, or the weight of protein synthesized, over a lifetime. Each of these parameters relates to the metabolic rate, which is quantified as the oxygen consumption per hour. Smaller animals generally have faster metabolic rates, to maintain their body temperature. If we compare the metabolic rates and maximum lifespans of different animals, the relationship is striking. A horse has a maximum lifespan of about 35 years and a basal metabolic rate of about 0.2 litres of oxygen per kilogram per hour. Over a lifetime, it consumes around 60 000 litres of oxygen per kilogram. In contrast, a squirrel lives for a maximum of seven years, and its basal metabolic rate is five times faster: about 1 litre of oxygen per kilogram per hour. In its lifetime, it too consumes around 60 000 litres of oxygen per kilogram. The correlation holds surprisingly well for most mammals. The 'constant' is known as the *lifetime energy potential.*

At first, the lifetime energy potential was thought of in terms of the rate of chemical reactions. Later, a connection was made between the rate of metabolism and the rate of free-radical production. The reasoning is as follows. A small proportion of the oxygen consumed in cellular metabolism (a few per cent) leaks out of the mitochondria in the form of superoxide radicals (see Chapter 6). Over a lifetime, such continuous leakage should account for a substantial production of superoxide radicals — perhaps as much as 2000 litres per kilogram. If a fixed proportion of respired oxygen escapes as free radicals, then the faster the oxygen consumption, the faster these radicals will be produced. In principle, therefore, a small animal that lives fast and dies young should have a high rate of free-radical production. This seems to be generally true. Across a spectrum of mammals, there is a strong inverse relationship between the rate of free radical production and the lifespan — the more free radicals, the shorter the life.[1]

---

[1] I am including molecules such as hydrogen peroxide under the general umbrella of free radicals, even though, strictly speaking, they are not (see Chapter 6).

The possibility that free radicals might play an important role in ageing was first suggested in 1956 by Denham Harman, then a young chemist at Berkeley. Harman had spent seven years at Shell Oil, working on the chemical properties of free radicals, before taking a course in biology at Stanford. He soon realized that the same reactions might underlie the process of ageing. The terms in which he summarized his argument in 1956 were a model of clarity, and can still be used today:

> Ageing and the degenerative diseases associated with it are attributed basically to the deleterious side attacks of free radicals on cell constituents and on the connective tissues. The free radicals probably arise largely through reactions involving molecular oxygen catalysed in the cell by oxidative enzymes and in the connective tissues by traces of metals such as iron, cobalt, and manganese.

The damage caused by free radicals to cell membranes, proteins and DNA has been measured with growing refinement over half a century. There is no question that free radicals are produced, or that they cause damage proportional to their rate of production. The problem with the free-radical theory, from the very beginning, was that correlations say nothing about causality. More free radicals are produced in species that age rapidly, but this could mean that free radicals cause ageing, or that they are a product of ageing, or even that they are a co-variable with no direct link to ageing. The best way of proving a causal relationship is to modulate it, for example by extending lifespan using antioxidants. Early experiments by Harman suggested that antioxidants could slow down ageing in mice, but these findings were not supported by later work.[2] As we saw in Chapter 9, there is still no evidence that antioxidant supplements increase maximum lifespan. Instead, a balanced diet probably corrects for vitamin deficiencies that might otherwise cut short our lives. Such failures have led many to dismiss the importance of free radicals.

There is a more general difficulty with the rate-of-living theory: it does not apply universally, even among warm-blooded vertebrates. That is why I had to limit my remarks 'to most mammals'. Birds and bats are at low risk of predation, as they can fly away; and they have evolved life-

[2] The mice in Harman's control group (which did not receive extra antioxidants) did not live as long as they should have done. One possible explanation is that the control mice died early because their standard diet was deficient in antioxidants, and Harman simply corrected for this. In fact this is quite likely because we are still uncertain about the *optimal* diet for most animals, either in the wild or in captivity.

spans out of all proportion to their metabolic rate. Bats live as long as 20 years, despite a basal metabolic rate equivalent to that of mice, which live for three or four years at most. Pigeons have a basal metabolic rate similar to rats, but live for 35 years, nearly ten times longer than the poor rat. The most extreme examples are the hummingbirds, which have heart rates of 300 to 1000 beats per minute and must visit thousands of flowers a day just to stave off coma. They 'should' survive for no more than a year or two, but can actually live for ten years or more, consuming 500 000 litres of oxygen per kilogram in this time. In general, if we multiply the oxygen consumption of birds by their longevity, we can calculate that their exposure to oxygen free radicals ought to be ten times that of short-lived mammals such as rats, and more than double that of humans. The fact that birds have evolved long lifespans, despite such high metabolic rates, is often cited as the death-knell of the rate-of-living theory. However, this is to interpret the theory rather rigidly, as if all living things really were granted a fixed number of heartbeats. In fact, the exceptions only go to prove the rule — or at least a slightly modified version of it.

Birds are an ideal test case for the hypothesis that it is the rate of free-radical production, rather than some other facet of metabolism, that is linked to longevity. If lifespan is linked to metabolism in a general sense, then we would expect free-radical production to vary with metabolic rate in all cases, as indeed it does in most mammals. On the other hand, if free radicals are responsible for ageing, the fact that birds consume so much oxygen can only mean that they have a very efficient mechanism for cutting down free-radical production. Put another way, if the free-radical theory is correct, birds must produce fewer free radicals than mammals, even though they consume far more oxygen.

Gustavo Barja, a biologist at the Complutense University in Madrid, spent most of the 1990s examining this question, steadily refining his measurements of hydrogen peroxide released from the mitochondria of birds and mammals, and the amount of damage done to mitochondrial and nuclear DNA. He found that isolated pigeon mitochondria consumed three times as much oxygen as rat mitochondria isolated from the equivalent tissues. Despite such high oxygen consumption, pigeon mitochondria produced only one third as much hydrogen peroxide under the same conditions. Barja concluded that the proportion of oxygen converted into free radicals in pigeons is nearly 10 per cent that of rats, corresponding to the nearly tenfold longer lifespan of pigeons. He went on to obtain analogous data from mice, canaries and parakeets; not yet formal proof, perhaps, but if not, an uncanny coincidence (Figure 11).

Why are bird mitochondria so efficient? Presumably, flight must have imposed a powerful selection pressure for energetic efficiency, regardless of lifespan — the power-to-weight ratio required for flight demands an efficient energy metabolism. The mammals have lagged far behind. Barja found that bird mitochondria are more oxygen-tight: relatively few free radicals leak from the mitochondria, and a correspondingly greater proportion of respired oxygen is converted into water. As a result, birds actually need fewer antioxidants to mop up the few escaping radicals. This explains one long-standing puzzle: the poor relationship between the antioxidant content and the lifespan of birds and mammals. The assumption that birds should need more antioxidants to live longer is falsified because they leak fewer free radicals in the first place. From an anthropocentric point of view, this is a blow: it is a much tougher proposition to redesign our genetically leaky mitochondria than it is to add more of the right antioxidant. Nonetheless, there are grounds for hope. Even if the example of the birds does not apply to us, Barja's findings do

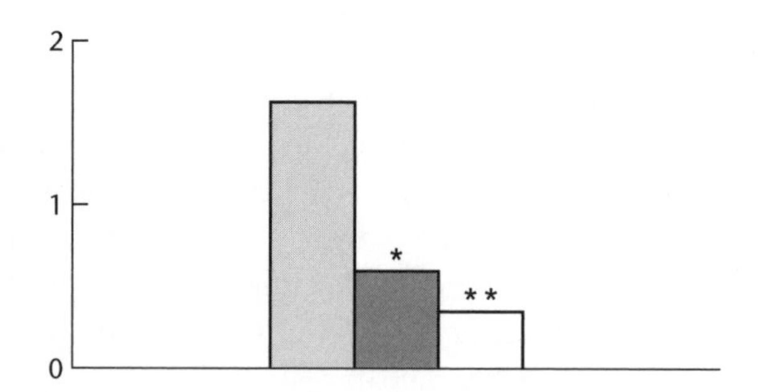

Figure 11: Hydrogen peroxide leakage from heart mitochondria in **(a)** mice (maximum lifespan 3.5 years), **(b)** parakeets (maximum lifespan 21 years) and **(c)** canaries (maximum lifespan 24 years). The rate of leakage is much lower in birds than in comparably sized mammals; the asterisks denote statistically significant differences between the groups ($p<0.05$ and $p<0.01$, respectively). All three animals are similarly sized, and have similar resting metabolic rates. Analogous relationships are found in pigeons and rats. The mitochondrial theory of ageing predicts that free-radical leakage should be lower from the mitochondria of bats than from mice (which have similar metabolic rates but live only a fifth as long), but this is as yet unknown. Reprinted with permission from Gustavo Barja and the New York Academy of Sciences.

corroborate the theory that free radicals limit lifespan. If we cannot emulate the birds, what can we do about free radicals?

We have seen that it is possible to extend the lifespan of nematode worms. We do not need to select for mitochondrial efficiency over generations: only a handful of master genes, such as *daf-16*, need to be activated. *daf-16* almost certainly controls the expression of a large number of subsidiary genes. Most of these have yet to be identified, but stress-resistance is a good pointer. Unable to wait for the systematic identification of all the genes controlled by *daf-16*, a few researchers have leapt in to measure changes in the expression of known stress proteins — like the police checking up on the whereabouts of known criminals before investigating the scene of the crime. We might predict, for example, that more SOD (superoxide dismutase) would be made in stress-resistant nematodes. True enough: in 1999, a Japanese team reported that long-lived mutant nematodes have dramatically more mitochondrial SOD (see Chapter 10) than normal adults. Mutations in *daf-16* blocked this rise, implying that the gene for mitochondrial SOD is indeed among those regulated by *daf-16*. Similarly, in 2001, a team at Manchester University reported that in stress-resistant nematodes metallothionein (another stress protein, also discussed in Chapter 10) reached seven times the normal levels.

There are good grounds for thinking that stepping up resistance to oxidative stress is a common means of slowing ageing in all animals, although the effects may be less pronounced than in worms. I shall outline three pieces of interlinked evidence that I find convincing: first, the effects of SOD and catalase; second, the effect of DNA-repair enzymes; and third, the apparent mechanism of calorie restriction.

In 1994, William Orr and Rajinder Sohal, at the Southern Methodist University in Dallas, reported in *Science* the first direct evidence that increased levels of antioxidants could slow down the rate of ageing. Orr and Sohal genetically engineered *Drosophila* so that they produced extra cytosolic SOD (see Chapter 10) and catalase. The genetically engineered (transgenic) flies lived up to a third longer than normal ones. Of particular importance, neither enzyme by itself had any effect on lifespan — the two work together and their production needs to be coordinated. When produced together, the extension in both mean and maximum lifespan was linked with a greater resistance to ionizing radiation, and less oxidative damage to DNA and proteins. The transgenic flies were also more active in old age, equating to a 30 per cent increase in their lifetime energy potential (equivalent to a greater number of heartbeats). The effect of increasing SOD and catalase levels is therefore not simply a matter of decreasing the rate of living: the transgenic flies lived at the same rate as normal flies, but for longer. More recent studies have achieved lifespan extensions of 50 per cent, using improved genetic engineering techniques.[3]

The requirement for SOD and catalase to work together hints at the importance of antioxidant networking; and of course SOD and catalase do not work in isolation. Stress-resistance is the product of many factors, including efficient protein turnover and DNA repair — our second example.

The importance of DNA repair in people is highlighted by a distressing example where it is defective — Werner's syndrome. This rare genetic disorder causes people to age at an accelerated rate. Their hair goes white and they suffer from various symptoms of premature ageing, including cataracts, muscle atrophy, bone loss, diabetes, atherosclerosis and cancer. Those afflicted usually die of age-related diseases such as heart disease and cancer by their early 40s. In the struggle to understand this intractable syndrome, scientists had hoped to learn something about the ageing process in general, and so help everyone; but, in fact, the spectrum of ailments is not really representative of normal ageing, and in the end frustrated researchers dismissed Werner's syndrome as a 'caricature of ageing'. Then came a big step forward: in 1997, the gene responsible for the syndrome was isolated. It encodes an unusual dual-function enzyme: one enzymatic function unwinds the DNA double helix (a helicase action), while the other excises and replaces any erroneous letters (an exonuclease action). The protein thus repairs errors in DNA caused by replication, recombination or spontaneous mutations — many of which are produced by oxygen free radicals.

Most people suffering from Werner's syndrome appear to have a

[3] In all these studies the *cytosolic* form of SOD was overproduced, rather than the mitochondrial form. This might be an important distinction. Mitochondrial SOD can protect both the inside of the mitochondria and the surounding cytosol from superoxide radicals, as most superoxide radicals escape from the mitochondria. In contrast, cytosolic SOD can only provide partial protection to the cytoplasm. To the best of my knowledge, transgenic *Drosophila*, engineered to overproduce mitochondrial SOD *and* catalase, have not yet been produced. I would be surprised if they did not have even longer life extensions.

mutation in the helicase part of the enzyme and cannot repair damaged DNA properly. Among other things, this mutation increases vulnerability to ultraviolet radiation, which damages DNA. Such vulnerability is the opposite of stress-resistant organisms, which are able to withstand high levels of ultraviolet radiation. We might predict, then, that DNA-repair enzymes should be among those whose manufacture is increased in long-lived mutant organisms, such as *daf-2* mutant nematodes. Although this has not been formally demonstrated, we do know that stress-resistance and longevity are associated with better DNA repair. If cells from animals with varying maximal lifespans are cultured, and then exposed to ultraviolet rays or other types of stress (such as hydrogen peroxide), we can measure the amount of DNA repair taking place. Such studies have generally shown a positive correlation between the maximum lifespan of animals and their ability to repair DNA.

These two examples suggest that lifespan is modulated by stress-resistance. Stress-resistance, in turn, is mediated (at least in part) by changes in the levels of stress proteins such as SOD, catalase, metallothionein and DNA-repair enzymes. The master-switch role of *daf-16* shows that the expression of these genes is coordinated in simple animals. My third example, calorie restriction, suggests that, even in complex animals, the stress response can be coordinated by a relatively simple switch. The profile of the response itself, however, is not always analogous to that in simple organisms.

We are only just beginning to unravel the mechanisms through which calorie restriction (see Chapter 11, page 229) extends lifespan. The mechanism certainly involves the number of calories, rather than a reduction in any particular source of calories, such as fats or carbohydrates. In general, calorie-restricted diets cut the overall intake of calories by about 30–40 per cent, while maintaining a balanced diet in other respects. It is therefore not the same as malnutrition or starvation (which for most people would involve a calorie restriction of between 50 and 60 per cent).

Ever since the effects of calorie restriction were first noticed in the 1930s, researchers have interpreted them in terms of the rate-of-living theory — eating less lowers the metabolic rate and oxygen consumption. In this light, calorie restriction has always seemed inherently futile, even though lifespan can be extended by 30 to 50 per cent in virtually all animals: who wants 50 per cent more life if the price we must pay is not just serious dieting, but also 50 per cent less energy? Even a couch potato might prefer to live fast and die young. Yet calorie restriction is turning out to be far more interesting than anyone had imagined. For a start, it does not necessarily reduce metabolic rate at all. When measured in terms of the metabolic rate per kilogram of *lean* weight, oxygen consumption may actually increase. Calorie restriction can therefore increase the lifetime energy potential — we get more heartbeats. For male rats, this increase is in the order of 50 per cent. The effects of calorie restriction are mediated by concerted changes in gene expression. The benefits are well worth queuing for. In all animals studied so far, calorie restriction delays ageing, not just the timing of death. This is true for more than 80 per cent of the 300 indices of ageing tested in rodents, including physical activity, behaviour, learning, immune responsiveness, enzyme activity, gene expression, hormonal action, protein synthesis and glucose tolerance.

The net effect of calorie restriction is to increase stress-resistance. Blood glucose levels fall, and this in turn lowers insulin levels. Metabolism is switched away from sex and towards bodily maintenance. Resistance to oxidative stress increases, especially in tissues where damage is normally highest, such as the brain, heart and skeletal muscle. Exactly *how* this effect is achieved is, curiously, an open question. Consistent changes in antioxidant enzymes are yet to be reported. We might predict that a spectrum of stress-related genes would be activated, including those for SOD, catalase, metallothionein and the DNA-repair enzymes, but this is not consistently the case.[4] One is tempted to say 'who cares', if the benefits apply to us; but again, we are not certain if they do. Any direct trial would take decades to complete.

In 1987, the National Institute of Ageing in Baltimore, Maryland and the Wisconsin Regional Primate Research Center in Madison, began two trials in primates — 200 rhesus and squirrel monkeys. In April 2001, the Wisconsin team, headed by Richard Weindruch, published an interim report on the effects of calorie restriction on gene expression, and thus the types and levels of proteins synthesized, in rhesus monkeys. To measure the effects, they determined which out of a selected sample of 7000 genes were switched on and which were switched off in the calorie-

[4] Although calorie restriction does not produce an ongoing stress response in rats, calorie-restricted rats are better able to mount a stress response to heat shock (where the animal is subjected briefly to a higher temperature than normal) than are normal ageing rats. In other words, in normal ageing, the ongoing stress response to mitochondrial leakage blunts the acute response to sudden stresses (such as heat shock), whereas calorie restriction reins back the chronic stress response and so sharpens the response to sudden stresses.

restricted monkeys and in a parallel group of fully fed monkeys of the same age. Their findings were surprising and intriguing. Although stress-resistance was increased in the calorie-restricted group, as expected, there was hardly any difference in the level of production of stress proteins in fully fed ageing monkeys, compared with middle-aged monkeys subjected to calorie restriction. Instead, calorie restriction had three big effects. First, it strengthened the internal structure of cells, more than doubling the rate of synthesis of almost all structural proteins. Second, it lowered the synthesis of proteins that promote inflammation, such as tumour necrosis factor (TNF-α) and the enzyme nitric oxide synthase. Third, it lowered the expression of genes responsible for oxygen respiration, in particular cytochrome *c* (to 1/23 of normal!). This last effect is consistent with a reduction in the metabolic rate — live slow, die old.[5]

Far from gearing up to combat stress, the shifts in gene expression in calorie-restricted monkeys almost seem to skirt the issue. If anything, the expression of stress proteins goes down rather than up. A possible explanation is as follows. We already live twice as long as the chimpanzees, which in turn live longer than rhesus monkeys. The mechanism underpinning our extra years is likely to be better stress-resistance. As we shall see in the next chapter, age-related diseases also produce a stress response, which draws on many of the same gene products. For example, ageing rhesus monkeys increase the synthesis of at least 18 stress proteins, including metallothionein and various DNA-repair enzymes. Unless calorie restriction is started very early in life, it is hard to see how the imposition of one stress response over another could extend our lives much further. Instead, in rhesus monkeys at least, the changes in gene expression brought about by calorie restriction in middle-age seem designed to lower the level of metabolic stress imposed on the system. In other words the important parameter is not stress-resistance *per se*, but the degree of stress on the system. This can be lowered by improving stress-resistance, or by lowering the level of stress imposed.

---

[5] There is another intriguing possibility here. Release of cytochrome *c* from mitochondria initiates apoptosis, or programmed cell death. It is conceivable that lower levels of cytochrome *c* in cells might cut the number of cells that undergo apoptosis in ageing organs. Function would then be maintained for longer. This is pure speculation, but I daresay Richard Weindruch will be looking into the possibility.

It is time to pause and take stock. The disposable soma theory argues that longevity is a trade-off between the resources committed to reproduction and those committed to bodily maintenance. We maintain our bodies in two ways — by preventing damage from happening in the first place, and by repairing any damage actually done. The amount of damage prevented depends, in large part, on the intrinsic rates of production and elimination of free radicals. The amount of damage repaired depends on the rate of turnover of DNA, membranes and proteins, in which damaged molecules are replaced new for old. The repair work can only be performed efficiently if the machinery is not itself damaged. Free radicals are undiscriminating, and damage the repair machinery, and the DNA encoding it, as easily as anything else. In the end, then, poor prevention leads to poor repair.

The rate of ageing is determined by the level of resources committed to prevention and repair. These resources are programmed genetically, but their deployment is influenced by environmental factors, such as the availability of food or the likelihood of sex. The switch between sex and longevity is conserved in nematodes, *Drosophila*, rats and humans, but the genetic response to the switch varies. In nematodes, longevity is apparently achieved by increasing production of stress proteins; in rhesus monkeys, by suppressing oxygen metabolism, and so metabolic rate. Given the parsimony of natural selection, the response elicited is always likely to be the most cost-effective, and so will depend on the level of stress-resistance already built into the system. Nematodes have low levels of a small number of different stress proteins, and so can easily up their levels. Rhesus monkeys, on the other hand, have much higher levels of numerous different stress proteins. Rather than having to make more of all of these, it is less costly to suppress metabolism instead. Regardless of the actual mechanism, the outcome in every case is to reduce the stress on the system, which enables animals to weather out the hard times and breed again when conditions improve. Thus, stress can be avoided either through the countering action of stress proteins, or by lowering the rate of respiration and the intensity of inflammation. Either way, the secret of a long life is low metabolic stress.

Similar mechanisms appear to have underpinned the evolution of longevity in species freed from predation and starvation. Lifespan reflects the rate of accumulation of damage, which varies with the metabolic rate, the production and elimination of free radicals, and the capacity for repair. Seen in this light, metabolic rate might even have been a factor in the evolution of gigantism, discussed in Chapter 5. Larger size permits a lower metabolic rate, hence a longer lifespan. In modern species, high oxygen levels reduce lifespan; in Carboniferous times, then, it is conceivable that greater size might have been a means of dealing with high atmospheric oxygen by lowering the metabolic rate. Be that as it may, the consistent factors underpinning lifespan extension in all cases are efficient prevention and repair of damage caused by free radicals. We can reasonably conclude that oxygen free radicals are a primary cause of ageing, and that ageing can be slowed down, in principle, by altering the expression of genes responsible for bodily maintenance.

If ageing really is caused by free radicals, we need to answer two difficult questions. First, how is ageing apparently deferred until after sexual maturation, a period, for us, of three decades or more? Second, how do some cells, such as bacteria, cancer cells and sex cells, avoid ageing? Indeed, it is not just single cells that avoid ageing: some animals, such as *Hydra* (a small, tentacled, freshwater relative of the sea anemone) appear to escape ageing altogether. They live in shallow, oxygenated waters and show no signs of senescence. How do they contrive to avoid the damaging effects of free radicals?

The first question — how is ageing deferred — can be answered by thinking about the peculiar nature of the mitochondria, and the way in which they operate in the cell. Recall that mitochondria were once free-living bacteria, which eventually evolved into the organelles responsible for oxygen metabolism in plants and animals alike. We saw in Chapter 8 that mitochondria have retained vestiges of their independent past, in particular their own DNA and their ancestral way of dividing, simply splitting in two by binary fission: an asexual process. Mitochondria are therefore asexual genetic systems that replicate themselves within a sexually reproducing organism. Their own DNA is critical to their function. If their DNA is damaged, mitochondria cannot work: it is impossible to build a mitochondrion from nuclear genes alone. Thus, all oxygen-breathing animals are totally dependent on the integrity of mitochondrial DNA. If damaged mitochondria are passed on to the next generation, the offspring will be compromised or die.

The mitochondrial theory of ageing was first proposed by Denham Harman, author of the free-radical theory, as a refinement of his original theory. His ideas were later developed by Jaime Miquel, at the Institute of Neurosciences in Alicante, Spain, and others. Essentially, the idea is as follows. Free radicals are formed continuously in the immediate vicinity of mitochondrial DNA. Mitochondrial DNA is naked — it is not coated in proteins — and so is exposed to attack. Even worse, repair of mitochondrial DNA is said to be rudimentary. As a result, errors accumulate quickly. Because mitochondria very rarely indulge in 'sex' by fusing together, these errors cannot be cleansed by recombination and so they persist. The persistence of mutations is confirmed by the rapid mutation rate of mitochondrial DNA, compared with nuclear DNA, over evolutionary time.[6] We are left in an odd situation, in which the most toxic compartment of the cell shelters the most vulnerable DNA. A vicious circle develops. Mutated mitochondrial genes direct the production of faulty respiratory proteins, which leak more free radicals, causing more DNA damage. The spiralling descent seems to lead inexorably to ageing and death. Indeed, it is astonishing that we survive as long as we do.

In 1988, Christoph Richter, Jeen-Woo Park and Bruce Ames, at Berkeley, measured the amount of damage to mitochondrial DNA compared with nuclear DNA (which is, of course, cordoned off behind its own membranes and wrapped in proteins, at a safe distance from the mitochondria). Their findings seemed to provide good support for the mitochondrial theory of ageing: the apparent load of oxidative damage to mitochondrial DNA was nearly 20 times that of nuclear DNA. During the 1990s, several research teams attempted to replicate these early results. Extrapolated from some 20 papers, the scatter of results almost defies belief. According to Bruce Ames and Kenneth Beckman, in a refreshingly honest reappraisal published in 1999 (it is always good to see scientists rising above their own theories) the range of estimates of oxidative damage spans more than 60 000-fold! There is no suggestion that anyone is fabricating data — it is simply that even the most sophisticated modern

[6] Mitochondrial DNA evolves over thousands of years, but is not recombined through sex. Thus, there is no 'mixing' of mitochondrial genes. If a Turkish woman marries a native American, *all* children born to them will have pure Turkish mitochondrial DNA. Because of the rate of evolution, races that diverged from each other thousands of years ago can be distinguished from each other through their mitochondrial DNA, whereas those within a group, all of whom inherited their mitochondrial DNA down the maternal line, share similar mitochondrial DNA. This is the basis of 'mitochondrial Eve', the mythical mother of mankind who passed her mitochondrial DNA to all those living on the planet today. According to Bryan Sykes, a specialist in mitochondrial DNA at Oxford, and author of *The Seven Daughters of Eve*, all modern Europeans are descended from seven mythical women, representatives of seven different tribes who migrated into Europe at different times.

techniques generate copious artefactual errors. Ames and Beckman concluded:

> In summary, despite considerable popularity and intuitive merit, the theory that mitochondrial DNA is more heavily damaged by oxidative damage than nuclear DNA does not stand on firm ground. Due to the variation between competing methods of analyzing oxidative damage, it must be concluded that the background level of oxidative damage of mitochondrial DNA is not yet known with certainty; nor, for that matter, does there exist a firm estimate of oxidative damage in nuclear DNA with which to compare it.

Do these messy experimental shenanigans signal the demise of the mitochondrial theory of ageing? In its original form, probably. There are a number of biological objections too. For example, even though there are fewer, larger, less-efficient mitochondria in ageing tissues, they are in some sort of working order: there are few signs of the catastrophic damage predicted by the mitochondrial theory of ageing. In relation to this, seriously damaged mitochondria ought to destabilize cells and set off the cell-suicide programme — apoptosis. But examination of ageing organs suggests that apoptosis does not take place on the scale predicted by the mitochondrial theory. So how do the leaky mitochondria maintain their integrity? Well, they have multiple copies of their genes, which are kept in functional clusters to ensure they have at least one working copy of each gene. Then, it seems, mitochondria are better at repairing their DNA than was once thought: an enzyme responsible for correcting oxidative damage to mitochondrial DNA was isolated in 1997. Mitochondria can also tolerate a large number of mutations — they apparently have a mechanism for editing erroneous RNA to make workable proteins. Finally, an evolutionary thought: if mitochondrial DNA is really so vulnerable, why did it persist there — why was it not all transferred to the nucleus? Genetic studies suggest that there is no physical reason why it should not have done, so there must be some benefit to DNA remaining in the mitochondria.[7] In sum, these considerations suggest that the mitochondrial theory, as originally stated, is biologically naive.

[7] One possibility put forward by John Allen (who we shall meet later in the chapter) is that mitochondrial genes allow a rapid response to sudden changes in oxygen level, nutrient supply or the presence of respiratory poisons. The energy status of the cell is so critical that cells need to respond swiftly and appropriately to sudden change. Having to rely on bureaucratic nuclear genes to do this is like waiting on the government to make decisions about the disposition of troops on the ground in a war. Mitochondrial DNA is thus a kind of front-

And yet . . . there is too strong a link between metabolism and ageing, across all species, to dismiss the mitochondrial theory out of hand. Mitochondrial DNA sequences certainly change relatively rapidly (over generations), which implies that their DNA does suffer more mutations than nuclear DNA. In addition, mitochondria from ageing tissues are unquestionably damaged to some extent, even if not catastrophically so. A more subtle version of the mitochondrial theory must be true. I favour a model put forward by Tom Kirkwood, working this time with the German biochemist Axel Kowald, and known as the MARS model (Mitochondria, Aberrant proteins, Radicals, Scavengers). The idea typifies Kirkwood's contribution to ageing research. Trained originally as a mathematician, he took a step back from the details of competing theories, and considered the broader network of interactions within cells. In particular, Kirkwood and Kowald asked what would happen to protein turnover if mitochondrial function declined only slightly. They made three assumptions: first that free radicals would escape from the mitochondria to damage other cellular components, such as the protein-synthesis apparatus; second, that prevention and repair is never 100 per cent efficient; and third, that mildly damaged, albeit functional, mitochondria produce less energy than their undamaged cousins, ultimately causing a cellular energy deficit (in other words, the cell cannot produce as much energy as it needs).

Kirkwood and Kowald built these three assumptions into a computer model, to see how well they could simulate the pace of the changes that occur during ageing. The detailed equations presented in their 1996 paper are enough to make most biochemists tear their hair, but their conclusions make good intuitive sense. A very slight mismatch between the rate of free-radical production and the ability of the cell to repair that damage, coupled with a growing energy deficit, leads to an insidious decline in mitochondrial function. The decline unfolds over many decades, until finally a threshold is crossed. At this point, the mitochondria probably resemble those isolated from old tissues. The emphasis now shifts from the mitochondria to the protein-synthesizing machinery. In a relatively

---

line rapid reaction unit, enabling the sensitive local control of critical gene expression and respiratory function. Mitochondrial DNA might be called 'altruistic', in that it serves the greater good; but in reality cells and organisms have a selective advantage if they retain DNA in their mitochondria, and so will survive better than cells that eliminate their 'altruistic' DNA. This reminds us that the 'selective unit' is always the organism (even if the organism is a single cell) and not individual genes.

short space of time, compared with the overall process, the cell's ability to maintain its biochemical equilibrium collapses. Once the equilibrium is lost, it is only a matter of time before the cell dies. This process matches both the timescale and the acceleration of ageing observed in real life. Critically, at no time does the model system actively lower the performance of its maintenance systems. The cell's resources were, in fact, insufficient from the beginning, and this allowed the gradual undermining of its integrity.

Although simultaneous equations inevitably simplify the real cell, I agree with Kirkwood and Kowald's conclusion that the model provides a plausible framework for understanding the process of ageing. It distinguishes between what is theoretically possible and what is improbable. In the absence of convincing experimental data, they seem to be barking up the right tree. If they are, then there is a big implication. Mitochondrial respiration will eventually undermine the integrity of cells. The speed at which this happens depends on the ability of cells to protect themselves, but no cell is 100 per cent efficient, so all creatures containing mitochondria should die. This returns us to the second difficult question: how do some cells, and even some simple animals, avoid ageing?

When August Weismann first distinguished between the mortal body and the immortal germ line at the end of the nineteenth century, he made a remarkable prediction: that all somatic (body) cells would have a finite lifespan. For much of the twentieth century, Weismann's prediction remained controversial. The debate was put on a more empirical footing in 1965 by the American biologist Leonard Hayflick, who finally proved that human fibroblasts (connective tissue cells involved in wound healing, and easily grown in culture) can divide no more than 50 to 70 times before succumbing to 'replicative senescence' and dying. Thus, unlike bacteria, fibroblasts cannot be cultured indefinitely: in the end, the entire population dies out, apparently of old age. The potential number of divisions that a single cell can make before dying (or more precisely, the number of population doublings) became known as the Hayflick limit. Different cell types have different Hayflick limits, but we now know that essentially all somatic cells senesce and finally die.

There are intriguing variations on this theme. Fibroblasts taken from short-lived species, such as mice, have a lower Hayflick limit than fibroblasts from long-lived species, such as humans (about 15 cell divisions compared with 70). This relationship is robust across all species tested. The Hayflick limit also varies with the age of the donor. If fibroblasts are cultured from an old donor, they divide fewer times before senescing and dying than those taken from a young individual. Presumably, they had already used up some part of their limit while dividing in the body, and so had fewer divisions left to them. Cells taken from people with Werner's syndrome, which causes accelerated ageing, also quickly curl up and die. The implication is startling: cells can count. When they have counted up to their limit, they die. The limit is encoded in the genes. Genetic diseases in which ageing is accelerated have a lower limit.

Cancer cells are an exception. They behave more like bacteria. Cancer cells somehow get around the Hayflick limit and continue to multiply indefinitely. The most famous example is the tumour of the unfortunate black American Henrietta Lacks, who died of cervical cancer in Baltimore in 1951. Doctors took a sample of her tumour in the 1940s, and cultured the cells to see what kind of a tumour it was. The cells, known as HeLa cells, were so vigorous that they are still being grown in research centres across the world 60 years later. They show no signs of senescence. In total, they now weigh more than 400 times Henrietta's own body weight.

The story of the Hayflick limit came to a head in 1990, when Cal Harley, founder of the Californian biotechnology company Geron Corporation, made a connection between the ability of cells to count and the length of their *telomeres* — the 'tips' at the ends of individual chromosomes. Telomeres are often said to resemble the ends of a shoelace — their purpose is to prevent 'fraying'; in other words, to preserve the integrity of the chromosome. They are also said to be the secret of eternal life. They are not, as we shall see.

Telomeres are a characteristic biological fudge: they are needed because our DNA replication machinery was inherited from bacterial ancestors with circular chromosomes, whereas those of all eukaryotes are linear. Because of the way the biochemical machinery for replicating DNA works, it is impossible to replicate the extreme ends of a linear DNA molecule. As a result, the chromosome gets shorter each time it is copied. The solution? A fudge. Evolution cannot whip new DNA-replicating machinery out of a hat, just like that, but it is easy enough to add a bit of extra non-coding DNA to each end of the chromosome, to which the enzymes can bind at the beginning and end of replication. The loss of this extra DNA does not matter, at least until its loss is complete: then the chromosomes start to fray and the cell can divide no longer.

These extra caps of non-coding DNA, then, are the telomeres. What Cal Harley showed was that they get steadily shorter in human fibroblasts growing in culture. Each time a cell divides it must replicate its DNA, so a little bit of telomere is lost with every cell division. Human fibroblasts lose all their telomeres after a maximum of about 70 cell divisions. Thus, the shortening telomeres function as a biological clock, which sets a limit on the number of times a cell can divide. This limit is determined by the original length of the telomere and the rate at which it is used up — but in general, the longer it was at the beginning, the more cell divisions are possible.

How do cancer cells escape? It seems they make use of an enzyme, called *telomerase*, which regenerates the telomeres, so that their length is not perpetually truncated. Telomerase is thought to be present in all cancer cells. These cells do not magic the enzyme out of thin air. The gene is present in all our cells, but is normally switched off. In our bodies, it is usually used only by *stem cells*, unspecialized cells that can divide and differentiate to produce new tissues, and by sex cells, whose *raison d'être* is reproduction. In 1997, scientists at Geron succeeded in cloning part of the gene for telomerase. When they introduced it into cultured human somatic cells, along with a promoter gene to make sure that the transfected telomerase was active, the new gene made the cells essentially immortal. The cell population was able to continue dividing indefinitely, but did not behave like cancer cells, which tend to form clumps similar to tumours, even in a petri dish. The findings were published in *Science* in 1998 and generated huge excitement — here was the secret of eternal youth! The product of a single gene could overcome ageing, or at least replicative senescence, in somatic cells.

The excitement over telomerase echoes the long human quest for immortality. Gilgamesh would have been enraptured. The gene-centric molecular biologists who advocate programmed ageing were triumphant. If our lifespan is set by the length of a piece of DNA, then the specific length must have been 'programmed' in some way to ensure that our lives are the right length, presumably for the good of the species. Evolutionary biologists took a different view. As we saw in the last chapter, if selection pressure falls with age, then there can be no unfolding programme for ageing. If this is the case, the telomeres must have a different significance.

Their apparent control of senescence in cell culture must be an artefact that is irrelevant to their role in the body.

These diametrically opposed interpretations of a central fact, which is not itself in dispute — that telomerase confers immortality on cells grown in culture — shows the importance of theory in science. Facts mean little in isolation unless they can be interpreted within the wider framework of a theory; and it is usually those needling little facts, which resist interpretation by any theory, that bring dogmas crashing down. In the case of telomerase, however, there is no need for a radical reinterpretation. Telomerase is necessary, but not in itself sufficient, for cells with straight chromosomes to go on replicating themselves indefinitely. It is almost irrelevant to the ageing of our bodies.

Many cells in the adult body do not divide, and therefore do not lose telomere length. They do not need telomerase because they do not have a problem with shrinking telomeres. The brain, heart, major arteries and the 'skeletal' muscles that enable us to move are composed largely of specialized cells that have a job to do, that do not divide and are not easily replaced. A centenarian's brain has nerve cells (neurons) that are 100 years old. Even though we do not understand the workings of the conscious mind, it clearly resides, in one sense or another, in the great network of connections formed between neurons throughout our lives. With the 100 billion neurons we start out with, we make some 200 million million connections. It is hard to imagine how this fantastic web of connections could be replicated by replacing old neurons with new ones, which would have to reproduce the exact spatial connections of their defunct predecessors. If they failed, our minds would change, our memories would be wiped or transfigured. Some songbirds that sing a new song each year are thought to replace certain neurons: something similar would surely be true for us. We might live forever, but unless we wrote it down we'd never know. The problem, then, is that the evolved structure of the human body is simply not compatible with eternal life, unless we can find a way of replacing worn-out neurons — and here we enter the realms of science fiction.

Cells that do divide regularly, such as stem cells and the cells that give rise to sperm, have active telomerase. They have no problem with shrinking telomeres either. Even circulating immune cells, which do not express telomerase when quiescent, reactivate it when stimulated to proliferate by bacteria. In other words, if our immune cells need to undergo many rounds of cell division, they have all the telomeres they need. All that remains are certain epithelial cell types, such as kidney cells and liver cells, and fibroblasts, which do divide in the body, but only when needed. These cells do not produce telomerase and potentially face the Hayflick limit, but it is questionable whether they ever reach it. Fibroblasts taken from elderly donors can usually still divide between 20 and 50 times before becoming senescent and dying: clearly they never reached their limit in the body. If they have no telomerase, it is because they do not need it.

A few other miscellaneous facts confirm the same story. There is a poor correlation between telomere length and the maximum lifespan of different species. Mice have far longer telomeres than humans, even though we live 25 times longer. Different species of mice, all with the same maximum lifespan, have very different telomere lengths. Remarkably, 'knock-out' mice lacking the gene for telomerase have normal lifespans until the third generation, when they do show signs of accelerated ageing, the significance of which is uncertain. Finally, the number of cell doublings needed to make a body does not relate to the subsequent rate of ageing. The cells of an elephant must divide many more times to produce an elephant than must those of a mouse to produce a mouse; yet elephants live far longer. In short, it seems fair to say that, for all the hullabaloo, telomerase does not hold the secret of eternal life. Without the enzyme, eternal cell replication is not possible in eukaryotes, due to a glitch in the DNA-replication machinery passed down by evolution. Telomerase thus facilitates cell division in the same way that a light switch facilitates lighting a room: it is technically helpful, but just as the light switch is not the source of the light, telomerase is not the spring of everlasting life. So why is telomerase switched off in epithelial cells? Some argue that a limit on the possible number of cell replications might protect against cancer, but this seems unlikely. The Hayflick limit is too high to be relevant: it is equivalent to the Chinese government decreeing that, to safeguard against population growth, it will impose a limit of 70 children per couple. The Hayflick limit is just as irrelevant to preventing cancer. The most likely answer is that, like most genes in most cells of the body, telomerase is switched off simply because it is not needed.

How is it possible, then, that normal cells can be transformed into immortal cells, just by adding the gene for telomerase? How do mitochondria fit into this story? I had my first inkling of this a few years ago, when I started growing kidney tubule cells in culture, and wasted weeks in the lab. I had taken a few lessons from people growing other sorts of cell, and had applied their methods to my own problem. Each time, my culture plates became overgrown with spidery looking cells, which I assumed were fibroblasts. Fibroblasts do very well in cell culture, and even a trifling contamination can lead to them taking over the plate. I threw away my fibroblasts and started again, more conscious this time of good technique. The same thing happened again and again. Finally I went to see a fibroblast specialist, who looked at my plates and laughed — "They're not fibroblasts", he said, "I don't know what they are, but they're absolutely not fibroblasts. They're probably your kidney cells!"

I was shocked. I had spent hours looking at kidney sections down the microscope, and I knew what tubule cells looked like: prolific brush borders, providing a massive surface area for reabsorbing solutes, and thousands of mitochondria, packed together like a Roman phalanx. My cultured cells had no brush border, and I could see no more than a handful of mitochondria. There was nothing for it but to turn to the textbooks and the original papers. I was in for another shock. My sad cells were exactly what kidney tubule cells were supposed to look like in culture! I had been planning experiments to see how vulnerable the cells would be to oxygen, and whether they could be protected by antioxidants, but now, after reading the small print, I realized that cultured kidney tubule cells do not require oxygen at all: they live quite happily by anaerobic respiration. In fact, the only way to get them to breathe oxygen is to deprive them of glucose in the culture medium, and to catch them in the act of growing, before they have completely covered the dish. I abandoned my experiments as irrelevant to real kidneys, chastened but a little wiser.

This pattern is characteristic of cells grown in culture: they don't need much energy, so they don't have many mitochondria. In fact, this is true not just of cells grown in culture, but of any cells that have a low energy expenditure. Perhaps surprisingly, these include actively dividing cells, such as stem cells and cancer cells: their energy requirement is much lower than that required for specialized metabolic work. Just think of the brain: it accounts for 20 per cent of resting oxygen consumption, but only 2 per cent of body weight. If the oxygen supply to the brain is cut off for more than a couple of minutes, we lose consciousness. Neurons do not divide, and the brain's support network, the glial cells, only divide occasionally — the brain needs all this oxygen for its normal metabolic work. Other metabolically active body tissues have a similarly grasping demand for oxygen. Liver, kidney and heart-muscle cells have as many as 2000 mitochondria each, so densely packed that it can be hard to see any cytoplasm at all. In contrast, stem cells, whose job it is to replenish cell populations with a continuous turnover, such as skin cells, have remarkably few mitochondria. Similarly, immune-system cells such as lymphocytes, which again divide frequently once activated, are virtually devoid of mitochondria.

In general, there is a striking relationship between the degree of cellular differentiation — the cell's commitment to a metabolic purpose — and the number of mitochondria. Specialized differentiated cells have large numbers of mitochondria and suffer the consequences — serious oxidative stress. Stressed cells gain the most benefit from improved stress-resistance. Recall, for example, that the protective effects of calorie restriction are most marked in the long-lived cells of tissues where oxidative stress is worst, such as the brain, heart and skeletal muscle. This, not telomerase, is what really confers immortality on populations. To survive, get rid of your mitochondria — throw them overboard like so much ballast. Cancer cells do. Cancer cells become less differentiated as they multiply, and lose their mitochondria in the process. They thrive on anaerobic respiration. Most tumours are a dense mass of tissue with a low requirement for oxygen. Indeed, oxygen is toxic to many tumours: radiotherapy is three or four times more effective when the tumour is oxygenated. As is so often the case, the exceptions prove the rule. Some cancer cells are rich in mitochondria. In particular, some glandular tumours (oncocytomas) and liver tumours (Novikoff hepatomas) have cells with huge numbers of mitochondria. In both cases, however, close biochemical inspection suggests that the mitochondria in the tumours are not actually functional. Thus, cells can reproduce indefinitely if they have active telomerase and a small number of relatively inactive mitochondria.

There is a second factor that helps to preserve rapidly dividing cells: their fast turnover. When a cell divides, it must reproduce its cytoplasm and mitochondria, as well as its DNA. This means that mitochondria replicate faster in rapidly dividing cells than in non-dividing cells, even if the latter are packed with mitochondria. In most cells, the mitochondrial population varies from good condition to completely shredded. Undamaged mitochondria replicate faster than damaged mitochondria. Each time a cell divides, then, the new pool of mitochondria is derived from the least-damaged survivors of the last pool, and this helps to replenish the population. In rapidly dividing cancer cells, then, we can predict that there should be relatively few mitochondria in relatively good condition.

In the case of non-dividing cells, where the rate of mitochondrial replication is much slower, the rate of mitochondrial breakdown becomes more important. In these cells, mitochondria are normally replaced every few weeks. In non-dividing cells, partially damaged mitochondria may be broken down more slowly than healthy mitochondria, and this discrepancy can lead to a takeover by damaged mitochondria.[8] The phenomenon is known as the 'survival of the slowest', or SOS, and may contribute to the demise of old differentiated cells.

We can conclude that the lifespan of a cell depends on the activity of its mitochondria and the efficiency of its damage-prevention and repair systems. Damage-prevention and repair are never 100 per cent efficient, so energetic cells will eventually accumulate defective mitochondria. In the end, these will undermine the integrity of the cell. This situation is exacerbated in non-dividing cells, which cannot replenish their mitochondrial populations by selecting for less-damaged mitochondria. We therefore have a spectrum of potential longevity, ranging from stem cells and cancer cells, which are virtually immortal, through to neurons, body muscle cells and heart muscle cells, which are doomed from the moment that they specialize in tasks that require large amounts of energy. In principle, the lifespan of these metabolically active cells can be extended by building up their ability to resist oxidative stress. However, all the energy directed at cell renewal is subtracted from tasks that the cell would normally perform. To protect neurons from free-radical attack, energy must be diverted away from thinking or coordinating the body — obviously to the detriment of performance. Longevity and biological fitness are therefore incompatible, and we must find an optimal trade-off. Can we live longer? Perhaps: some tortoises can live for 200 years, but their success does not depend on quick movement and sharp wits. Their shells confer a different kind of protection, enabling them to be less metabolically active. They have a different trade-off.

[8] The idea, proposed by Aubrey de Gray at the University of Cambridge, is that damage to mitochondrial DNA may prevent oxygen metabolism, and this would paradoxically lower the burden of free radicals. As a result, the mitochondrial membranes would be *less* damaged than normal, and so their turnover would be lower. This may sound implausible, but it is supported by empirical data.

Two broad conclusions emerge from this analysis. First, Weismann was wrong again: there is no fundamental distinction between germ cells and somatic cells. Some somatic cells, such as cancer cells, achieve immortality by losing mitochondria and replicating quickly. This is how the simple, tentacled *Hydra* courts immortality: it has a large pool of stem cells, which can develop into any one of the mature cells in its body. It continually replaces its worn-out cells. The price for this is a simple body plan, allowing cells to be replaced without affecting the function of whole organs. Our own stem cells might possess similar powers of regeneration — think only of cloning — but our body structure is fundamentally different: as noted earlier, we cannot replace the neurons in our brain while at the same time maintaining a sense of continuity and experience. As one body system begins to wear out, the repercussions are felt by others. If the ageing pituitary gland in the brain starts producing fewer hormones, the vitality of stem cells in the skin will inevitably be affected. Until we find a way around this problem, we will never outlive our neurons.

Second, the disposable soma theory is not just about sex. Sex steals resources that we would otherwise use for staying alive, but then so does being human. If we are to think, run, create, interact — anything that makes us human — we sell ourselves a short life. Perhaps Raymond Pearl was right after all — perhaps laziness does pay, so long as we don't eat or drink our way to an early death. The idea is supported by a recent best-selling book, *The Okinawa Way*, written by a Japanese cardiologist and two American colleagues. Based on a 25-year study, the book argues that the secret of the Okinawans (the inhabitants of a Japanese island with more centenarians than anywhere else in the world) goes beyond genes, diet and exercise to their relaxed lifestyle and low levels of stress. The Okinawans have a word for it, *tege*, which means 'half-done': forget timetables, forget finishing today things that can be done tomorrow. I suspect they are probably right.

We are left with a final problem, which still has the power to make or break the arguments so far. I have argued that mitochondrial respiration will be the death of us. In killing cells, mitochondria first damage their own integrity. But if *all* mitochondria damage themselves, how do mitochondria-bearing organisms evade decay over generations? How is it that babies are born young?

The situation reminds me of the decline and fall of the Byzantine Empire. If we believe the eighteenth-century historian, Edward Gibbon, the empire was in a state of continuous decline for 1000 years. Some emperors succeeded in reviving its flagging fortunes temporarily, but the "corrupting spirit" of the Greeks meant it was only ever a matter of time before the empire fell. Gibbon's 'corrupting spirit' meets its match in mitochondria: it should be only a matter of time before mitochondria corrupt their hosts, even if it takes 1000 generations. Yet the fall of Constantinople has no echo in nature: how have we avoided the corrupting spirit?

Let me define the problem. Undamaged mitochondrial DNA is *necessary* for the function of any organism. Sex cells must pass on newly minted mitochondria, so the mitochondrial DNA must somehow be rejuvenated. The problem is that mitochondria replicate their DNA asexually. As we have seen, sex is reinvigorating for genes, but without sex it is hard to see how the mitochondrial genome can regenerate itself. How can it reset its biological clock to zero in a newborn infant? Free-living asexual organisms such as bacteria preserve genetic integrity over generations by combining rapid reproduction with heavy natural selection. Bacterial-style selection for mitochondria is out of the question, however. They would have to replicate themselves at the same rate as cancer cells, and we would turn into mitochondrial tumours. This paradox — how mitochondria regenerate themselves without either sexual recombination or heavy selection — is known as Muller's ratchet and seems insurmountable; but clearly it is not. So how do they do it?

To understand the solution to this conundrum, we must think about the fate of mitochondria following the act of sex, in particular the fate of sperm mitochondria. Wriggling human sperm are a familiar sight to television audiences: we all know that their power and endurance is extraordinary. Surprisingly, there is some confusion about exactly how they power their performance. There is a common misconception that sperm are too small to contain mitochondria. In fact, sperm contain about 40 to 60 mitochondria, encased in the midpiece. The sperm's mitochondria *enter* the fertilized egg, along with the midpiece, but may not survive there for long. What becomes of them is uncertain, but essentially *all* our mitochondria are inherited from our mothers. This is true not just for us, but also for the great majority of sexual organisms, including plants.

Why male mitochondria are not passed on to the next generation has taxed the minds of some of the finest biologists. The most widely accepted general explanation was articulated well by John Maynard Smith and Eörs Szathmáry in their book *The Origins of Life*. Essentially, if mitochondria are inherited from both parents, the stage is set for the evolution of 'selfish' organelles. The argument is as follows. When a cell divides, all its nuclear DNA is replicated, with half going to each daughter cell: the two daughters have identical sets of genes, so there is no unequal competition. This is not true of mitochondria, which have their own DNA and use it to replicate independently. The overall make-up of the mitochondrial population in a cell therefore depends on the speed at which individual mitochondria replicate (or break down), and this makes the cell vulnerable to abuse. Any mutation in *mitochondrial* DNA that increases the speed of a mitochondrion's replication will lead to its progeny taking over the entire cell and its descendants, even if the same mutation makes them worse at oxygen respiration (indeed, especially if it makes them worse at respiration, as they will damage themselves less). If the mutant mitochondrion is in a sex cell, then the entire organism will become compromised as it grows. According to the selfishness theory, the proliferation of selfish mitochondria is prevented by the device of *uniparental inheritance,* in which only one parent provides all the mitochondria. Instead of mixing unrelated mitochondria from the merger of two similar (but otherwise unrelated) sex cells, one sex specializes in providing all the mitochondria, while the other specializes in providing none at all. Thus, gender grew out of an evolutionary trick to exclude selfish mitochondria.

This theory is almost certainly true in some instances, but there are two objections to it as a general explanation. First, a mutation that causes uniparental inheritance is only advantageous if the selfish mitochondria are waiting in the wings to take advantage. This is improbable: any organism that harbours selfish mitochondria is less likely to survive and reproduce than a robustly fit organism. It is equivalent to pitting a broken old man, with his broken mitochondria, against a heroic youth, to win the favours of a lady. In fact, the chances are that an organism with defective mitochondria would not even get through development. Think of the difficulties many couples experience in getting pregnant. Some evidence suggests that a large proportion of embryos fail to develop past the early stages of pregnancy because of a problem with their mitochondria. Such problems may also account for the high failure rate of cloning experiments.

Second, a number of species do not rely on uniparental inheritance at all — their mitochondria are inherited from both parents. They circumvent the selfishness problem somehow. As we have seen, the same may even be true of us to a lesser extent — the fate of sperm mitochondria is uncertain. Some researchers put their apparent disappearance down to a simple dilution effect. This has not been ruled out. A sperm has 40 to 60 mitochondria, while a human egg cell is believed to contain more than 100 000. The dilution factor is therefore at least 1000, which is in fact below the detection limit of many techniques for detecting mitochondrial DNA. Some studies in mice, using more sensitive techniques, suggest that male mitochondrial DNA is present at a frequency of between 1 in 1000 and 1 in 10 000, relative to the maternal contribution: very close to what one would expect from dilution alone. The issue is still unresolved, although we shall see in a moment that recent work does suggest a solution.

These two objections to the selfish mitochondria theory question its validity as an explanation for the evolution of sexes; but the fact remains that some animals go to bizarre lengths to exclude the male mitochondria. A few species of *Drosophila* apparently sequester the sperm mitochondria in the gut of the larvae during development, and defecate them out soon after hatching. Something peculiar is going on. At the level of the sex cells, mitochondrial transmission is virtually the defining difference between the sexes. Why should this be? The answer, in my view, was spelled out in 1996 by John Allen, a biologist at Lund University in Sweden, in the *Journal of Theoretical Biology* — a journal always crammed with impressive erudition and lofty ideas, from the sublime to the ridiculous. Allen draws on the logic of the mitochondrial theory of ageing to explain the evolution of two sexes. In essence, he argues that male mitochondria are not passed on to the next generation because they are time bombs: they have been fatally damaged by oxygen, and if passed on would cause the birth of prematurely aged babies. Breathing oxygen dictates the need for two sexes. If so, then oxygen is the ultimate gender bender.

Allen's fundamental idea is as follows: if mitochondria damage their own DNA by respiring oxygen, and cannot systematically cleanse their genome of error by either sex or binary fission, then the only way to prevent mitochondria from passing on damaged DNA to the next generation is to stop them from respiring at all. In other words, the only way of maintaining mitochondrial integrity is to switch them off. This proposition leads to a number of predictions, many of which are undoubtedly true, and all of which are testable. If these predictions each turn out to be true, then we can use the detailed process of sexual reproduction to confirm the validity of the mitochondrial theory of ageing.

To follow Allen's ideas through, we need to go back to one of the fundamental problems facing sexual reproduction — how to find an appropriate partner. The problem affects single cells as well as lonely-hearts, and the solution is somewhat similar. To have two people searching for one another is no more effective than having one person stay put and the other doing the searching: this is the idea behind dating agencies. For sex cells, one cell *must* move around in its quest for a suitable partner, but the probability of meeting the cell of choice is no greater if both cells move around. One cell can stay put, as long as it signals its presence or availability. In our case, and in many other animals, the sperm are motile, while the eggs are immobile. Indeed, the word 'male' is conventionally defined as the sex which produces a large number of small, mobile gametes, while 'female' is defined as the sex that produces a small number of large, immobile gametes.

Motility, of course, requires active mitochondrial respiration, and this damages mitochondrial DNA. Since the objective is *not* to pass on damaged mitochondria, we can predict that sperm will not pass on their mitochondria to the next generation. If the reason that they do not is indeed because of damage, then we can also predict that the sperm's mitochondria should be damaged and as a result destroyed. There is some evidence that this is the case. Peter Sutovsky and his team at Oregon Health Sciences University, Oregon, published a paper in *Nature* in 1999 showing that in cattle the male mitochondria become tagged with the protein ubiquitin. This tag is normally used as a marker of damaged proteins, consigning them to breakdown and turnover. The implication is that the sperm's mitochondria are spotted as defective and destroyed in the early stages of embryonic development. Sutovsky's more recent work confirms this mechanism, in cattle at least. Thus, discrimination between male and female mitochondria seems to be achieved on the basis of *damage*, as predicted by Allen's theory.[9]

[9] This may also explain how some species can receive mitochondria from both sexes. We might predict that *either:* only undamaged mitochondria survive, that is, only those not tagged with ubiquitin; *or* that there are subpopulations of mitochondria within both sex cells that are switched off, as discussed above. We might also guess that the motility of sex cells would be limited in both cases. Pollen, for example, requires little output of energy to fertilize a flower.

A second prediction relates to the timing of sex-cell production. Because the recombination of chromosomes during sex cleanses the nuclear gene-line, and the new combinations are subjected to selection for viability, then it should not matter exactly *when* the new sex cells are produced. There is no obvious reason why both types of sex cell should not be produced continuously throughout life. So why is it, then, that sperm *are* produced through life, but eggs are only produced early in development and then last half a lifetime? Well, think about the mitochondria. Sperm mitochondria are not passed on to the next generation. It does not matter if these mitochondria are damaged, provided that they are still functional enough to get the sperm to the egg. In our bodies, this is the average state of mitochondria for most of our lives: damaged but functional. Thus, there is no reason why sperm should not be produced continuously throughout our lives. The only proviso is that the *nuclear* DNA must be shielded against escaping mitochondrial free radicals by antioxidant defences. This is indeed the case. The midpiece of sperm, containing the mitochondria, is encapsulated in selenium-containing proteins. Sperm contain a higher concentration of selenium than any other cell type in the body. Dietary selenium deficiency is a common cause of infertility in some parts of the world. One of the selenium proteins is a form of glutathione peroxidase, which disposes of hydrogen peroxide. Glutathione peroxidase would probably not protect the mitochondria from damage, but would prevent hydrogen peroxide from diffusing into the nucleus, where it could react with iron to produce hydroxyl radicals.

What about the egg? In this case the mitochondria *are* passed on to the next generation. If the eggs are formed throughout life, then their mitochondria will become progressively more damaged as time goes by. Nuclear DNA can be rejuvenated by sex but mitochondrial DNA cannot. One solution is to cordon off undamaged mitochondria very early in life, switch them off, enclose them in an egg, and then maintain the egg in a dormant state until it is needed. This is very close to what does happen, and brings us to our third prediction: that the mitochondria in the egg should be switched off.

The easiest way of switching off mitochondria is to halt the production of respiratory proteins. Imagine a room full of dominoes standing in line: the simplest way of preventing the whole line from toppling is to remove a domino or two, so that a falling domino cannot touch the next in line. So it is with the chain of respiratory proteins in the mitochondria: if a few strategic proteins are omitted from the chain, then respiration cannot take place. The strategic proteins omitted are those encoded by the mitochondrial genes, certainly in mice and the African clawed frog, *Xenopus laevis*. In the mouse, the mitochondrial genome is largely inactive in the egg and the early embryo. In *Xenopus*, DNA-binding proteins are known to inhibit mitochondrial gene transcription. Thus, in a few known cases at least, the mitochondria are indeed 'switched off' in the egg.

If this sort of mitochondrial inhibition turns out to be generally the case, as we would predict, then the egg cells would be unable to provide all their own energy by respiration. This leads to a final predication from Allen: that the follicle cells surrounding the developing egg should provide the egg with energy in the form of ATP. Whether or not this is true is unknown, but the morphological structure of the follicles suggests that it may well be.

Overall, then, the facts fit the theory. Passing mitochondria from one generation to the next is a liability that requires extraordinary measures to make it possible at all. These measures probably contributed to the evolution of two specialized types of sex cell, or *anisogamy*. Anisogamy, in turn, is equated with the origins of sexes: once the two types of sex cell have become mutually dependent, there is no way back, so the only path is towards the increasing specialization of sexual traits. Breathing oxygen is thus intimately linked with both ageing and the origins of gender.

Considering the situation from the other extreme of life, I feel that the elaborate precautions required to reset the mitochondrial clock to zero in a new generation confirm the main tenets of the mitochondrial theory of ageing. If so, then we have reached a watershed conclusion: there is indeed a process of ageing that is independent of age-related disease. Ageing is not simply the accumulation of late-acting mutations, as argued by the theory of antagonistic pleiotropy (see Chapter 12, page 239). Even without succumbing to genetic disease, we will eventually die of mitochondrial wear and tear. It is quite plausible that the time required for mitochondrial burn-out in long-lived cells, such as neurons, heart and skeletal muscle cells, is close to the maximum human lifespan of 115 to 120 years.

Few people live out their maximum lifespan: even in the Western world, most people die of some disease, usually with a genetic basis, in their seventies or eighties. It is no use working out how to prolong our maximum potential lifespan if only a handful of people live that long. The question for the next chapter, then, is how do age-related diseases, such as cancer and heart disease, fit in with the mitochondria story? Are they completely unrelated, or might it be possible to postpone the onset of such diseases by delaying the underlying process of ageing?

If it is possible to postpone the onset of *particular* diseases through a *general* mechanism, then the present emphasis of medical research, on pinpointing the genetic causes of disease, is wrong. With the excitement surrounding the human genome project, and our focus as a society on individual rights, pharmaceutical research is heading towards the individualization of treatment. Great weight is placed on tiny genetic differences between individuals, such as single-nucleotide polymorphisms — differences of just one letter in a given stretch of DNA. I suspect we may be losing our way in the detail. If slowing the process of ageing can postpone the onset of age-related disease in species as diverse as nematode worms, *Drosophila*, rats, monkeys, and perhaps ourselves, then we should be looking for commonalities, not particulars. In the next chapter, we shall see that there are good grounds for thinking that the whole thrust of gene-searching for drug treatment is misdirected.

CHAPTER FOURTEEN

Beyond Genes and Destiny

The Double-Agent Theory of Ageing and Disease

OEDIPUS KILLS HIS FATHER AND SLEEPS WITH HIS MOTHER. He does all this in ignorance: he had been left for dead at birth and raised in another land. He returns unknowingly to his homeland and becomes a good and noble king, only to be cut down by the machinations of fate. His terrible future is revealed by the old sage Tiresias: "Blind from having sight and beggared from high fortune, with a staff in stranger lands he shall feel forth his way; Shown living with the children of his loins, their brother and their sire, and to the womb that bare him, husband-son, and, to his father, parricide and co-rival."

When I first read Sophocles' great tragedy, I was amazed at how un-Freudian the story was. When he discovers the true nature of his actions, Oedipus tears out his eyes and condemns himself to a wandering exile, thus fulfilling the prophecy; hardly the action of one who desires his own mother. Curiously, his wife-mother, Jocasta, is more ambiguous. She is the first to grasp what has happened, and she tries to prevent the truth from emerging. Only when she sees that Oedipus is set on the truth does she damn him and hang herself. One wonders if she would have continued as before, had the truth not been revealed; but if Sophocles intended a subplot here, he paid little attention to it. The most striking element of *King Oedipus*, and indeed so much of Greek tragedy, is the implacable role of fate. The characters, for all their eloquence, are just puppets. Motive scarcely matters. Jocasta's attempt to turn a blind eye to the workings of fate merely illustrates the impossibility of her task, and the penalty for anyone who tries.

Today, millions of people enjoy reading astrology columns in daily newspapers, and some no doubt believe them, but the sense of ineluctable fate went out with Christianity. When Adam and Eve ate of the apple of knowledge, humanity was freed to suffer or prosper of their own free will. The concept of sin is a foundation of Christianity, yet must have been alien to the ancient Greeks — how can Oedipus be said to have sinned, he who was condemned by an oracle before his own birth? For Christians, sin is a choice, and we are judged on the choices we make. The difference is clear in tragedy. The Greek sense of tragedy is quite unlike Shakespeare's. Hamlet is faced with choices throughout the play, notably the ultimate question, "To be or not to be?" The terrible final scene is the outcome of a series of contingencies. The tragedy of Hamlet lies in the fact that it could all have been averted. One can imagine a satirical reworking, in which a peace-broker brings the two sides together to mediate a solution. The mediator would have failed with Oedipus. Indeed, there was a mediator, Jocasta, and she did fail. What a tragic breed we are! The tragedy of Oedipus lies in its inevitability, the tragedy of Hamlet in its evitability. After two millennia of Christian choices, it is the inevitability of Greek tragedy that shocks us today.

For the first time since the ancients, a sense of implacable fate is returning. The certainties of Greek theatre have been superseded by the certainties of modern genetics, which at times seem just as disturbing. We read about genes 'for' heart disease, cancer or Alzheimer's disease. Few people, even the scientists working on them, have a clear idea of exactly what these genes do, but we eye them with mistrust. We resist the intrusion of insurance companies who wish to pry into our genetic makeup — to read our oracles — yet our resistance owes more to a sense of personal infringement than a questioning of the veracity of genetics. We seem to accept that if we have the gene 'for' multiple sclerosis, then we will go on to develop the disease. We accept the inevitability of genetics in the same way that the Greeks accepted the inevitability of fate. The analogy is sharpened by our powerlessness to alter the course of many diseases. Many people prefer not to know what they cannot change. Tiresias put it well 2500 years ago: "Ah! How terrible is knowledge to the man whom knowledge profits not."

Many writers have riled against the idea of genes 'for' diseases. No gene is 'for' a disease any more than an aeroplane is 'for' crashing. Genes, however, like aeroplanes, do go wrong. Historically, the attitude of medicine has been that this is a mischance, a part of the human lot. The human body is tremendously complex, so there are many ways in which it can go wrong. Genes are one of these ways. A gene goes 'wrong' and the result is havoc. Cancer is the classic example. A handful of chance mutations lead to that most terrible of human fates. These mutations only need to happen in one cell out of 15 million million. There is no 'reason', beyond such unsatisfying explanations as mischance, environmental toxicity or genetic susceptibility.

The spirit guiding the human genome project is the apotheosis of this view: genes go wrong and cause disease. Therefore, to cure the disease, find the gene and put it right. Today this might not be possible, but in the future we will no doubt perfect gene therapy. All we need to do is excise the faulty gene and replace it with a nice new one: replace the carburettor and the engine will work again. Many single-gene disorders, such as haemophilia or muscular dystrophy, are in principle amenable to this approach. In the case of haemophilia, the gene that codes for a blood-clotting protein, factor VIII, is mutated, so the protein is absent. The protein can be replaced by transfusion, or ultimately the gene can be fixed by gene therapy. There are many practical obstacles to overcome, but in conceptual terms the only subtlety is to ensure that the right amount of factor VIII is present at the right time.

The trouble is that single-gene disorders are rare. For the vast majority of diseases, especially the diseases of old age, a whole assortment of genes increase our susceptibility to disease. There is typically no genetic 'defect' as such. The word is too black and white — there are as many shades of grey between a working gene and a broken gene as there are between good and evil. Consider: a gene codes for a protein. If the sequence of the gene changes in the course of evolution, the structure of the protein changes. Sometimes the new protein may not work at all — in which case, if it is important, it will be eliminated by natural selection along with its bearer. Sometimes the change will have no effect on the function of the protein: it will simply be slightly different.[1] Then there

[1] The sequence of the same gene in different species may vary in almost every letter, without affecting function. The differences are due to 'evolutionary drift', in which mutations that do not affect the function of the protein are passed on and species drift apart over time (see Chapter 8). The evolutionary relationship is often betrayed by conservation of purpose,

may be several other versions that work to some degree or other. Given a particular set of environmental conditions, one of these may work best — but that is not to say that the others are 'broken'. Change the conditions and a different form may well work better. In the same way, a tractor is not really cut out for the city, but comes into its own in the countryside. If you move with your tractor from the countryside to the city, and cannot afford to buy a car, you may not be as well adapted as before, but you are still better off than if you had to walk. The tractor is not broken.

The different working versions of a gene are known as *polymorphic alleles*. It is hard to overstate their importance: they are the molecular units of variation and adaptation, the very essence of the individual. The genetic differences between people do not lie in different genes but in ever-so-slightly different versions of the same genes. On average, our DNA has between one and ten variant letters in every thousand, which are known as single-nucleotide polymorphisms, or SNPs (pronounced 'snips'). These are being catalogued exhaustively, although we have a long way to go: there are expected to be a million SNPs in the human genome. When they are shuffled and recombined in sex, these SNPs account for our endless genetic variety. For exactly the same reasons, they also influence our susceptibility to both diseases and treatments.

Some polymorphic genes — particular SNP configurations — may come to predominate within a population as a result of evolutionary selective pressures. Selective pressures can blur the distinction between a pathological process and an evolutionary trade-off. Our genes must make the best of a bad job. In previous chapters we have noted several examples of diseases that are not really pathological. Insulin-resistance in diabetes, for instance, is a genetic response to hard times, selected for over many generations. It is only pathological if a high-energy Western diet is superimposed over a 'thrifty' genotype. Similarly, sickle-cell anaemia and the thalassaemias protect against malaria through small changes in the structure of haemoglobin. These anaemias are maintained at a high frequency in areas where malaria is endemic because the carriers do not suffer from anaemia, but are protected against malaria. How many other human diseases are maintained in the gene pool because they offer a hidden benefit is anybody's guess.

---

shown by the three-dimensional structure of the protein and the preservation of particular amino acids near the active site. The proteins may work equally well despite these differences. I suppose it would be possible to find a million different working versions of the same gene in a million different species.

We are left with a curious situation, in which our genes are held responsible for disease, even though there is nothing actually wrong with them. They are simply variable. To treat a disease on the basis of genetic polymorphism is to say that all individuals are different and should be treated as such. This is very close to what leading figures in the pharmaceutical industry *are* actually saying. There is a revolution in healthcare, we are told by commentators as distinguished as Sir Richard Sykes, the ex-chairman of Glaxo Wellcome. We are misguided if we think there is such a thing as Alzheimer's disease: in reality it is a kaleidoscope of deceptive conditions, a hall of mirrors, caused by unique combinations of polymorphic genes. These combinations produce a spectrum of diseases that 'look' superficially similar — they look like Alzheimer's disease — but are in fact quite different, and may respond differently to treatment. This, we are told, is why we have had so little success in curing the disease: we dilute successful responses with less successful responses, in people whose genes were inappropriate for that particular treatment. We used to search for particular genes that predisposed us to disease, now we must consider whole genotypes. Treatments will become ever more specialized as we understand and begin to target individual genotypes. Blockbuster drugs will give way to genetic therapies tailored to individuals.

This is the rising field of *pharmacogenomics* and woe betide anyone who says it is misguided. It is, though. Particular genes, or even whole genotypes, may predispose us to the common diseases of old age, but in a wider sense this is irrelevant. Imagine you are crossing a road. You have a chance of being knocked over and killed. Your behaviour influences your chance of survival: if you step out into a busy road, without pausing to look, you have a far better chance of dying than if you wait patiently at a zebra crossing for the traffic to stop. We can whittle away at the statistics of deaths on the roads by introducing speed limits, sleeping policemen and better road markings, or by building bridges and subways, or by educating the public, or by clamping down on drink driving. If all these small changes were controlled by genes, then targeting each gene would have a small but incremental effect on the number of traffic accidents. However, we would only have a significant impact on mortality if we targeted all the 'genes' simultaneously; and even then we could be sure there would still be people killed. Ultimately, the only way to prevent traffic accidents altogether is to ban cars, impractical as this may be. Similarly, in the case of diseases, we can fiddle with predisposing genes, and change our risk profile slightly, but in the end the only way of preventing the diseases of old age is to *prevent old age*. Is this aim as ludicrous as banning cars, or can it be done?

With this question, we return to the link between ageing and age-related disease. We saw in the last chapter that there almost certainly *is* a process of ageing, which is independent of age-related disease: mitochondrial respiration undermines the integrity of cells and organs regardless of whether we suffer a disease or not. We saw that mitochondrial respiration may set an upper limit on our lifespan of perhaps 115 to 120 years; but what about the reverse case? If ageing is independent of age-related diseases, are these diseases necessarily independent of ageing? In other words, would we suffer from dementia or heart disease if we did not age? Is there something inherent about being old that increases our risk of disease? The idea sounds intuitively reasonable, but the implications are far-reaching. Banish ageing and we banish many diseases, regardless of whether we carry susceptibility genes or not.

If our risk of disease increases with our age, then the question we should ask is not *why* does a particular variant of a gene predispose us to Alzheimer's disease, but *why are its effects delayed until old age?* This question is rarely addressed in medicine, which must try to cure people who are already riddled with specific ailments, but has been answered by the evolutionary biologists. As we get older, our risk of accidental death accumulates, so there is less evolutionary pressure to maintain physiological function in an older person than in a younger person. Thus natural selection cannot eliminate a gene that causes Alzheimer's disease at 140, because none of us lives to that age. Selection pressure has fallen to zero. The consensus is that age-related diseases are caused by the detrimental late effects of genes that are maintained in the gene pool because their late effects are counterbalanced by beneficial effects earlier in life. There is a trade-off between early advantages and late disadvantages. This is the idea of antagonistic pleiotropy, which we met in Chapter 12. We parked the idea there, noting that it was not a good explanation of ageing (because it could not account for the swift and flexible changes in lifespan observed in nature) but that it was potentially a good explanation of age-related diseases.

A common view of antagonistic pleiotropy is that our genes are out of step with our lifestyle. We spent half a million years evolving as hunter-gatherers. Restless wandering was combined with an ability to subsist on a meagre diet for weeks or months at a time. Then, a few thousand years ago, we became farmers. Food was plentiful, but the staple diets were far less varied, and courted malnutrition. Rice, for instance, is a good source of carbohydrates and some proteins, but a poor source of other proteins and a number of vitamins. Health deteriorated. Skeletal remains show that the first farmers were less healthy than their hunter-gatherer forebears. Even so, the sheer quantity of food could support much larger populations. People lived together in towns and cities. Contagious diseases became rife. Entire cities were wiped out by plagues. For the next few thousand years, infections became the strongest selection pressure on the human genome. The genotypes of peoples living across whole continents were shaped by diseases such as malaria. The high incidence of sickle-cell anaemia in Africa and Asia is a direct result. Perhaps fewer people starved in the age of farming, but many died young from infections instead.

In the past few hundred years, all this has begun to change. Better hygiene, better nutrition and advances in medicine have created a brave new world, in which most of us can expect to live out our three score years and ten, and more. Two hundred years is just ten generations — presumably too short to adapt to our cushy new lives. We sit around and overeat. Our genes adapted to meagreness for half a million years, and infection for a few thousand, but are caught reeling by this new onslaught. We are genetically geared to extract as much as possible from an impoverished environment, and have been transplanted into the midst of riches. In our youth, we have no problem. As we age, the abuse catches up with us. The theory of antagonistic pleiotropy says this is too bad: selection pressure is low once we are past 40 or 50. Until conditions such as obesity begin to shape the reproductive population, there is next to no selective pressure for change. Thus, our genes condemn us to rot in a world of plenty. What a depressing scenario.

There is more than a grain of truth in this pessimistic view of disease, but also some problems with it. For a start, age-related diseases have always been with us, among the lucky few who survived to old age: they did not just appear in the past couple of centuries or even millennia. More important, they are also found in ageing animals — and not just in captive animals, which might be overfed, but also in wild animals shielded from predation. Old mice suffer from the same sort of ailments as old people. Their joints stiffen, their skin wrinkles, they lose their ability to remember and learn, their immune system degenerates, and they have a rising incidence of heart disease and cancer. If we take a single parameter, such as the number of cross-links between collagen fibres in the skin (which cause wrinkles) there is little difference between old mice and old men. In each respect, the way that we age is strikingly similar. The difference lies in the rate. Mice and rats pass through the sequence of age-related changes in four years, we take 70.

Similar patterns apply to other animals: the spectrum of age-related changes is analogous, but the rate of ageing is different. Tiny nematode worms live just a few weeks, yet still age in a way that we can recognize — they move and feed more slowly, they become infertile, their outer cuticle becomes wrinkled, and they accumulate the fluorescent age pigment lipofuscin, just as we do in our neurons and muscle cells. At the other extreme, many birds, some of which live for well over a hundred years, also suffer equivalent degenerative conditions to mammals, including stiffening joints, congestive heart failure, atherosclerosis, cataracts and a variety of cancers. The entire animal world cannot be out of step with its environment! There must be more to age-related disease than just a mismatch between genes and environment.

We do not have to be out of step with our environment, of course, to suffer from the effects of antagonistic pleiotropy. In Chapter 12, we noted that genetic conditions such as Huntington's disease are examples of pleiotropy in action: a barely measurable increase in fecundity in youth is enough to offset the most dreadful stripping away of faculties later in life. Diet is irrelevant: the effect is written in a single gene. If we carry the gene for Huntington's disease, we will get the disease whatever we eat. Something similar may be true of other diseases. Some variants of polymorphic genes, such as the *ApoE4* allele of the *ApoE* gene, increase our susceptibility to Alzheimer's disease.[2] A quarter of the population inherits a single copy of the *ApoE4* gene, increasing the risk of dementia fourfold. Two per cent of the population inherits a double dose, increasing the risk of dementia eightfold. For a gene to be this frequent in the population, we might suspect a hidden benefit earlier in life. What this putative benefit might be in the case of *ApoE4* is unknown. The point is that the extra risk of dementia is not enough to rid us of the *ApoE4* allele. One may well wonder how many other diseases of old age, almost all of which have

[2] There are three common *ApoE* alleles in the population — *ApoE2*, *ApoE3* and *ApoE4*. They code for different versions of a protein called apolipoprotein E, which helps deliver lipids and cholesterol to cells around the body. For this reason the *ApoE* genes also affect our risk of heart disease and stroke. How they are involved in Alzheimer's disease is a mystery, although apolipoprotein E is thought to assist neuronal repair in some way. The *ApoE4* product seems to exacerbate the deposition of amyloid, the main component of senile plaques found in the brains of people with Alzheimer's disease.

a genetic component, are similar to Alzheimer's disease in this respect.

But wait a moment. Earlier in this chapter I made a strong assertion: targeting susceptibility genes is not the way to cure Alzheimer's disease, or any other age-related disease. Instead, we must try to slow the whole ageing process. The secret to this lies in the theory of antagonistic pleiotropy. The idea of antagonistic pleiotropy sounds simple enough, but there is a quandary at its heart: *when* is a late effect? At what point in our lives do genes start to have a negative effect instead of a positive effect? Should we measure this 'time to negative effect' in years, or in some other kind of unit? If the units are years, then the effect of antagonistic pleiotropy is as defined as the fate of Oedipus. If we have two copies of the *ApoE4* gene, we shall succumb to dementia at the hour of our appointed fate, and have little more chance of stopping it than we do of stopping time. But if the effects are dependent on *age*, not on time, then the tragedy of Alzheimer's disease is contingent *on being old*, on having crossed an age threshold, rather than the time that elapsed before we reached the threshold. Like Hamlet, our fate is then a matter of historical contingency, of having crossed the threshold, not an Oedipal certainty.

In the case of Alzheimer's disease, an age threshold may account for the wide variation seen in the age of onset. *ApoE4* shifts the risk of Alzheimer's disease to a younger age, so that people with two *ApoE4* genes are more likely to succumb to Alzheimer's disease by the age of 65. Yet having two copies of *ApoE4* does not exacerbate the severity of dementia, or noticeably change its pathology, or speed up the clinical course. The disease is similar in every respect, except that it happens earlier. In this sense, *ApoE4* does not 'cause' the disease so much as shift a condition that would happen anyway into an earlier time frame. This implies that there is a threshold: the disease develops in the same way once the threshold has been crossed, regardless of which *ApoE* allele you have. The chronological age at which the threshold is crossed may vary between 60 and 140.[3] As Einstein said, time is relative; but in the case of ageing, relative to what?

We all know people who have aged well and others who have aged badly. There may be a discrepancy between our biological age and our chronological age. The average life expectancy of 75 years conceals a huge amount of variation. It is not uncommon for people in their 50s to die of

[3] Sir Richard Sykes, in answer to a question about risk factors for Alzheimer's disease, drew a laugh from his audience when he said that people at lowest risk of dementia — people with two *ApoE2* alleles — would still get dementia by the age of 140.

an age-related disease, such as a heart attack or cancer, nor is it uncommon these days to live until over 100. It is questionable whether age in years is as useful an indicator of life expectancy as biological age. There are numerous ways of thinking about biological age, but a reliable way of quantifying it is in terms of the *oxidative damage* accruing to individual cells and organs. People who reach the age of 100 in good health often have a similar accumulation of damage to their DNA, lipids and proteins as people in poor health at the age of 50.

To visualize the difference in simple terms, consider a population of cells exposed to radiation. Imagine that an average cell dies after it has taken 100 'hits'. If we now double the radiation intensity, the cells will accumulate 100 hits in half the time. They 'age' at twice the rate. Time is not an appropriate measurement of their age: the number of hits is far more relevant. In this instance, the number of hits reflects the biological age.

In this chapter, I will argue that biological age is central to our risk of disease. Our biological age equates to the number of 'hits' we have taken. This in turn depends on how we handle oxygen, or, more particularly, oxidative stress. In other words, old age is not a function of time, but a function of oxidative stress, which tends to rise over time. Thus, we ought to be able to prevent degenerative diseases if we can prevent oxidative stress. To find a cure for dementia, we should forget about the genes that increase our susceptibility to dementia, and look instead for genes — or other factors — that can protect us against oxidative stress. In so doing, we stand not only to prevent dementia, but at the same time to ward off other age-related diseases such as cancer and diabetes.

In an age of healthcare rationing, governments and pharmaceutical companies are spending billions of pounds a year on research and development, to create designer drugs tailored to individuals. We are in danger of becoming obsessed with details and dismissive of important platitudes: we are all getting older in a rather similar way. The challenge of slowing ageing need be no more intractable than that of curing dementia, and there are some good reasons to think it may be more tractable.

The idea that age-related diseases are linked with being old rather than the number of years lived — with crossing an age threshold rather than a length of time — is sustained by a wide range of observations. We have already noted that different species age at completely different rates, but still suffer from the same diseases. Similar but smaller variations take place within a single species. Radiation poisoning or smoking speeds up the rate of ageing, as well as our likelihood of suffering from age-related diseases such as cancer. Accelerated ageing syndromes, such as Werner's syndrome, are associated with an early deterioration, including cataracts, muscle atrophy, bone loss, diabetes, atherosclerosis and cancer. Those afflicted usually die of age-related diseases, such as heart disease and cancer, by their early 40s. Conversely, for most people, a healthy diet lowers the risk of many age-related diseases, including heart disease, cancer and dementia. So too does calorie restriction, which we have seen slows the decline, in rodents at least, in physical activity, behaviour, learning, immune response, enzyme activity, gene transcription, hormone action, protein synthesis and glucose tolerance. We also saw that increasing production of enzymes such as SOD (superoxide dismutase) and catalase slows ageing in *Drosophila*, improving the insects' activity in old age.

There are two critical points to take home from this. First, age-related diseases are connected with being old, regardless of how much time has elapsed. Second, factors that change the underlying rate of ageing affect *when* we suffer from disease. If ageing is slowed down, age-related diseases are postponed; if ageing is sped up, the diseases are upon us in middle age. In other words, the diseases are similar in all cases, but the time taken to succumb to them varies. Going a step further, it seems to be easier, from an evolutionary point of view, to change the underlying rate of ageing than it is to rid ourselves of age-related diseases altogether: animals have different lifespans but similar diseases. This view is the antithesis of modern medical research. The distinction was highlighted by Tom Kirkwood in his final BBC Reith lecture of 2001:

> The idea that science should aim to postpone disabling conditions like Alzheimer's disease, without necessarily extending life itself, has become something of a mantra. It is a mantra that goes by the name of 'compression of morbidity'. The aim is to squeeze the bad things that happen to us at the end of life into as short a period as possible. Another way of putting this is that we want to extend the health span, while leaving the life span as it is. On the whole, people seem to find this more reassuring than that scientists want to make us live longer. The trouble is, however, that compression of morbidity makes assumptions about the extent to which we can decouple ageing and disease.

If we accept that ageing and age-related diseases go hand in hand in nature, then we begin to see why it is proving so dishearteningly difficult to decouple them in medicine. It is like trying to separate the squeal from the pig, or the mind from the brain. Having said that, it is not enough just to denounce the mantra as wrong. If our attempt to extend the 'health span' while maintaining the lifespan is misguided, what might we offer instead that is practical and likely to succeed? After all, if nothing else, we know that some genes increase our susceptibility to age-related diseases. We have had some sort of success in postponing conditions like Alzheimer's disease, diabetes and cancer by a few years. If we drop this approach, what should we replace it with?

The answer is concealed in an ever-growing pile of data. Three large bodies of work provide the clues, and we have touched on all three in earlier chapters. The problem is that these three fields are largely independent. Few researchers are comfortable in crossing the boundaries of their own field. As medical research grows ever more specialized, it becomes at once less acceptable and more necessary to transgress the limits of personal expertise. Perhaps this is the most valuable role of the science writer: writers must transgress their own expertise as a matter of routine, and flights of fancy can at least be brought down to earth by experts. In my own case, everything I have to say is based on the work of others, but I cannot find a synthesis in the literature that straddles the three fields. I shall therefore run the risk of error or inadvertent plagiarism and put forward my own ideas.

The first clue comes from the mitochondrial theory of ageing: oxidative stress rises gradually through our lives, especially in the mitochondria, and this rise is the *causal* basis of ageing (if not of age-related diseases). There are two objections to this claim. First, the rise in oxidative stress is hard to measure in practice, and some researchers dispute that it happens at all. I hope I convinced you in Chapter 13 that mitochondria alone do cause a rise in oxidative stress, and that this is the causal basis of ageing. We have yet to establish whether mitochondria can also cause age-related diseases. The second objection is more treacherous: we have had little success in blocking the putative rise in oxidative stress using antioxidant supplements. This failure is often taken as evidence that there is no rise in oxidative stress, or that the rise is inconsequential. This is facile logic. It might simply be that dietary antioxidants are not up to the job — indeed *cannot* be up to the job. In the last pages of Chapter 10 I noted that dietary antioxidants are far from a panacea. Antioxidants may even be counterproductive in that they can *suppress* the powerful genetic response to oxidative stress that is produced by proteins such as haem oxygenase and metallothionein. Thus, clue 1 is double: oxidative stress rises as we get older, but we can only gain limited protection from dietary antioxidants.

Clue 2 actually derives from the failure of antioxidant supplements to extend life, and comes from the field of cellular signalling. Signals are as critical to the behaviour of cells as electronic communications are to modern society. Chemical signals control the expression of genes — switching them on or off — in the same way that information constrains our decisions as individuals in society. Whether cells divide or mature into neurons, die or become cancerous, secrete hormones or absorb salt, does not depend on their genes. All our cells have the same genes. The behaviour of a cell depends on *which* genes are active at any one time, and this depends on the signals it has received. The signals are converted into an appropriate response through the action of transcription factors, regulatory proteins that bind to DNA and direct the transcription of particular genes into proteins. We saw in Chapter 10 that the activity of several important transcription factors depends on their oxidation state. Many types of physiological stress, such as infection, radiation poisoning and inflammation, cause an *increase in oxidative stress*. Transcription factors such as NFκB and Nrf-2 become oxidized and migrate to the nucleus, where they bind to DNA and coordinate the transcription of 'stress' genes.[4] The products of these genes muster resistance to the threat. Thus, clue 2 stands opposed to clue 1: some sort of oxidative stress is a *necessary* signal for cells to marshal their genetic response to physiological stress. If we block oxidative stress, we may make ourselves more vulnerable to infection. Seen in this light, it is quite conceivable that we are 'refractory' to large doses of dietary antioxidants because they interfere with our response to stress.

Clue 3 comes from the evolutionary theory of antagonistic pleiotropy — the idea of a trade-off between the detrimental effects of genes in old age and their beneficial effects in youth. The theory was first proposed by George C. Williams in 1957, and has since been accepted by most evolutionary biologists, though it is not in common currency in medical

[4] The migration of NFκB to the nucleus requires its oxidation. Once in the nucleus, however, it must be restored to its unoxidized state before it can bind to DNA. Thus, the action of NFκB is carefully stage-managed, and it only exerts an effect if the cell retains control of the oxidation state of the nucleus. If both nucleus and cytoplasm are oxidized, it is more likely that the cell will fail to muster a response, and will die instead.

science. The role of pleiotropy in the diseases of old age was discussed by Williams and the physician Randolph Nesse in an enlightening book on the "new science of Darwinian Medicine", *Evolution and Healing*, first published in 1994. Much of this book is packed with vivid examples, but I found the section on pleiotropy disappointing. We are offered the examples of Alzheimer's disease and haemochromatosis (in which heavy iron absorption overcomes the risk of anaemia in youth, but courts misfortune in middle age; see Chapter 10), and one or two others. Tantalizingly, they cite Paul Turke, an evolutionary anthropologist and physician who argues that the whole immune system is age-biased, in that the noxious oxidants released by immune cells to kill invading microbes can also damage the body; but Turke, though quite right, did not close the circle. Such a theory cannot, in itself, explain why mice, which have elaborate immune systems of their own, should die of age-related diseases after four years while we take 70. In a more recent textbook on evolutionary medicine, based on a conference in Switzerland in 1997, the concluding discussion notes tersely that "At this time, few examples of tradeoffs have been established." Thus, clue 3 is that, despite its theoretical utility, few concrete examples of pleiotropic trade-offs have ever been described. Could something be missing from the equation?

I suggest that there is a trade-off between oxidative stress as a signalling pathway that musters our defences against infection, and oxidative stress as a cause of ageing. In effect, the diseases of old age are the price we pay for the way in which we are set up to handle infections and other forms of stress in our youth. In both cases, the shadowy agent pulling the strings is oxidative stress. The outcomes are diametrically opposed: resistance to disease in youth and vulnerability to disease in old age. The duplicitous role of oxidative stress is central to both, in what I shall call the *'double-agent'* theory of ageing and disease (Figure 12).

The problem is that oxidative stress is a *necessary* part of our response to infection: without it, we are unable to mount a genetic defence against pathogens. Oxidative stress activates transcription factors such as NFκB, which coordinate the broader genetic response promoting inflammation and stress resistance.[5] Unfortunately, the rise in oxidative stress during ageing also activates NFκB. The pressure to eliminate infection in youth is

[5] NFκB is not the only transcription factor that responds to oxidative stress; Nrf-2, AP-1, Rel-1, P53 and several others do too. However, in medical circles NFκB has become virtually synonymous with the stress response, so I shall focus on it here.

far higher than the pressure to eliminate inflammation in old age. We cannot get rid of NFκB (we would succumb to infection) and yet its effects in promoting inflammation change the whole balance of the body as we get older. It is this shift in balance, rather than the time that has elapsed, that is responsible for the negative pleiotropic effects of other genes.

Exactly *why* infections should produce a rise in oxidative stress is uncertain, but it does seem to be the case that they do. In many cases,

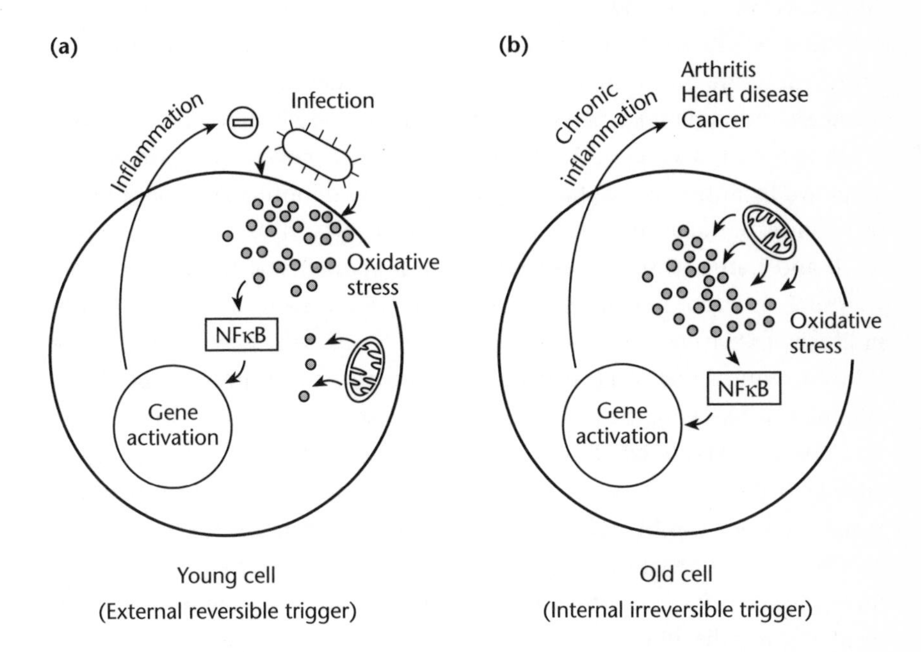

Figure 12: Schematic representation of the 'double-agent' theory of ageing. In the young cell **(a)**, infections (an external reversible trigger) cause a rise in oxidative stress within the cell. Oxidative stress activates NFκB, which migrates to the nucleus, where it orchestrates the transcription of 'retaliatory' genes. These genes code for stress proteins and proteins that mediate inflammation, such as tumour necrosis factor and nitric oxide synthase. Active inflammation resolves the infection. Once the trigger is removed, oxidative stress returns to normal. In the old cell **(b)**, an equivalent rise in oxidative stress is brought about by leaky mitochondria (an internal irreversible trigger), which also activate NFκB and the inflammatory response. In this case, it is impossible to resolve the trigger, so the inflammatory response becomes chronic. This contributes to the diseases of old age, and also attenuates the acute response to infections and other physical stresses. Because oxidative stress is pivotal to our recovery from infections in youth, and therefore affects our likelihood of surviving to have children, it is positively selected for by natural selection to our own detriment in old age.

oxidative stress is exacerbated by activated immune cells, such as neutrophils (which produce potent oxidants to kill the pathogen); but cell-culture experiments in the absence of immune cells suggest that the mechanism is more subtle and fundamental than this. This is a critical point: infections increase oxidative stress regardless of whether the immune system is involved. In the case of influenza, for example, Heike Pahl and Patrick Baeuerle, at the University of Freiburg in Germany, have shown that a single viral protein — haemagglutinin — induces oxidative stress in cultured cells.[6] Pahl and Baeuerle showed that this rise in oxidative stress activated NFκB, which in turn coordinated the genetic response of the organism to the infection. Conversely, if the oxidative stress was abolished using an antioxidant such as dithiothreitol, neither NFκB nor its subordinate genes were activated. A similar dependency on oxidative stress, and abolition by antioxidants, has been demonstrated in infections with many other viruses, including human immunodeficiency virus 1 (HIV-1), hepatitis B and herpes simplex, as well as by components of the bacterial cell wall such as endotoxin and lipopolysaccharide. In each case, the infection produces oxidative stress, which activates NFκB, which orchestrates the transcription of numerous other genes. The entire response can be short-circuited by blocking the rise in oxidative stress using antioxidants.

The response to NFκB is generally two-pronged: a bolstering of resistance to inflammation — a stress response — coupled with an inflammatory attack on the invading microbe.[7] The inflammatory attack can be severe: fever is part of the host's reaction to infection; it helps us clear the infection. Yet fever is obviously detrimental to our long-term health, insofar as it is disabling. We cannot expect to persist in this state for a prolonged period. Similarly, our response to endotoxin or malaria can be too violent — if we have a serious infection, the ferocity of our own immune response may push us over the brink into septic shock or cerebral malaria, which might well kill us. Some pathogens have learned to modulate the inflammatory response, or even take advantage of it. (For example, HIV has several defensive genes that are activated by NFκB, and it uses inflam-

[6] Pahl and Baeuerle suggest that the mechanism may depend on the sheer volume of viral protein accumulating in the cell's export pathway, the endoplasmic reticulum. Overload of the endoplasmic reticulum releases calcium, which in turn activates enzymes such as cyclo-oxygenase and lipoxygenase, which increase the production of oxygen free radicals.

[7] There are, in fact, at least seven related forms of NFκB, which are activated in slightly different ways, and which activate different selections of genes. One form, for example, increases stress-resistance but does not promote inflammation. When I talk about NFκB here I mean the most common, pro-inflammatory form.

mation as a signal for proliferation.) In general, though, inflammation is a positive force that has been selected for by evolution because it helps to *resolve* infections. Once the infection has been resolved, the inflammatory attack dissipates and we regain health. In other words, once the pathogen has gone, oxidative stress falls and NFκB is switched off. As a result, the genes controlling the stress and inflammatory responses are switched off. Normal housekeeping genes are switched back on. The body reverts to its 'peace-time' routine. The whole process is reversible.

Now think what happens during ageing. Our mitochondria leak free radicals, bringing about an insidious rise in oxidative stress as we get older. There comes a point when the oxidative stress is severe enough to activate transcription factors like NFκB. We begin to go through a low-grade stress response and inflammation. Virtually all diseases of old age are characterized by a chronic activation of stress proteins and persistent inflammation. Because broken mitochondria cannot be repaired, the situation is self-perpetuating — or worse, self-exacerbating. Inflammation damages cells and bodily structures, and so gives the jittery immune system a 'real' target. Proteins that are normally hidden inside cells, or behind barriers such as the blood–brain barrier, are exposed to immune surveillance and attack. Any concomitant conditions are exacerbated. We might be able to dampen this attack by using anti-inflammatory drugs, but, unlike infection, we cannot remove the primary cause: we cannot mend broken mitochondria. Nor can dietary antioxidants prevent the mitochondrial leak, so they too cannot counter the oxidation of the cell. On the contrary, as we have seen, antioxidants may even weaken the stress response — which is, after all, a response to genuine physiological stress.

No gene is an island, any more than a man is an island. If a gene becomes more or less active, the effects are felt by other genes. The activity of all genes depends on their immediate environment, in other words, the chemical balance inside cells. Oxidative stress shifts the spectrum of active genes, regardless of the exact cause of the stress. The rise in oxidative stress over a lifetime means that many genes that are active when we are 20 are less active when we are 70, and vice versa. Other genes keep working throughout our lives, but their effects shift because their environment changes. In the same way, a song accompanied by solo violin in an intimate setting differs from the same song chanted in a football stadium with a backing rock band. As we age, a pleiotropic gene exerts its negative effects because its environment has become oxidized and pro-inflammatory, not because a particular period of time has elapsed. If we wish to overcome the negative effects of pleiotropy, we must therefore prevent the oxidation of cells and tissues with age.

Before we consider some particular examples, we should ask if there is any empirical evidence for concerted changes in gene expression with age. Do cells and tissues really become more oxidized? If so, does this really change the pattern of genes that are switched on? If rhesus monkeys are anything to go by (which share 95–98 per cent of their genes with humans), the answer is almost certainly *yes*. In Chapter 13, we discussed a study of rhesus monkeys, carried out by Richard Weindruch and his team at the Wisconsin Regional Primate Center, Madison, and published in 2001. In one arm of the study, the team compared the activity of 7000 genes in young animals (aged 8 years) with that of the same genes in ageing animals (aged 26 years; the maximum lifespan of rhesus monkeys is 40 years). The similarities and differences were striking. Of the 7000 genes, about 6 per cent changed their activity by twofold or more in the course of 18 years, some becoming more active (300 genes) and others less active (149 genes).[8] Many of the genes that became *more* active with age were concerned with inflammation and oxidative stress (including a rise in the activity of the NFκB gene). The genes that became *less* active with age were concerned mostly with mitochondrial respiration and cell growth.

Weindruch and his colleagues argued that the shift in gene expression was *caused* by a rise in oxidative stress from damaged mitochondria. Although the changes measured were correlative — so a causal relationship was not proved — the notion that mitochondria were damaged was supported by the declining activity of respiratory genes (presumably they are not transcribed if they are not needed) and by high levels of oxidative damage to mitochondrial DNA, proteins and lipids. The level of oxidative damage correlated with the activity of pro-inflammatory and stress genes. Again, a causal relationship was not proved; but the most reasonable conclusion is that failing mitochondria brought about a rise in oxidative stress, which altered the spectrum of genes switched on. If so, then oxidative stress does indeed shift the genetic balance in old age towards stress-resistance and inflammation, as predicted.

Thus, our argument — the double-agent theory — is as follows. Infectious diseases cause a rise in oxidative stress, which is largely *responsible*

[8] Lest these numbers seem small, we should remember that a change in activity of this magnitude in a single gene, such as that for haem oxygenase, can mean the difference between life and death while fighting off an infectious disease. Changes of this magnitude in 450 genes, therefore, are presumably serious.

for coordinating our genetic response to the infection. As we age, mitochondrial respiration also causes a rise in oxidative stress, which activates essentially the same genes through a common mechanism that involves transcription factors like NFκB. Unlike infections, however, ageing is not easily reversed: mitochondrial damage accumulates continuously. The stress response and inflammation therefore persist, and this creates a harsh environment for the expression of 'normal' genes. The expression of normal genes in an oxidized environment is the basis of their negative pleiotropic effects in old age (Figure 12).

We are faced with two components to oxidative stress in the cell: mitochondrial leakage and non-mitochondrial factors such as infection. Let us assume that the overall degree of oxidative stress is additive. If the stress from old mitochondria is added to the stress from infection, the combination might easily be overwhelming — perhaps this helps to explain why old people are more prone to die from infectious diseases such as influenza and pneumonia. Earlier in life, the stress from mitochondrial leakage is less serious, but other factors may 'top up' the overall stress to a higher level. In the case of infections, the rise in oxidative stress is normally constrained and reversible, but other factors, such as smoking or high blood glucose, may be more persistent. Any factor that brings about a general rise in oxidative stress should exert similar effects through the same genes. If oxidative stress is persistent, it might have the effect of simulating old age, at least in some tissues or organs. Factors that cause a 'premature' rise in oxidative stress would thus be expected to cause premature ageing, and a greater risk of other diseases.

In the rest of this chapter we will look at how this works out in practice, using the example of Alzheimer's disease. Dementia illustrates many of the points we have discussed, such as misleadingly complex genetics. There are several well-established genes that increase the risk of Alzheimer's disease, which seem to have no association with oxidative stress. This disease is therefore a good example of how susceptibility genes that seem unrelated to oxygen are in fact influenced by mitochondria, oxidative stress, and inflammation.

Why do some people with no genetic susceptibility still get Alzheimer's disease? This question should be as revealing as the opposing question — why do some genetic mutations increase the risk of dementia? After all, more than half of the people who develop Alzheimer's disease have no known genetic risk factors. Following the simplistic nature–nurture interpretation, if the reason is not genetic, it must be environmental. For 30 years, researchers have looked in vain for evidence that aluminium or mercury provoke Alzheimer's disease, but unequivocal links have never been proved; if there are links, they must be very weak. What about a virus? Many people are infected with the herpes simplex virus (the virus that causes cold sores), which typically localizes in the brain to the same regions that degenerate in Alzheimer's disease. There is a link, as we shall see; but still, only half the people who are infected with herpes simplex get dementia, so this cannot be the only answer. In recent years, most serious research interest has centred on genetic factors, if only because the clues seem to lead somewhere. The fact that the clues do *not* point to most people who get Alzheimer's disease is held in abeyance until our understanding improves. Surely they must be linked in some way!

The pathology of Alzheimer's disease is characterized by two striking features, which were originally described by Alois Alzheimer himself in 1906: tangles and plaques. The tangles consist of twisted fibrils of a protein called *tau*. These fibrils are remnants of the extensive network of tubules that normally maintains the structure and function of neurons. As the tangles form, the neurons die off around them, finally exposing the tangled fibrils like bones in an excavated graveyard. In contrast, the plaques are formed outside the neurons. They consist of dense deposits of a protein fragment called *amyloid*, mixed up with inflammatory cells (some types of glial cell and invading white blood cells) and detritus. In the venerable tradition of scientific controversy, the plaques and tangles have each attracted a dedicated tribe of researchers, convinced that their preferred pathology is the primary cause of dementia; and ne'er the twain shall meet. Most researchers, though, are open-minded enough to admit that there are problems relating the two features together. The premise that one of them must come first, and somehow produce the other, is difficult to verify. Biopsies of brain tissue can only be taken at autopsy, so most pathological data relate to the late stages of the disease. Animal models offer a potential escape from this problem, but so far no animal model of Alzheimer's disease parallels the human syndrome in all its details.

The theory that amyloid toxicity is central to Alzheimer's disease gained ground in the mid 1990s as virtually all the mutations that make us prone to the disease promote the deposition of amyloid. Amyloid is a fragment of a larger protein called the *amyloid precursor protein* (APP), which straddles the external membrane of neurons. In healthy brains, APP is cleaved to produce the soluble fragment, amyloid, which then circulates in the cerebrospinal fluid. What it does there is uncertain, but some evidence suggests that it is necessary for normal neuronal function. Amyloid only becomes toxic when precipitated into dense clumps. The first clue to this came from people who had inherited very rare mutations in the *APP* gene, which cause dementia in early middle age (familial Alzheimer's disease). The mutations shift the cleavage point of the amyloid precursor protein, so that the amyloid fragments are longer and 'stickier'. The sticky fragments clump together to form plaques more easily. In 1995, two further genes were discovered, *presenilin 1* and *presenilin 2*, which also cause dementia in middle age when mutated. The presenilin genes code for proteins that are thought to help in processing APP. Again, people with these mutations generate long, sticky amyloid fragments, which precipitate readily. Apolipoprotein E4, the protein product of the *ApoE4* allele, also exacerbates amyloid deposition, but exactly how is not clear — we will return to this. Regardless of the exact mechanisms involved, all established genetic factors point to amyloid deposition as the primary pathology in Alzheimer's disease.

There are two problems with this interpretation of the disease. First, tangles are often formed *before* the amyloid plaques, and indeed some people with classic symptoms of Alzheimer's disease never develop amyloid plaques. In general, the onset of dementia corresponds to the loss of neurons, rather than the quantity of amyloid in the brain. Second, transgenic mice with mutations in the *APP* or the *presenilin* genes produce amyloid plaques, but only form tangles in old animals. Once the tangles have formed, the senile mice begin to lose neurons, and show signs of dementia, insofar as that can be measured in mice. Something similar is true of rhesus monkeys injected with amyloid — they develop tangles and lose neurons *only* in old age. Amyloid alone, it seems, is insufficient to produce Alzheimer's disease, despite the fact that all known genetic mutations point to it as the prime suspect. Perhaps this should not be entirely surprising. After all, even people with *APP* or *presenilin* mutations do not develop Alzheimer's disease until middle age — well past their childhood, and therefore unlike other single-gene disorders such as haemophilia. Something seems to be missing from the equation: might this be the same something that produces dementia in people with no genetic susceptibility?

The secret of plaques and tangles is not to be found in genetics, but in chemistry. Let us think about amyloid first. People with no genetic susceptibility to amyloid deposition still deposit amyloid in their brains. This is because the deposition of normal amyloid (and perhaps even the sticky variety) depends on its oxidation: it clumps together when oxidized. In the dense amyloid plaques, amyloid is invariably oxidized. Amyloid is more likely to become oxidized later in life, as oxidative stress in the brain rises. As we have seen, oxidative stress rises in everyone in the end, regardless of their genetic make-up, because mitochondrial respiration inevitably damages neurons. But does oxidative stress rise first and then *cause* amyloid deposition? We do not know for sure, but the question can be recast more practically: if oxidative stress occurs earlier in life for some reason, does it bring about a correspondingly early deposition of amyloid and onset of Alzheimer's disease?

One clue to the relationship between oxidative stress and dementia is provided by people with Down syndrome, who often get Alzheimer's disease in early middle age. Again, the disease almost seems to have 'moved forward' to an earlier time slot. We saw in Chapter 10 that people with Down syndrome suffer from oxidative stress, as a result of an imbalance in antioxidant enzymes.[9] Could a rise in oxidative stress underlie the early onset of dementia in people with Down syndrome? Quite probably, according to pathologists George Perry and Mark Smith, and their team at Case Western Reserve University, Cleveland, in a study published in 2000. This team measured the oxidation of proteins and DNA in people with Down syndrome, and found that a marked rise in oxidative stress always preceded the deposition of amyloid. Oxidized proteins and DNA began to build up in their late teens and 20s, with amyloid deposition occurring by their 30s. Thus, it seems that a rise in oxidative stress *does* foreshadow an increased risk of Alzheimer's disease, regardless of age.

What about the tau protein, the chief component of the tangles, the other pathological trait of Alzheimer's disease? In a 1995 study, Olaf Schweers and his colleagues at the Max Planck Unit for Structural Molecular Biology in Hamburg showed that the tau protein *only* coagulates when

[9] People with Down syndrome inherit an extra copy of chromosome 21, which includes the gene for SOD among others. SOD eliminates superoxide radicals, but in doing this produces hydrogen peroxide. Unless the hydrogen peroxide is eliminated by catalase, the extra SOD *increases* oxidative stress. Other genes on chromosome 21 include the *APP* gene, so people with Down syndrome also overexpress the amyloid precursor protein and presumably amyloid; however, the amyloid fragment is normal, not the long, sticky version.

oxidized. Conversely, if oxidation is prevented by antioxidants, tau will not coagulate.[10] In other words, tangles, like plaques, usually form under conditions of oxidative stress. This is presumably why tangles do not form in transgenic mice or rhesus monkeys, despite extensive amyloid deposition: the pathology must 'wait' for a more general rise in oxidative stress.

Thus, for most people with Alzheimer's disease, oxidative stress is the earliest pathological change, and is responsible for producing both main features, the tangles and the plaques. This idea is borne out by the effects of apolipoprotein E. Recall that the *ApoE* gene is polymorphic, which means that there are several different versions of the same gene. None of these versions is a mutant, in the sense that none is broken: all have been maintained by evolution, so all must have positive benefits. One of these benefits seems to be some degree of antioxidant activity. In old age, though, we have seen that *ApoE4* raises our susceptibility to Alzheimer's disease. The phrase 'raises susceptibility' is misleading. If *ApoE4* is beneficial earlier in life, then it makes more sense to see it as beneficial later on too — but perhaps less so than its cousins. In other words, rather than raising risk, it might be less effective at suppressing risk. Why this should be is not known, but the *ApoE4* protein *is* known to be more sensitive to free-radical attack than its cousins: it is therefore plausible that the *ApoE4* protein is 'lost' preferentially with age. Early in life, when selection pressure is high, this difference is not felt because oxidative stress is low. Later in life, as oxidative stress rises, the benefits of *ApoE4* (whether antioxidant effects or more general effects on cholesterol transport) are gradually lost, as more and more *ApoE4* proteins are disabled by free radicals.

If this reading is correct, then the loss of *ApoE4* proteins with rising oxidative stress accounts for two findings that otherwise resist interpretation. First, we noted that infection with herpes simplex virus increases the risk of Alzheimer's disease. This effect is marked in people with two *ApoE4* genes, but barely noticeable in people with *ApoE3* or *ApoE2*. Activation of herpes simplex in the brain produces oxidative stress and inflammation. As the *ApoE4* protein is sensitive to raised oxidative stress, its beneficial

[10]The most obvious alteration in tau is its reaction with phosphate: it becomes abnormally highly phosphorylated in Alzheimer's disease. Researchers had tacitly assumed that phosphorylation was necessary for the tau fibrils to coagulate into tangles. Schweers and his colleagues contradicted this view: phosphorylation is *not* necessary for tau deposition, and tau fibrils can form even in the complete absence of phosphate. More recent studies have confirmed their findings.

effects are lost preferentially in people who have both *ApoE4* and herpes simplex. In other words, people with *ApoE4* are sensitive to oxidative stress anyway, and if they happen to harbour the herpes simplex virus, they are more likely to develop oxidative stress. They are therefore more likely to succumb to Alzheimer's disease.

Second, and on the brighter side, people with two *ApoE4* genes gain most from antioxidant therapies. This is demonstrably the case in all conditions for which *ApoE4* is a risk factor, including dementia, heart disease and stroke. In one sense, this is a puzzle: as we have seen, antioxidants have little effect on mitochondrial respiration, and by suppressing the genetic stress response might even exacerbate oxidative stress inside cells. However, they *can* help protect the *ApoE4* protein from free-radical attack outside cells, as the external fluids are at once more accessible to antioxidants and less tightly controlled by gene activity than the insides of cells. Antioxidants may therefore shield the *ApoE4* protein from oxidation, or supplement its own failing antioxidant actions. This in turn postpones the onset, or slows the progression, of dementia. One antioxidant proved to delay the onset of Alzheimer's disease is vitamin E. If you know you have two *ApoE4* genes ask your doctor about taking vitamin E supplements. If you have other *ApoE* genes, you may not gain much from taking extra vitamin E; but neither will you lose much (as long as you don't overdo it).

Once formed, the tangles and plaques exacerbate oxidative stress both directly and indirectly. Direct amyloid toxicity is dependent on the binding of metal ions, such as iron and copper, which can catalyse the formation of free radicals. Such metals undoubtedly bind to amyloid plaques in the brains of people with Alzheimer's disease. If amyloid is added to cells grown in culture, its toxicity depends on free-radical formation in this way. Conversely, amyloid toxicity is abolished, in the same simple system, by free-radical scavengers or metal chelators (which block the action of iron or copper). Thus, amyloid plaques are *formed* through the action of free radicals on amyloid, and then *exert* their toxic effects by producing more free radicals. They are free-radical amplifiers. In the brain, it seems likely that amyloid damages the neurons surrounding the plaques in this way, though it is unlikely to injure more distant neurons.

The indirect toxicity of both tangles and plaques almost certainly results from inflammation. Plaques and tangles are recognized as alien by the brain's resident inflammatory cells, the microglial cells, which attempt to engulf the 'invaders' and attack them with chemicals, including free radicals. The plaques, especially, are indigestible and fester away. The jumpy microglial cells pump out inflammatory messengers that recruit and activate immune cells from elsewhere in the brain, as well as the blood stream. The entire brain is put on perpetual red alert. The chemical balance of the brain shifts inexorably towards oxidative stress, and susceptible neurons begin to die off. The most vulnerable neurones often die of 'excito-toxicity', in which they are provoked into a frenzy of electrical firing, and finally sink exhausted into an early grave. Thus, brain inflammation promotes the formation of more and more tangles and plaques, and finally the loss of neurons on a huge scale. By the time that Alzheimer's disease can be diagnosed by standard clinical criteria, a quarter of the brain's neurons — 25 billion of them — are dead. This vicious circle of inflammation is so important that the Canadian researchers Patrick and Edith McGeer have described Alzheimer's disease as arthritis of the brain.

Inflammation again! We may be sure that NFκB, the inflammatory transcription factor, will be involved, and indeed it is. NFκB is a source of anguish among researchers. The chemistry of life is so appallingly complex that researchers are quick to regress into a childlike sense of good and evil, pitting the 'good guys' against the 'bad guys'. This naive view of molecules is willingly embraced by the pharmaceutical industry, which strives to target the 'bad guys' — a molecule that is good and bad by turns is a dreadfully shifting target for a drug. This is unfortunate. As we saw in Chapter 9, even icons of goodness such as vitamin C can exert an unpredictable mixture of 'good' and 'bad' effects. Applying a simplistic moral order — are you with us or against us — invariably results in a 'paradox', a massively overused word in academic journals. A quick search for the word 'paradox' on Medline (the main database of medical abstracts) yields nearly 4000 articles in which it is a key term: 'Another Calcium Paradox'; 'The Paradox of Antioxidants and Cancer' or 'Beta-Carotene: Friend or Foe?' One might think that biochemists were fond of solving paradoxes, but few are solved in these articles. Instead, the weight of evidence for and against is stacked into heaps and left for posterity to judge. For all the detailed scientific analysis of data, confusion radiates from many articles. My own favourite is the title 'Does Growth Hormone Prevent or Accelerate Ageing?' It is hard to avoid the impression that, for all the astonishing advances in medicine, some fundamental questions remain unanswered even in outline.

The trouble stems from the traditional approach of medical research, which takes a snapshot in time — sometimes a fraction of a second, before quenching a biochemical reaction — and then tries to piece together the frozen relationship between molecules. This approach is analogous to forensics, which scrutinizes clues scientifically to solve a murder mystery, but omits to consider motive. For all the science in the world, a proper understanding will only come from motive. Motive is often rooted in historical accident, such as a humiliating experience years before. The same is true of the way in which our bodies are built, right down to the molecular level. Our bodies are historical accidents of evolution and ultimately can only be understood from an evolutionary perspective: how things got to be the way they are. From this point of view, a good guys–bad guys philosophy is a woefully inadequate way of thinking about molecules as complex as NFκB. Even so, this is the norm. NFκB is usually portrayed as Janus-faced, capable of abrupt swings from the good to the bad and the ugly. Sometimes it destroys neurons, sometimes it protects them. It is important, but profoundly unreliable as a drug target.

Seen in the light of infections, though, the behaviour of NFκB is consistency itself. Activation of NFκB in Alzheimer's disease, as in an infection, has two complementary effects: it fans the fires of inflammation, and at the same time shields our healthy cells from the same flames. In infections, the rationale is obvious: the immune system attacks the invader with free radicals that might just as easily harm our own cells. To prevent damage to our own cells, their genetic resistance to oxidative stress is tuned up. A shock that might normally kill our body's cells, or make them commit suicide (apoptosis) is now weathered out until the storm is over. Some cells are stimulated to divide, to repopulate tissues that were less prepared for the storm, and which suffered accordingly.

The difference in Alzheimer's disease is that the storm is never-ending, albeit less violent. The inflammatory glial cells in the brain are incited by chemical messengers to attack the plaques and tangles, but healthy neurons are threatened with collateral damage. They respond by stepping up their own resistance to the onslaught. Powerful protectors, such as haem oxygenase and SOD, are produced like so many sandbags (see Chapter 10) to bulwark the neurons. Yet even such powerful protectors cannot help indefinitely. They impose a curfew on the cell and, like real curfews during a continuous bombardment, there is a limit to how long cells can bear the strain. This is not a paradox: it is a clockwork response to oxidative stress, which is strained by its own duration.

If NFκB is 'switched off', healthy neurons are *more* vulnerable to damage — of course, they no longer have their sandbags to protect them — but then the inflammation may at least come to an end, so there is less need of protection. The balance is delicate and unpredictable. A practical solution is to block the activation of NFκB only in inflammatory cells. To some extent this can be achieved using aspirin or non-steroidal anti-inflammatory drugs (NSAIDs).[11] Several studies have shown that people prescribed aspirin or NSAIDs over a number of years to control rheumatic pain have less than *half* the risk of dementia, compared with their contemporaries. Conversely, people who have other sources of brain inflammation, such as a stroke, traumatic brain injury and viral infections, have several times the risk of dementia. I imagine that such vulnerable groups would benefit most from aspirin or antioxidants, though I am not aware of any systematic studies to prove the point.

Before drawing this chapter to a close with some parallels in other age-related diseases, what have we learnt from Alzheimer's disease? First, the known genetic mutations affect a small fraction of people with Alzheimer's disease and their effects are delayed until middle age. This delay implies that, as in mice and monkeys, oxidative stress must cross a threshold before neurons die *en masse* and dementia can be diagnosed clinically. Second, all other known risk factors for Alzheimer's disease, including Down syndrome, *ApoE4* and herpes simplex infection, are associated with a rise in oxidative stress. Third, oxidative stress alone is sufficient to cause dementia in old age in people with no known risk factors (about half the people who succumb to dementia in old age). Fourth, factors that lower oxidative stress, such as aspirin and vitamin E, can postpone the onset of dementia by a few years, if not indefinitely.

[11]High or continuous doses of aspirin and NSAIDs can cause gastrointestinal bleeding and ulcers, and thousands of people are hospitalized for side-effects each year (although about 97 per cent of people can tolerate moderate doses of aspirin without problems, the other 3 per cent totals many thousands). New, more specific versions of aspirin, known as COX-2 inhibitors, are now on the market and hold the prospect of similar potency with fewer side-effects. At low doses, however, COX-2 inhibitors and aspirin inhibit the enzyme cyclo-oxygenase, and so have only a limited effect on other proteins whose production is controlled by NFκB, such as tumour necrosis factor or nitric oxide synthase. At high doses, aspirin (but not the COX-2 inhibitors) inhibits NFκB, which might account for some of its previously unexplained effects — a startling finding published in *Science* in 1994 by Elizabeth Kopp and Sankar Ghosh at Yale University. Glucocorticoids suppress the activity of NFκB even more strongly, accounting for their potent immunosuppressant effects. When taken in medicinal doses, however, glucocorticoids have many unpleasant side-effects, including weight gain, chronic infections, bone loss and glandular atrophy.

In conclusion, Alzheimer's disease is linked with age *because age is a function of oxidative stress*. Factors that exacerbate oxidative stress early in life accelerate the onset of dementia, while factors that alleviate oxidative stress postpone dementia. However, unless we abolish oxidative stress we can never get rid of Alzheimer's disease. The difficulty is that we cannot abolish oxidative stress, because it is necessary to coordinate our resistance to infections and other physical stresses; but we can probably modulate it with a little more subtlety. At the beginning of this section on Alzheimer's disease, I posed a question: why do people without known risk factors still get dementia? I believe we have answered the question. They still get oxidative stress. Another permutation of the same question may throw a more practical light on the prevention of dementia. Why do some people *with* known genetic risk factors (such as *ApoE4*) *not* get Alzheimer's disease? What do they have that is protecting them? These are questions we will touch on in the final chapter.

I have argued that ageing and age-related diseases are degenerative conditions brought about by the combination of mitochondrial leakage, oxidative stress and chronic inflammation. Some genes, infections and environmental factors exacerbate oxidative stress at an earlier age, and this speeds up the ageing process, in some organs at least. We have seen that amyloid deposition, apolipoprotein E4, Down syndrome and reactivation of latent viral infections all exacerbate oxidative stress. So too do smoking, high blood glucose and various environmental toxins.

Nicotine is blamed for many things but, although addictive, it is not responsible for the deadly diseases caused by smoking. This is why nicotine gums and patches offer a safe way of quitting smoking. Cigarette *smoke* is dangerous because it is the most dastardly free-radical generator known (I enjoy smoking, though I intend to quit when I finish this book). Many chemicals in cigarette smoke, including semiquinones, polyphenols and carbonyl sulphide, react with oxygen to form superoxide radicals, hydroxyl radicals and hydrogen peroxide, as well as nitric oxide and peroxynitrite. A single puff of cigarette smoke is said to contain 10^{15} (a million billion) free radicals. The mind boggles. If this were not enough, cigarette smoke activates inflammatory cells, which add their own toxins to the brew. The result is oxidative stress, especially in the lungs and the walls of blood vessels. Cellular glutathione levels are suppressed (quitting smoking raises blood glutathione levels by 20 per cent in three weeks) and this activates transcription factors like NFκB. Smokers 'turn over' antioxidants such as vitamin C much faster, and so should take more dietary antioxidants to counter the threat. Most do not. Smoking thus provokes inflammation, and this is the chief reason for the high risk of both heart disease and cancer.

Too much glucose is another modern killer, for surprisingly similar reasons. Poor control of blood glucose levels is the hallmark of diabetes (see Chapter 12), and in diabetics glucose may reach very high levels after meals. Glucose reacts in a complex manner with proteins to form brownish caramels that accumulate with age, known as *advanced glycation end-products*, or AGEs (this is also why meat browns when it is cooked). Such caramels account for the clouding of the lens of the eye in a cataract. Caramelization of proteins is accelerated by oxygen, and most AGEs are really oxidation products. Not surprisingly, caramelization blocks the function of proteins, but worse follows. AGEs, like amyloid, are free-radical amplifiers: they are mostly formed by free radicals and then exert their toxic effects by producing more free radicals, causing oxidative stress and so inflammation. Because glucose is delivered to cells via the blood stream, the vessel walls are worst affected. In diabetes, small blood vessels in the eyes, kidneys and limbs become damaged and blocked, causing blindness and kidney failure, and all too often necessitating amputations. This whole process is speeded up in diabetes, but it happens at a slower speed in everyone, as AGEs accumulate with age as a result of mitochondrial leakage in all tissues. For this reason, diabetes is often referred to as a form of accelerated ageing; it is really an accelerated form of oxidative stress.

Inflammation of blood vessel walls induces cellular proliferation, oxidation and deposition of cholesterol, and the development of atherosclerosis. These are ideal culture conditions for some tough bacteria, such as *Chlamydia pneumoniae,* which infect damaged arterial walls, and from their safe haven antagonize immune cells. Up to 80 per cent of people with cardiovascular disease are infected with *Chlamydia,* but whether this is a cause or an effect of heart disease is still disputed. It is simplest to say it could be both: the *cause* of atherosclerosis is oxidative stress, and the process can be started, or perpetuated, by smoking, AGEs, oxidized cholesterol, *ApoE4,* infections, or just old age. Any one of these factors makes the others more likely. All are united by oxidative stress, and converted into the common currency of inflammation by NFκB and its kin.

Because many of these factors are 'external' (not produced within our own cells by mitochondrial leakage) they are more responsive to antioxidant therapies than is mitochondrial ageing, which cannot be easily reversed. This is why a healthy diet, or possibly antioxidant supplements, can postpone the onset or progression of heart disease, but do not ultimately prevent ageing.

Cancer is also provoked by oxidative stress and inflammation. We have seen that the effects of radiation are mediated by the production of oxygen free radicals from water (see Chapter 6), which then attack DNA, proteins and lipids. Because oxygen itself produces the same radicals, it is in fact a carcinogen (or more technically, a pro-carcinogen). The more air we breathe, the more likely we are to get cancer, hence the strong association between cancer and age. Many carcinogens, including benzene, quinones, imines and metals, also act by generating free radicals. The importance of oxygen radicals in this process is borne out by tell-tale chemical signatures of hydroxyl radical attack on DNA, such as 8-hydroxydeoxyguanosine (8-OHdG; see Chapter 6, page 124), which are excreted in the urine. Smoking can increase the excretion of 8-OHdG by 35–50 per cent. For the most part, the oxidized fragments are excised from the DNA and replaced with new, correct, letters; but mismatches do occur (for example, guanine (G) is replaced with thymine (T)). When cells divide, the mistakes are passed on in the genetic code as mutations. Both oxidative damage and mutations accumulate with age. By the time a rat is old (two years) it has about a million lesions in DNA per cell, twice as many as a young rat.

The frequent mutations in cancer cells are usually considered to be the cause of cancer, but in fact it is not known whether such mutations occur first, and then stimulate cancer cells to proliferate, or whether the mutations accumulate in cancer cells that are already proliferating. Certainly, tumours 'evolve' and accumulate more mutations over time. There is good evidence to suggest that oxidative stress and inflammation create an environment conducive to cell division in the first place. Apart from the oxidative stress associated with irradiation, smoking, carcinogens and ageing, a third of cancers worldwide (notably in the developing world) are caused by chronic infections, such as hepatitis B and C and schistosomiasis. Not surprisingly, given the ubiquity of oxidative stress, NFκB is involved. High levels of activated NFκB are found in most cancers and may be necessary for the transformation of a normal cell into a cancer cell. Why should this be? Again, the normal response to infection provides the clue. First, NFκB strengthens the cell's resistance to oxidative stress: it makes cancer cells tough. Second, NFκB stimulates cell proliferation. In an infection, the rationale is to replace damaged tissue with new cells, but in cancer, activation of NFκB simply makes proliferation more likely. Thus, perpetual activation of NFκB toughens cancer cells and stimulates their proliferation. Apart from anything else, this makes tumours more resistant to treatments such as chemotherapy and radiotherapy. Switching off NFκB, if possible, makes tumours more sensitive to treatment, and is a promising line of cancer drug development.[12]

There are hundreds of diseases in which oxidative stress is known to play a role. I hope these few examples are enough to establish my general point, the 'double-agent' theory of ageing: oxidative stress rises with age and activates the genes responsible for fighting off infections by way of transcription factors such as NFκB. These genes were never intended to be switched on for months or years at a time: their purpose is to improve our chance of surviving infection in youth, so that we may recover, go forth and multiply. In terms of selective pressure, or pleiotropic trade-offs, the importance of this task outweighs the personal misfortune of ageing and age-related disease. Even so, the message of this chapter is positive. We are not the victims of a thousand random genetic muggers, intent on 'doing us in'. Quite the contrary. The behaviour of our genes depends on oxygen and oxidative stress. When we learn how to modulate oxidative stress with more finesse, then, and only then, can we go beyond our genes and destiny.

[12]In cancer, the gene for NFκB is often mutated to be continuously active. One possible problem with blocking the activity of NFκB is that this causes immune suppression. Immunosuppression makes the progression of cancer more likely, as the immune system normally targets and eliminates cancer cells. Such involuted links are painfully difficult if not intractable.

CHAPTER FIFTEEN

Life, Death and Oxygen

Lessons From Evolution on the Future of Ageing

WHICH CAME FIRST, THE CHICKEN OR THE EGG? This question symbolizes our fascination with cause-and-effect problems. It might be rephrased, did the egg 'cause' the chicken, or did the chicken 'cause' the egg? Given that the one follows the other in an apparently endless succession, the question seems impossible to answer: it is the kind of infinite regression that philosophers love. Some people see such regressions as evidence of a prime mover, who created both the chicken and egg simultaneously. Then there are the pedants, who insist on answering the question. The tiresome truth of it is that the pedants are right: there is an answer. We will look into this answer briefly, because it throws light on the more important problems of life, death and oxygen.

The answer is not logical but historical: we confuse an infinite regression with an incomprehensibly long time and the contingencies of history. There were not always chickens and eggs: they evolved. More than this, they evolved in a particular way, by sex and natural selection. In a sexual species, the tiny changes in genes, which accumulate generation after generation, are only passed on through the sex cells. The genes in the sex cells are largely unchanged by the experience of the organism: we can mutate them by smoking or irradiation, but if we develop large muscles through working out in the gym, we cannot pass them on to our children (though we may pass on our frame or propensity for working out). This is the distinction between the competing theories of Darwin and the French naturalist Jean Baptiste Pierre Antoine de Monet, better known as the chevalier de Lamarck. Lamarck believed in the inheritance of acquired traits, a theory that was popular in the Soviet Union during the Stalin years. Guided by the Marxist pseudoscience of Lysenko, Stalin hoped that imposing communism for a few generations would imprint the 'genes' for communism on the Russian people.

If Lamarck had been right, then a bird could become more like a chicken as it grew, just as a Russian might become a better communist. The bird would then pass on its newly fledged chicken-genes to its offspring. If this were the case, the chicken would come before the egg. There is nothing illogical about this scenario — it is in fact what bacteria do. Bacteria do not generate sex cells. When they divide, they pass on any new traits they have acquired to both daughter cells. As it happens, this is not how sexual species pass on their genes. In sex, the body is thrown away as a genetic dead end, while the sex cells contain all the inheritable genes. The genetic changes that led to the evolution of a chicken therefore took place in one or other of the sex cells — or both — and came together in the egg on fertilization. This means the first chicken must have hatched from an egg laid by a bird that was not a chicken. Clearly the egg came first.

The first chicken, of course, did not appear suddenly: there was a gradual transition from non-chickens (actually, the red jungle fowl *Gallus gallus*) to domestic chickens. The eggs of earlier birds therefore evolved before chickens. Eggs with hard shells were in fact invented by the reptiles, around 250 million years ago. After the Carboniferous, the climate grew cooler and drier, and the great coal swamps dried up. The first reptiles developed scales and shelled eggs to escape the constraints suffered by amphibians, which depended on water. Eggs with shells could be laid on land and did not dry out. This was beginning of the 'age of reptiles', which lasted until the demise of the dinosaurs, 65 million years ago. As a related historical accident, the hard shell made copulation necessary. The shell forms before the egg is laid, so fertilization has to take place internally. Thus, all reptiles copulate, and passed on this trait to their descendants — the birds and mammals. A little understanding of the history of life, then, tells us that both copulation and eggs came before chickens — and the historical narrative makes the idea of infinite regression seem absurd.

The role of oxygen and free radicals in ageing and disease presents a similar problem: which comes first, the radicals or the disease? In the 1950s, Rebeca Gerschman, Daniel Gilbert and Denham Harman argued that the reactive intermediates of oxygen respiration *caused* ageing and disease. Such big claims have not been proved experimentally even today. Even so, many people, including some eminent researchers, cling to the belief that antioxidants are a miracle cure. Most researchers in the field, though, would concur with the characteristically trenchant view of John Gutteridge and Barry Halliwell: "By the 1990s it was clear that antioxidants are not a panacea for ageing and disease, and only fringe medicine still peddles this notion."

Few scientific fields hold a greater promise of fame — a cure for ageing! — than the free-radical field, and none has suffered so many reversals. Much of the excitement that followed the discovery of superoxide dismutase (SOD) in the late 1970s (see Chapter 10) dissipated as pharmaceutical drugs failed to deliver the anticipated miracle. High doses of dietary antioxidants likewise failed to impress. The field too easily became a forum for bad science, in which claims were unsupported by hard evidence. We examined an instance of this in Chapter 9: the claim that plasma levels of vitamin C correlate inversely with mortality. They do, it is true, but the unspoken implication — the reason this study was published in *The Lancet* — is that if we eat more vitamin C we will be less likely to die. Perhaps this too is true, but the study came nowhere near proof, as the authors themselves were first to admit: if anything, it proved the contrary, as people taking vitamin C supplements gained no extra benefit. The claim is equivalent to saying that the number of hours we spend on our feet correlates inversely with mortality, so if we stand up more often we will live longer.

Not surprisingly, other fields of medicine have become suspicious of the boy who cried wolf too many times. This feeling was conveyed well by one of the reviewers of my proposal for this book:

> I confess to some prejudice against the free-radical field, which is complex and messy and seems at the same time to attract messianic types, who think that free radicals explain all diseases, not to mention ageing. (So if we eat enough free-radical scavengers we shall all live forever.) Of course this is not to say that free radicals are unimportant or uninteresting, only that it is difficult to separate the science from the hype.

Apart from the hype, there are some genuine scientific difficulties here. I wonder if it is even possible, using direct experimental methods, to prove that free radicals *cause* disease. The problem is that most free radicals are present at tiny concentrations and for fleeting periods: no sooner are they formed than they transform into something else. The only known way of measuring free radicals directly is a technique called electron spin resonance (ESR), which can detect tiny magnetic signals deriving from the spins of unpaired electrons in free radicals (see Chapter 6). Unfortunately, these transient signals are easily lost against background noise, and the method is not sensitive enough to detect radicals as reactive as the hydroxyl radical, which disappear in billionths of a second. There are ways around this difficulty, but they generate their own problems of interpretation.

The easiest way to skirt these issues is to measure free radicals indirectly, by quantifying the build-up or excretion of their end-products — in particular, oxidized DNA, proteins and lipids. Now the problem is one of attribution: do the oxidized products really reflect free-radical attack in the body? For example, one oxidized breakdown product of DNA is 8-hydroxydeoxyguanosine (8-OHdG). We have seen that 8-OHdG is formed when hydroxyl radicals attack DNA, but some 8-OHdG is formed as an artefact, and some may be formed by enzymes. Estimates of total DNA oxidation by hydroxyl radicals are in reality a best guess. While it would be perverse to ignore the large body of evidence which *suggests* that free radicals actually cause disease, definitive claims are no more than hype. The same applies to other measures of free-radical formation, including standard tests for the products of lipid and protein oxidation. We cannot infer the definite involvement of free radicals from such tests any more than we can infer deliberate arson from the smoking remains of a building. If we accept that oxidized proteins, DNA and lipids *are* the tell-tale footprints of free radicals, we still do not know whether free radicals cause disease. We have little idea about timing or causality. The signs of oxidation are often concomitant with the signs of disease, but this is not to say that one causes the other. The simplest way to prove that free radicals *do* cause disease is to block their action using antioxidants. As we have seen, antioxidants rarely cure diseases, let alone ageing. Of the many possible explanations for this — perhaps they are not potent enough, or do not get to the right place in the right amount at the right time — the most inherently believable is that free radicals are only part of the problem. Even when antioxidants do help, it is not easy to prove that they do so by working as an antioxidant. In the case of vitamin C, many of its actions have nothing to do with antioxidant activity: it may act by stimulating the synthesis of carnitine, or the production of peptide hormones and neurotransmitters. Short of a methodological breakthrough, it is difficult to progress beyond this point with present experimental techniques.[1]

Set against the potential dead end of experimental research is the intuitive explanatory power of free radicals as a cause of ageing and disease. The fact is that free radicals *are* detected in virtually every disease known to man, and that in principle they *can* explain the progression of ageing and the rising incidence of age-related diseases. The massive accumulation of data is at least suggestive that free radicals have a *causal* role in many conditions. Many other facts line up with this interpretation. To take just one example: if free radicals are produced by mitochondria, we would expect DNA in the mitochondria to be damaged more than DNA in the nucleus. In Chapter 13, we saw that there are practical problems in measuring this predicted difference — estimates vary by a factor of 60 000. However, a high rate of damage should lead to a high mutation rate, and the mutation rate of mitochondrial genes *is* higher than the mutation rate of nuclear genes by an order of magnitude. The free-radical explanation is therefore supported by evidence from other quarters.

I think that we can best understand the importance of oxygen free radicals by looking at their place in a bigger picture. We cannot prove experimentally that free radicals cause disease any more than we can prove logically that eggs cause chickens, but we *can* see how far free radicals fit into an evolutionary framework by telling the story of oxygen. The story we have pieced together in this book goes some way towards answering intractable experimental questions. The answers hold significant implications for the future of medicine. Before we peer into future possible worlds, let me recap the story, bringing out the elements most important to our own lives and deaths.

In the beginning there was no oxygen, but there was ultraviolet radiation and water. Without an ozone layer, the intensity of ultraviolet radiation

[1] One such breakthrough has already happened: the advent of 'knock-out' mice, in which the genes for enzymes like SOD are disabled so that the protein is not produced. Such experiments are valuable, and illustrate the importance of SOD in newborn mice; but the fact that SOD is necessary in baby mice does not mean it is important in ageing people. The extension of lifespan in *Drosophila* by the overproduction of SOD and catalase is more revealing, but limited, so far, to *Drosophila*.

in the air and surface oceans was at least 30 times greater than today. Radiation splits water to produce the same reactive oxygen intermediates that we generate when we breathe — hydroxyl radicals, superoxide radicals and hydrogen peroxide. These unstable intermediates reacted together, and with water, to generate hydrogen and oxygen. Hydrogen was light enough to seep away into space. Oxygen reacted with iron in the rocks and with sulphurous gases emanating from volcanoes, and was trapped in the crust. In the thin, dry air of Mars, the oxygen intermediates were petrified, literally, as the red iron oxides that lend the planet its colour today.

On Earth, something different happened. Life adapted to the surface oceans. LUCA, the Last Universal Common Ancestor (Chapter 8), had already evolved antioxidant enzymes that could protect her against the reactive oxygen intermediates derived from radiation. Genetic studies suggest that LUCA possessed antioxidant enzymes, including SOD, catalase and peroxiredoxins. More than this, LUCA had a sophisticated metabolism. She could trap oxygen using a form of haemoglobin, and generate energy from it, using the enzyme cytochrome oxidase — the grand ancestor of the enzyme that continues to do the same job for us. LUCA could do all this as long as 3.85 billion years ago, soon after the end of the meteorite bombardment that cratered the Moon and the Earth. Even though free oxygen had not yet accumulated in the air, the earliest ancestor that we know about was already generating energy from oxygen respiration, and was resistant to oxidative stress.

The oxygen intermediates formed by radiation reacted with dissolved iron salts and hydrogen sulphide, gradually depleting these substances from the shallow seas and lakes. Both were early raw materials for photosynthesis, so their depletion raised the selective pressure to find an alternative. In such sheltered environments hydrogen peroxide was relatively abundant, and it was a practical alternative as it could be split by the antioxidant enzyme catalase. Catalase thus doubled as a photosynthetic enzyme. As multiple enzymes clustered around the photosynthetic reaction centres, two catalase units became lashed together to form an 'oxygen-evolving complex'. This complex could harness the energy of sunlight to split water and release oxygen. Water-splitting, oxygen-producing photosynthesis evolved only *once* on Earth. In a thought-provoking quirk of fate, *all* life on Earth that uses water as a raw material for photosynthesis has inherited a water-splitting complex based on catalase units. This could never have happened if life had not learnt how to tolerate radiation first. Perhaps it would never have happened without catalase; and it almost certainly never did happen on Mars. The ostensible sterility of Mars might be put down to this detail alone.

On Earth, photosynthetic cyanobacteria injected oxygen into the air, and *fast*: faster than the volcanoes could spew out sulphurous gases, faster than erosion could expose virgin rocks to the air. The crust oxidized, but still there was some oxygen left over. When radiation split water, the hydrogen could no longer escape into outer space: instead it reacted with the excess oxygen to form water again. As oxygen built up, an ozone layer formed, which blocked the penetration of ultraviolet radiation into the lower atmosphere. The loss of oceans slowed to a trickle on Earth but continued apace on Mars and Venus, where no oxygen buffer had formed. Loss of water in this way may have cost Mars and Venus their oceans.

Water retention was the first gift of photosynthesis. The second was oxygen itself. At several times in the long Precambrian era, catastrophic geological upheavals — snowball glaciations and bouts of mountain building — overturned the lengthy periods of evolutionary stasis, and buried so much organic matter that excess oxygen was injected into the air. Each time, life leapt forward. In the first injection, 2.7 billion years ago, our own ancestors, the eukaryotes, made their first wispy appearance, leaving behind tell-tale molecular fingerprints — sterols similar to cholesterol. In the next, larger, injection, which followed the snowball Earth and mountain-building episodes around 2.3 to 2.2 billion years ago, the eukaryotes made their first robust appearance in the fossil record. Soon after, the fossil record began to show signs of multicellular algae; but little else happened in the next billion years. Then came the greatest of all geological roller-coaster rides: a succession of at least two snowball Earths, in which ice shrouded the Earth episodically over a 160 million-year period and finally pushed atmospheric oxygen up to modern levels.

Afterwards, as the ice retreated and the dust settled, the first large animals floated on-stage — jellyfish-like bags of protoplasm, the vegan Vendobionts. With their bulk they had guts and made faeces. Their heavy faecal pellets sank and were buried in the ocean depths, depleting the organic matter in the ocean. Not only this, but faecal burial prevented the breakdown of organic matter by respiration (and therefore the consumption of oxygen), and so contributed to the oxygenation of the overlying oceans. The bacteria were forced to retreat, taking their stinking, sulphurous underworld with them. A new, oxygenated, ecosystem yawned open, a vast blank canvas awaiting the hand of nature. With oxygen came a fourfold leap in organisms' ability to extract energy from their food, and with that came predation. Now, for the first time, it really paid to eat, and extended food chains became possible. Life burst into the vacant eco-space in the Cambrian explosion, 543 million years ago. The oceans filled up with scampering, armour-plated monsters, the hunters and the hunted. With predation came evolutionary arms races, in which hunters and hunted competed for size (which depended on oxygen for energy and to build structural support). With size came the complex adaptations that enabled the colonization of the land.

Behind the scenes, oxygen was puppeteer. The first single-celled eukaryotes were degenerates, scavengers in a world of assiduous bacteria. They had lost the metabolic prowess of LUCA and scraped a living by fermenting organic remains or engulfing bacteria. They needed oxygen to make their membranes, but couldn't bear too much of it. Then one day a eukaryotic cell happened to swallow an oxygen-guzzling purple bacterium. Suddenly it could swim with apparent impunity through the shallow seas, protected from oxygen by its internal vacuum cleaner. The insider deal blossomed into a Mephistophelean pact: as the purple bacteria turned into modern mitochondria, they exchanged their surplus energy for a life spent dicing with death. Oxygen mutated DNA, forcing genes to change and evolve. It must have been one of the factors that drove the evolution of the most efficient of all genetic cleansing mechanisms, sexual recombination. But mitochondria presented a unique and profound problem of their own: they retained some genes that were necessary for the function of the eukaryotic cell as a whole. Stranded in the belly of the furnace, and prevented by their host cell from dividing as fast as bacteria, mitochondrial genes could not be rejuvenated by sex nor by bacterial-style selection. They could only degenerate. The solution was not sex but sexes. Given two sex cells that would combine to form the next generation, one cell could be powered by mitochondria destined for disposal, like the fuel tanks of a rocket, while the other could maintain its mitochondrial population in a dormant state, like hibernating astronauts, ready for duty on arrival in the next body. Early in embryonic development, the pool of sleeping mitochondria would be siphoned off and kept 'on ice' for the next generation.

With sex and gender came redundancy. When only some genes are passed down to the next generation, all other attributes are subsidiary to the transmission of these genes. Redundancy allows the specialization of support cells and ultimately of whole bodies. Bodies fundamentally became redundant machines for passing genes from one generation to the next.[2] The body protects the sex cells from injury, starvation and mutation, or from being eaten or infected, and advertises the quality of its genes by parading the protein products in public: a survival machine and display counter. How much the genes invest in bodily maintenance depends on their own likelihood of transmission — and this in turn depends on two main factors: fecundity (the number of offspring produced per unit time) and the time available.

The balance between sex and survival, between passing genes on to the next generation and surviving long enough to do so, underlies the evolution of an optimal lifespan. Reproduction must fit into the time window available before death becomes a statistical likelihood. It may pay to reproduce slowly and raise our children if we have 70 years at our disposal, but if we are guaranteed fodder for a sabre-toothed tiger within ten years of birth, we will not survive as a species unless we compress our reproductive cycle into the ten years available. For opossums, death by predation is likely within three years, and so this is how long they live. If the threat of predation is lifted, opossums can evolve to live longer. They commit more resources to survival — to keeping the body going for longer — and divert resources away from sex. Litter sizes and fecundity fall. Even so, the new-age opossums continue to produce litters for longer — their fecundity falls on a unit-time basis, not over a lifetime (they commit fewer resources at any one moment but have more moments).

In all animals studied, senescence is postponed by shifting resources away from sex, towards the prevention and repair of damage at a molecular level. This shift may take place over generations (in which case the changes are inherited as fixed differences in lifespan) or within a single lifetime (in which case the expression of existing genes is altered). Either way, the genes that prolong lifespan are similar: their effect is always to restrict molecular damage. The extent to which they do so depends on their mobilization and efficiency rather than their nature, just as a standing army and a conscript army differ in training and discipline rather than in methods.

The allocation of resources within a lifetime is controlled by a hormonal switch — insulin and the insulin-like growth factors — which

[2] This is a restatement of Richard Dawkins' 'selfish gene' theory. Most biologists do not see the concept as a radical idea but as a helpful way of viewing natural selection. Of course our bodies are more than gene machines, as Dawkins is the first to admit; but many harsh facts of life, such as illness, ageing and death, can best be understood rationally in these terms.

responds to the availability of food and the possibility of sex. The choice is simple: sex now, or defer and survive in the meantime. Calorie restriction simulates the physiological response to famine — survive now, sex later — and extends maximum lifespan in species as diverse as nematode worms and rats. Calorie restriction works at the genetic level, by changing the expression of genes responsible for maintaining the integrity of the body. The overall effect of these genetic changes is to *reduce metabolic stress* (the threat to the health of the cell caused by leakage of free radicals from mitochondria) for the duration of the famine.

Long-lived species can restrict metabolic stress throughout their lives, not just in periods of environmental hardship. The importance of metabolic stress to lifespan is borne out by comparisons between species. In many cases, lifespan varies simply with metabolic rate — the slower the metabolic rate, the longer the life. This idea is often criticized for being riddled with exceptions, but in fact the exceptions prove the rule. The metabolic rate is a proxy for free-radical production by mitochondria. The more free radicals that escape to react with cellular components, the sooner we will die. Lifespan therefore varies according to the rate of free-radical production and the degree of protection against their effects.[3] Birds have high metabolic rates, but live a long time because they leak relatively few free radicals from their mitochondria (good *primary prevention*) and have good repair systems. We also live a long time, despite leaking more free radicals than birds, because we have invested more in antioxidant defences (good *secondary prevention*) and, like birds, have good repair systems. Rats live a short time because they have a high metabolic rate, leaky mitochondria, poor antioxidant defences, and rudimentary repair systems.

Secondary defences are less effective than primary defences because the defences themselves can be damaged by free radicals. In the same way, in society, it is better to prevent an outbreak of violence before it happens than have to forcibly restrain a riot already in full swing. Even so, our secondary defences should be sufficient to help us live out our maximal lifespan of about 120 years. The fact that most of us do not, and

[3] Think of this in terms of radiation poisoning. The greater the intensity of radiation, the more likely we are to die. If we are shielded from the full intensity of radiation by a screen, then we are less likely to die, even though the radiation intensity is not altered. Our chances of death vary with the intensity of 'external' radiation (which equates to the metabolic rate) and the thickness of the screen (which equates to antioxidant protection); or more simply, with the intensity of penetrating radiation.

die of age-related diseases rather than old age, betrays a fundamental tension at the heart of life itself, a tension between the two cardinal properties of living things — reproduction and metabolism. If we want to rid ourselves of ageing and diseases, we must come to terms with this most ancient of tensions.

Imagine you are LUCA, floating in shallow seas under the boiling Sun. You need energy and you need to reproduce. If you don't reproduce, you will be torn to pieces one day — no physical matter can survive indefinitely without replication. But if you do replicate, if you succeed in cloning yourself, then you will survive in one sense or another. To clone yourself you need energy and you need a template. The best source of energy is the Sun. Even in the beginning, in the shallow seas, sunlight is splitting water and you are catching oxygen. In just a few hundred million years, your descendants will mimic this trick in photosynthesis, and release masses of oxygen into the air. Other descendants will become living machines that suck up this oxygen to fuel a spectacular leap in the potential of life: size, movement, strength, predation, consciousness, and mind.

The important point is this. From the very beginning, LUCA was using sunlight as an energy source. This had a downside: solar radiation produces a flux of free radicals, and the more energy, the more free radicals. Free radicals can destroy the all-important DNA template. Somehow, LUCA had to venture as close as she could to the sunlight for energy, without getting so close that her DNA template was jeopardized by free-radical attack. The health of the first cells *must* have depended on getting this balance just right. This, in turn, *must* have meant detecting free radicals and responding somehow if too many (or too few) were present. How could this have been done? Detection and response might have been coordinated by means of proteins that change their function when oxidized by free radicals. Certainly, representatives of all three domains of life have proteins that respond to oxidation. In Chapter 10 we met haemoglobin, SoxRS, OxyR, NFκB, Nrf-2, AP-1 and P53, and more are being recognized all the time. Today, when these proteins become oxidized, the cell responds by correcting the free-radical balance, either by swimming away from the source of danger, or by stepping up the antioxidant defences and repair mechanisms.

Free radicals are therefore dynamic indicators of the energy levels and general 'health' of the cell. They should have been among the earliest and most important indicators of cellular health, as they are a unique chemical bridge between the most basic traits of life — metabolism and reproduction. The 'right' number of free radicals indicates the 'right' balance between energy and replication (Figure 13). This is a critical point. Evolution works by building on existing systems in the same way that the Spanish conquistadors built baroque cathedrals on the solid Inca walls of Cuzco. In biology, older foundations are rarely obliterated completely.

If it is true that the earliest 'health-sensor' of living cells operated by detecting free radicals, we would expect this system to underpin more recent innovations, such as the immune system. I think that it does. I have often wondered why oxidative stress should be a common denominator of stressful physiological states, from radiation and heavy-metal poisoning to infection and ageing, but from this perspective the parallels

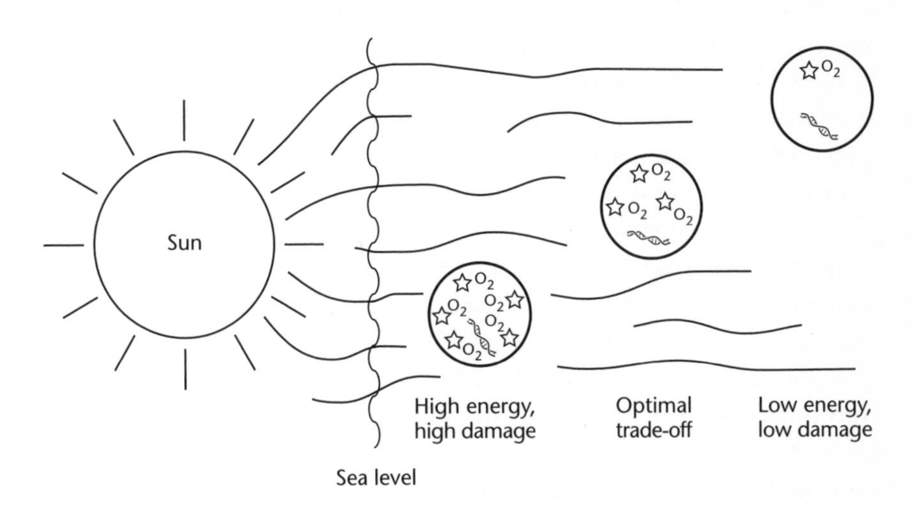

Figure 13: The primordial connection between metabolism and reproduction. LUCA (the Last Universal Common Ancestor) could generate energy using, indirectly, the energy of ultraviolet radiation, which splits water to produce oxygen (via free-radical intermediates). LUCA could trap the oxygen with haemoglobin and generate energy from its reduction using cytochrome oxidase. The greater the intensity of ultraviolet radiation, the greater the production of oxygen, hence the more energy available for successful reproduction. However, free-radical intermediates damage DNA and lower the chances of successful reproduction. The viability of primordial cells therefore depended on sensing the number of free radicals produced, and responding accordingly (optimal trade-off). This original 'health-sensor' system still underpins more sophisticated defences, such as our own immune system.

begin to make sense. While our responses to different kinds of threat have become more sophisticated and autonomous as they evolved, all depend on oxidative stress in the same way that the emergency services depend on a 999 operator to verify that it *is* an emergency, and route the caller to the appropriate service. To see how this works in the body, think of the immune system, a bafflingly complex system, capable of recognizing and destroying a billion different antigens, most of which represent hypothetical microbes that our immune system will never meet. Yet simple drugs can suppress this entire network, making operations like organ transplantation feasible. They do so by interfering with the free-radical 'health-sensors' — the 999 operators — that still underpin the immune system. Blocking the activity of one sensor, NFκB, with glucocorticoids or cyclosporin hinders the rejection of transplanted organs — a profound stress for the body — for months or years.[4]

These ideas underlie what I called the 'double-agent' theory of ageing in the last chapter. Far from being simply a pathological state, oxidative stress is a vital signalling mechanism that underpins the cell's genetic response to all kinds of injury. In particular, oxidative stress marshals our resistance to infection. This takes the form of an aggressive inflammatory attack (which eliminates the invader) combined with a stress response (which bolsters our own cells against the attack). The importance of this mechanism to our chances of sexual success — we have a good chance of recovery when young — outweighs the downside, which is postponed until old age and so has little impact on our reproductive success. In old age, oxidative stress rises as our mitochondria leak free radicals into the cell. The body perceives this as a threat and responds accordingly. Unlike infections, however, the new threat cannot be eliminated: there is no cure for broken mitochondria. Instead, the chronic inflammatory response is perpetuated indefinitely and contributes to our physical and mental demise.

Guided by this evolutionary perspective, we can conclude that oxygen free radicals *are* an underlying cause of both ageing and age-related diseases. The 'double-agent' theory may explain why antioxidant supple-

[4] Glucocorticoids, such as prednisolone, block NFκB by stimulating the synthesis of its natural inhibitor, IκB. At high doses, prednisolone has such a potent and general immunosuppressive effect that many early transplant recipients died from infections and cancers. Cyclosporin, the drug that led to a breakthrough in transplant medicine in the 1980s, is more selective for T lymphocytes, and so has a less catastrophic effect on overall health. It inhibits the enzyme calcineurin, which among other things blocks the activity of NFκB in T lymphocytes, but not in most other immune cells.

ments have had so little effect on lifespan: they cannot halt mitochondrial leakage, and cells are refractory to overloading with antioxidants, lest they smother the powerful genetic response to injury. Thus, oxidative stress rises to cause age-related diseases, and antioxidants cannot reverse this rise, although they may slow it to some extent.

Is there anything we can do to prevent the diseases of old age? Targeting 'susceptibility' genes is misguided, as there is nothing wrong with them: their negative effects are unveiled by oxidative stress, and in principle the best way of restoring their positive function is to alleviate oxidative stress. Despite the problems with antioxidants, this ought to be possible. Lifespan is flexible in nature. In Carboniferous times, life gave every sign of coping with higher oxygen levels (which must have increased oxidative stress). Healthy centenarians show that disease is not an inevitable feature of human ageing. How do they escape from age-related disease? If the double-agent theory is correct, there are two places we might look for help: infectious diseases, and the mitochondria themselves.

Malaria illustrates both the possibilities and the drawbacks of infectious disease as a solution to the problems of old age. To our shame, given the lack of real political will to eradicate it, malaria still affects half a billion people every year. The most feared complication, cerebral malaria, is caused by inflammation of the tiny blood vessels in the brain, which leads to fever, convulsions, coma and death in more than a million people each year. As we have seen, inflammation and fever are part of the reaction of the *host* to infection. If someone dies of cerebral malaria, they die more through the violence of their own immune system's counter-attack than through the virulence of the parasite.

The selection pressure exerted by malaria is strong enough to maintain sickle-cell anaemia and the thalassaemias at a high frequency in populations across Africa and Asia. These conditions are not isolated curiosities, but part of a spectrum of adaptations that enable people to survive in areas where malaria is endemic. Among the most important of these adaptations is *malarial tolerance*. Tolerance develops after infections in early childhood and lasts a lifetime. It is not the result of a heightened ability to kill parasites (as in vaccination) but the triumph of *realpolitik* —

a live-and-let-live policy in which the immune system is *suppressed* so that its attacks on the parasite do not damage the body. People who are tolerant to malaria sometimes harbour massive numbers of parasites in their blood — enough to kill ordinary people — yet show few or no symptoms of illness.

We know from the experience of transplantation and AIDS that immunosuppression has very serious side-effects: if the immune system is crippled we are far more susceptible to other infections, which may kill us, and are at high risk of some (but not all) cancers. Even today, transplant recipients have nearly 5 per cent chance of developing cancer within a few years of their operation, a 100-fold increase in risk over the general population. On the other hand, transplant immunosuppression has improved over the past 20 years and will no doubt continue to do so, while in AIDS, HIV infects the immune cells themselves, and so induces a purely pathological change in the immune system. In contrast, in malarial tolerance the immune system is regulated in a physiological manner, which has a lasting effect on susceptibility to malaria and presumably other aspects of health.

In 1968, Brian Greenwood, now at the School of Hygiene and Tropical Medicine in London, drew attention to the rarity of autoimmune diseases (such as multiple sclerosis, rheumatoid arthritis and lupus) in tropical Africa, but not in Africans living in North America. He suggested that the difference might relate to the frequency of parasitic infections in childhood, notably malaria. Over the following three decades, mounting evidence has confirmed that the low incidence of autoimmune diseases in Africa is indeed related to malarial tolerance.

I wonder how far we can take Greenwood's ideas. Where should we draw the line between an autoimmune disease and other forms of illness? An autoimmune disease is a condition in which our own immune system mistakenly attacks components of our own body. If ageing is essentially a chronic inflammatory response to oxidative stress, in which immune cells attack components of our own body, should we define the diseases of old age as autoimmune diseases? Not in a conventional sense, perhaps, but it is constructive to think of Alzheimer's disease as an autoimmune disease. If so, for example, we would expect the incidence of Alzheimer's disease to be low in tropical Africa; and dementia *is* rare in Africa. This is not just because more Africans die from infections before they reach old age, or because social pressures lead to the real incidence being concealed by relatives. In 2001, Hugh Hendrie and his colleagues at Indiana University in the United States and the University of Ibadan in Nigeria, reported the results of a five-year study of dementia in Nigeria and the United States. They tracked the fortunes of nearly 5000 Nigerian Africans and African Americans, living in the towns of Ibadan (an area where malaria is endemic) and Indianapolis. All were 65 or over, and none had overt dementia at the start of the study. After five years, the proportion of the study population diagnosed with dementia was significantly lower in Ibadan than in Indianapolis: 1.35 per cent compared with 3.24 per cent, or somewhat less than half the risk.

The report attracted interest, and Lindsay Farrer, at Boston University, wrote an editorial in the same issue of the *Journal of the American Medical Association*, in which she advocated a "global approach to bad gene hunting". Both articles discussed a curious finding: in Ibadan, there was *no* association between *ApoE4* alleles and the risk of Alzheimer's disease. Various possible genetic and environmental reasons were explored, in particular high blood pressure and other vascular risk factors for dementia; but rather surprisingly, neither article mentioned malarial immunosuppression in Africa. This seems to me the most probable explanation. We noted in the last chapter that *ApoE4* proteins are easily damaged by oxidative stress, and that this probably explains the link with Alzheimer's disease. We also saw that people with two *ApoE4* alleles gain more from antioxidants and anti-inflammatory drugs. In Ibadan, it is plausible that malarial immunosuppression blunted the severity of cerebral inflammation and so lowered the risk of Alzheimer's disease.

The Indianapolis–Ibadan Dementia Project also demonstrated the reverse side of the coin, and this too was passed off with little comment: the mortality rate in the African cohort was nearly double that of the American cohort, despite their better cardiovascular health. What they died of is not stated. I imagine many must have died from infections or cancers. Studies in Tanzania, where malaria is also endemic, showed that the death rate fell substantially in regions where malaria had been controlled by draining swamps. The scale of this effect was larger than could be attributed directly to malaria, and prompted research into the 'hidden morbidity of malaria'. This research has confirmed the suspicions: immunosuppression in areas where malaria is endemic perpetuates opportunistic infections, leading to the spread of diseases such as tuberculosis. In addition, cancers such as Burkitt's lymphoma (a malignant cancer of B cells) are common and linked with malarial immunosuppression, probably through infections in childhood with the Epstein-Barr virus, which are not properly cleared and persist.

Immunosuppression clearly influences health in old age, but at a potentially serious cost. Even so, there are grounds for hope, particularly if we gain the upper hand in the battle against infectious disease: we know that it is *possible* to lower the probability of age-related diseases by modulating the immune system. Whether or not this is a practical goal depends on exactly how it is done.

The mechanism of malarial immunosuppression is not well understood, and there are various competing theories. Scientists are people, and bring their own expertise, experience and biases to research problems. This is a far cry from how philosophers tell us that science 'should' work, but the idea of the scientific method as an inductive procedure in which facts emerge from accumulating data is as misguided as can be. In reality, experiments are conceived and interpreted in terms of particular hypotheses, so data are generated on particular aspects of a problem rather than the whole problem. I will therefore put forward the ideas I think are the most intriguing and perhaps the most likely to be right.

In terms of malarial immunosuppression, I am struck by a paper in *Laboratory Investigation* in 2000, by Donatella Taramelli and her colleagues at the University of Milan, in which they studied the behaviour of isolated immune cells. Feeding malarial pigment to the immune cells brought about a rise in oxidative stress, which stimulated a fivefold rise in activity of the stress protein haem oxygenase. When the same immune cells were challenged for a second time, they did not respond normally, by pumping out inflammatory messengers, but instead had a 'depressing' effect on neighbouring cells, which became glum and unresponsive. Extrapolating to people, Taramelli argued that frequent malarial infections in childhood might produce a swing in the behaviour of the whole immune system, from activation to depression, by means of a continuous activation of haem oxygenase.

Let me try to put Taramelli's findings in a broader medical context. If we grow up in an area where malaria is endemic, we can expect to be infected regularly in childhood. For the first few times, we will become ill, and we might die of cerebral malaria. If we survive, however, we adapt, or at least our immune system adapts. Instead of responding vigorously to infection, the immune system is restrained to some degree. The molecular detail is far from clear, but it seems probable that a new balance is established, in which NFκB is inhibited, and stress or antioxidant genes such as haem oxygenase are activated (probably via the counterbalancing transcription factor Nrf-2). The continuous activation of haem oxygenase prevents the immune system from causing too much collateral damage to our own body. If true, this scenario is all the more important because the mechanism is not unique to malaria: haem oxygenase and other stress proteins seem to underpin tolerance to common bacterial infections, and can provide almost complete protection against septic shock.[5] It is therefore plausible that frequent childhood infections could bring about a persistent immunosuppression later in life. As with malaria, toning down our immune reaction to infections should make us at once more vulnerable to infections, and less vulnerable to autoimmune diseases and the diseases of old age.

There are three considerations that make me think this is true. First, haem oxygenase appears to be necessary for our normal health, even though it is a stress protein and supposedly 'switched off' in normal circumstances. Recall from Chapter 10 that an unfortunate six-year-old boy diagnosed with haem oxygenase deficiency suffered from vascular inflammation, severe growth retardation, abnormal blood coagulation, haemolytic anaemia and serious renal injury. He died at the age of seven. Clearly, a regular dosing with haem oxygenase is necessary to temper inflammation. This idea is corroborated by studies of knock-out mice, in which the genes for haem oxygenase are mutated so that the protein is not produced. We saw that these mice have symptoms similar to those of people with chronic inflammatory diseases like haemochromatosis, including liver fibrosis, joint inflammation, restricted movement, weight loss, shrunken gonads and early death. Thus, haem oxygenase deficiency causes chronic inflammation and a short lifespan in both mice and men, whereas additional haem oxygenase suppresses the immune system and might potentially prolong lifespan.

[5] Haem oxygenase has a powerful immunosuppressive effect. Overproduction can block the rejection of hearts transplanted from mice into rats, and even graft-versus-host disease in mice injected with spleen cells from a different species. Whether haem oxygenase directly inhibits NFκB activation is not known, but the broader stress response *does* block the activation of NFκB and haem oxygenase is one of the most prominent players in the stress response. Hector Wong and his colleagues at the University of Cincinnati have shown that stress elicits a rise in IκB, the natural inhibitor of NF-κB, which suppresses the activation of NFκB. They also found that previous stress protects against a subsequent septic shock. Incidentally, psychological stress, which is well known to induce immunosuppression, may also operate through a stress response involving haem oxygenase.

Second, the incidence of autoimmune diseases, such as insulin-dependent diabetes, Crohn's disease and rheumatoid arthritis, is rising throughout the world, especially in Westernized countries. In Europe and the United States, the incidence of insulin-dependent diabetes has risen by an estimated 3–5 per cent a year in the past two decades. Hypersensitivity reactions, in which the immune system correctly recognizes foreign antigens, but then over-reacts to them, are also on the increase. The incidence of asthma and allergies has doubled in the last decade. Of the various possible reasons for this rise, one theory is gaining ground — the 'hygiene hypothesis'. Simply put, too much cleanliness in childhood is bad: we need regular infections for our immune systems to develop properly, just as we must use our eyes to develop a visual understanding of the world.

A number of studies have shown that frequent infections in childhood are linked with a lower incidence of allergies and autoimmune diseases later in life, and vice versa. The assumption is that the immune system needs 'house training' in infancy, and if deprived of appropriate stimuli behaves like a bull in a china shop at the slightest provocation. However, it is also plausible that regular infections in childhood could produce a persistent immunosuppressive effect, as in malaria. I am not aware of any clinical data to support this interpretation, but one animal study is intriguing: mice that are deficient in the transcription factor Nrf-2, and thus in haem oxygenase and other stress proteins, go on to develop an autoimmune disease similar to lupus, leading to kidney failure. In other words, if the balance skews to inflammation, and away from immunosuppression, there is a higher risk of autoimmune disease.

My final consideration supports the relationship between infections in childhood and lifespan, and is based on the work of the cell biologist Giovanna De Benedictis at the University of Calabria in Italy, and her collaborators, including the demographer Anatoli Yashin at the Max Planck Institute for Demographic Research in Rostock, Germany. Yashin and De Benedictis spent the 1990s searching for 'longevity' genes in centenarians. Their basic idea is simple: some genes raise our chances of reaching a ripe old age, while others have a negative or neutral effect. The genes that prolong survival are most likely to be found in the people who did survive, so the best place to look for them is in centenarians. We imagine that the genes which prolong survival should make us 'more robust' in some sense. This is certainly true of some genes (which we will come to shortly), but Yashin and De Benedictis found that the situation is in reality more complex. A surprising number of 'longevity' genes turned out to be linked with *frailty* (or susceptibility to disease) earlier in life. In other words, people who are ill a lot in their youth are more likely than most to survive to a ripe old age, as long as they don't die first. Yashin and De Benedictis attributed this durability of the weak to adaptation, or as Nietzsche put it, what doesn't kill us makes us stronger. So long as we avoid really serious illness, a weak disposition might perhaps lend itself to persistent immunosuppression, which reaps its reward in old age.

What can we conclude from infectious diseases? We will probably learn how to modulate the immune system with more subtlety than we know at present, and this should improve our health in old age. I suspect that stress proteins like haem oxygenase will hold the key, and we might even be able to modulate their levels by diet. Plants produce toxins to safeguard them against being eaten. Spices such as curcumin are known to stimulate the activity of haem oxygenase and other stress proteins (and show potential as anti-cancer agents). The trouble with curcumin is its bioavailability: when we eat it, very little is absorbed into the blood stream. How many other plant toxins, with better bioavailability, might stimulate the activity of stress proteins is anybody's guess. As I suggested in Chapter 10, it is feasible that the benefits of a diet rich in fruit and vegetables go beyond their antioxidant content. Plant toxins, if palatable (and we have adapted to many over evolution), are likely to have beneficial effects on our immune system. I think this may help to explain why plants are clearly beneficial to our health, while antioxidant supplements are much less so.[6]

Even so, there is a dilemma at the heart of immune modulation, however refined it is: the benefits are always part of a trade-off between susceptibility to infections, on the one hand, and to age-related diseases on the other. Any benefits will depend on a delicate balancing act in which genes, diet, environment, behaviour and luck all have a role. I can see no systematic way of delaying ageing or preventing age-related diseases here. The only way this might be done 'scientifically' is to prevent

[6] Curiously, a study by Chris Bulpitt and his colleagues at Imperial College, London (published in the *Postgraduate Medical Journal* in 2001) found that women who looked older than they really were had low levels of bilirubin in their blood stream, and vice versa. Bilirubin is an end-product of haem oxygenase. The implication is that high haem oxygenase activity makes women 'look' younger. In men, the strongest connection was with high levels of haemoglobin (which is, of course, broken down by haem oxygenase). Again, the implication is that high haem oxygenase activity makes men 'look' younger.

the root cause of inflammation by targeting the mitochondria, so blocking the rise in oxidative stress in the first place.

The question is, how can we be more like birds? Mankind has always envied birds their power of flight, but now it seems we should envy them their mitochondria too. Bird mitochondria hardly leak any free radicals. Why? The simple answer is that we don't know, though there are some clues that might point us in the right direction. But before we speculate on these clues, are there people with 'bird-like' mitochondria, and do they live for longer? Again, the place to look is the centenarians.

The answer is hidden in a short research letter published in *The Lancet* in 1998 by Masashi Tanaka and his colleagues at the Gifu International Institute of Biotechnology in Gifu, Japan. In less than two columns, the Japanese group laid out a formidable series of studies on mitochondrial DNA in hundreds of centenarians, healthy volunteers and hospital patients. Their results injected new life into longevity-gene hunting the world over. What Tanaka and his colleagues found was that more people with one particular variant of a mitochondrial gene survived into old age: 62 per cent of centenarians carried the variant, known as Mt5178A, compared with 45 per cent of a random sample of healthy blood donors. Equally important, in a separate group of inpatients and outpatients at Nagoya University Hospital, only one third of patients older than 45 had the variant gene, whereas two thirds had the normal version. In other words, more older people with the 'normal' gene ended up in hospital, presumably because they were more susceptible to age-related diseases. This discrepancy did not apply to younger patients, who shared both versions in roughly equal measure. The implication is that the normal gene does not affect health earlier in life, so the spectrum of younger people in hospital reflects the genetic mix of the population as a whole. Taken together, these results suggest that people with the Mt5178A variant are more likely to survive to a hundred and less likely to suffer from age-related diseases than people with the normal version.

There are two points I want to make about this study. First, nearly half the random sample of healthy blood donors in Japan carried the mitochondrial variant Mt5178A. Elsewhere in the world, this variant is much rarer. In one study, for example, only five Asians and one European carried the variant out of 147 samples. Thus, the majority of Japanese centenarians who carry the mitochondrial variant are the select survivors of a population in which it is already common. The frequency of the Mt5178A variant in the population as a whole may help to explain the long life expectancy of the Japanese: currently 84 at birth for women, and 77 for men. The less fortunate slight majority of the Japanese population who do not have the variant are *nearly twice as likely* to end up in hospital with age-related diseases. There could hardly be a clearer link between mitochondrial health and general health in old age.

The second point I want to make concerns the variant itself: it is a single-letter substitution in a mitochondrial gene (a C is replaced with an A). On what a slender thread hangs fate! We have about 35 000 genes, of which a mere 13 protein-coding genes are in the mitochondria instead of the nucleus. Of these 13 genes, a single-letter change in one of them is enough to *halve* our risk of getting any age-related disease, and virtually double our chance of living to a hundred. What on earth does this change in letter do? Well, at an arcane level, it causes a change of one amino acid in the protein encoded by the gene: a leucine is replaced with methionine. Why this should make a difference is not known, but I suspect the real significance lies in the protein itself. The protein is a component of the respiratory chain, the long chain of proteins responsible for passing electrons to oxygen to generate energy. It is not just any component, but part of the first functional complex of the chain, *complex 1* (NADH dehydrogenase). Complex 1 is a notoriously weak point in the chain, and the source of almost all escaping oxygen free radicals. I know of no studies to prove the point, but would be surprised if the single-letter change did not have an inordinately large effect on free-radical leakage from mitochondria. I will go further: this is exactly the kind of evolutionary change we would expect to find in bird mitochondria, making them more leak-proof. The pressure to select such changes in birds is much higher than in people, because flight itself demands very efficient energy production per gram body weight (the flight muscle needs to be lightweight and powerful — efficient — to enable flight at all).

Mt5178A is not the only mitochondrial variant to be linked with ageing and disease in people. Several others have been identified, although their effects are less pervasive. We get a sense of their overall importance from a looser relationship: the maternal inheritance of longevity. Mitochondria, as we have seen, are only passed on in the egg, so all 13 mitochondrial genes come from our mothers. If these genes really do influence lifespan, and we can only inherit them from our mothers, then our own lifespan should reflect that of our mothers but not our fathers. This seems to be the case, despite the many other factors that impinge on survival, and was recognized as long ago as the nineteenth century by the American physician, poet and humorist, Oliver Wendell Holmes. In one of his famous 'breakfast-table' essays, Holmes wrote that to achieve longevity one should not only choose one's parents wisely, but "especially let the mother come from a race in which octogenarians and nonagenarians are very common phenomena."

All this is very well, but what can we do about it? A few researchers talk glibly of transplanting foetal mitochondria into adult cells (or more technically 'gene therapy by mitochondrial transfer') but the idea is absurd as a 'cure' for ageing. We have an average of 100 mitochondria in each cell, so each of us harbours around 1.5 million billion mitochondria. I find it hard to imagine that we could make a real difference to such a large population simply by injecting a few new mitochondria. On the other hand, it is quite feasible to inject mitochondria into an egg cell. This has already been done as a fertility treatment, by injecting the contents of an egg from a fertile woman, along with a donor sperm, into the egg of an infertile woman — a procedure known as ooplasmic transfer. At least 30 babies have already been born using this technique, of whom the eldest celebrated his fourth birthday in June 2001. Even given the personal happiness that fertility treatments can bring, however, I find it hard to welcome 'reprogenetic technologies to shape future children', let alone to shape the elderly.

Setting aside the ethical objections to ooplasmic transfer there are still some difficult technical considerations. Egg cells are subject to natural selection. Of the 7 million eggs that develop in the female foetus, only a few hundred ever come close to ovulation in sexual maturity: one in 20 000. The basis of this selection is shrouded in mystery, but there seems to be a sophisticated cross-talk between the nucleus and the mitochondria, which is even influenced by the spatial distribution of mitochondria within the egg. Essentially, if the mitochondria aren't right, the egg never makes it.

If an egg is forced to develop artificially, the offspring frequently suffer from bioenergetic diseases. This problem may go some way towards explaining the disturbingly high failure rate of cloning, in which an alien nucleus is inserted into an egg from which the nucleus has been removed, and development is stimulated by an electric shock. John Allen, whom we met in Chapter 13, and his wife Carol develop this argument. They attribute the premature ageing of cloned animals like Dolly the sheep to a contamination of the egg cell with mitochondria. Dolly was cloned by fusing a whole somatic cell, including its mitochondria, with an enucleate egg. According to this argument, Dolly is ageing prematurely (she developed arthritis, for example, by the age of five) because many of her mitochondria came from a cell taken from a sheep that was already six years old. Dolly is therefore mutton dressed as lamb. Her biological age is probably closer to 11 than to 5. The Allens spelled out many practical ways of testing this theory in a paper published in 1999 (see Further Reading).

The amazing fact is that ooplasmic transfer and cloning ever work at all. No doubt many of the technical problems can be ironed out in time, although as far as preventing ageing is concerned we must ask ourselves, as a society, whether we would even wish to try. But turning away from such genetic manipulations, what else *can* we do? We are learning all the time about how mitochondria differ between species, and how our own mitochondria change in the course of our lives. Such differences are controlled not just by genes, but also by diet, activity and hormones.

One difference between species that correlates well with longevity is the lipid composition of mitochondrial membranes. All biological membranes are composed of a lipid bilayer, in which the water-hating tails of the lipid molecules in both layers point to the inside of the membrane. The bilayer is studded with proteins, which float like islands of pumice in a fatty sea. The inner mitochondrial membrane is especially rich in proteins — these make up the hundreds of respiratory chains that generate energy for the cell. Sixty per cent of the mitochondrial membrane is made of protein. As in an engine, the function of the respiratory chains depends on their 'lubrication'. This is provided by the lipid components of the membrane. The exact composition of these lipids has a profound effect on the function of mitochondria, just as the oil alters the behaviour of an engine. If the lubricant is not effective, the mitochondria leak more free radicals and generate less energy, which contributes to cell damage and metabolic insufficiency. In mitochondria, the lubricant of choice is called *cardiolipin*.

Each cardiolipin molecule incorporates four fatty acids, which can be unsaturated (containing double bonds) or saturated (not containing double bonds). Unsaturated fats keep the membrane fluid (in the same way that unsaturated oils are more fluid than saturated lards). This is because the double bonds kink the fatty-acid chains, which prevents them from lining up in neat arrays (making it harder for them to set). There is a price for fluidity, however: double bonds are easily oxidized. Some sort of compromise is necessary. The best compromise varies according to the kind of performance required. For example, a high metabolic rate requires a fluid membrane, while a long life demands resistance to oxidation.

With this in mind, Reinald Pamplona, Gustavo Barja, and their colleagues at the University of Lleida, in Spain, compared the fatty acid composition of mitochondria from different species, from rats to horses and pigeons to parakeets. They found a striking relationship. Animals with long lifespans had low levels of highly unsaturated fatty acids, such as docosahexanoic acid (with six double bonds) and arachidonic acid (with four double bonds) but much higher levels of slightly unsaturated fatty acids with two or three double bonds, such as linoleic acid. In other words, the longer the lifespan, the lower the level of unsaturation. The exact lipid composition varies somewhat with diet, but is largely refractory to change: animals convert one fatty acid into another to meet the requirements of their mitochondria. For example, the staple diet of laboratory mice contains no docosahexanoic acid (it is easily oxidized), yet their mitochondria contain 8 per cent. In contrast, horse fodder is rich in the precursors of docosahexanoic acid, but horse mitochondria contain only 0.4 per cent. We are left with a problem: the composition of mitochondria affects their function and our lifespan, but is not easy to alter by diet.

Worse follows. Animals get 'more unsaturated' as they age. Old rats double their content of highly unsaturated fatty acids, while the proportion of less-unsaturated fatty acids falls correspondingly. As a result, mitochondria become more vulnerable to oxidation with age, and lose their lubricant, cardiolipin. The cardiolipin content of rat mitochondria halves by old age. Similar changes probably take place in us. Thus, for a long life we must restrict the proportion of highly unsaturated lipids in our mitochondria, yet as we get older the proportion, contrarily, increases. If diet can only help a little, is there anything else we can do about it?

The answer is almost certainly yes. The composition of mitochondria is only partly influenced by diet, but equally, the changes that take place as we age are only partly controlled by changes in our genes. By this, I mean that the *sequence* of genes often remains inviolate, but their activity — whether or not they are expressed, or how much they are expressed — almost invariably changes. To reverse the changes that take place as we age, we need to reverse the changes in gene expression, and this is much easier than altering the sequence of the genes themselves. In rats, for example, calorie restriction can reverse the age-related changes in mitochondrial composition and function, making mitochondria less vulnerable to oxidation. In other words, the inexorable decline in mitochondrial function with age is partly physiological, and not purely pathological. Whether calorie restriction can orchestrate similar changes in people is unknown, but I see no reason why not.

Intriguingly, carnitine may exert similar effects. We met carnitine in Chapter 9 in relation to vitamin C. We need it to shuttle fats into the mitochondria for use as fuel, and to remove the left-over organic acids. We can synthesize carnitine ourselves, using vitamin C, but we also eat some in our food. One of the symptoms of scurvy is general lassitude, which may be explained by carnitine deficiency. Carnitine supplements have been used for many years (with regulatory approval) as energy-boosters, and to protect against heart weakness and muscle wasting. Its effects go beyond a shuttle-bus service: carnitine alters the lipid composition of mitochondrial membranes, restoring the cardiolipin content to youthful levels. These effects are not just cosmetic: old rats gain energy and are twice as active when fed carnitine.

Carnitine is no panacea, however: it also increases free-radical leakage and oxidative stress. This may help explain its disappointing record in age-related diseases such as Alzheimer's disease. Even so, the pro-oxidant effects can be suppressed using antioxidants such as lipoic acid, and this particular combination holds promise. In a series of papers published in the *Proceedings of the National Academy of Sciences of the USA* in February 2002, Bruce Ames and his colleagues at the University of California, Berkeley, reported that carnitine, given together with lipoic acid, improved the mitochondrial function and integrity of old rats, and boosted their energy levels. As Bruce Ames put it, "These old rats got up and did the Macarena." The rats also performed better in various tests of memory and intelligence. How much benefit we might gain from carnitine is another open question, but is at least beginning to attract serious research interest, and clinical trials are now underway. Presumably, high-dose vitamin C might boost carnitine synthesis in old age too, although surprisingly little is known about this: perhaps we have focused too tightly on its antioxidant properties.

Exercise itself benefits mitochondria. We saw in Chapter 13 that the health of a population of mitochondria reflects the rates of replication and breakdown. Damaged mitochondria are broken down more slowly than healthy mitochondria in old tissues. Because the rate of mitochondrial replication is very slow in such tissues, the damaged mitochondria ultimately take over. This vicious circle can be broken by gentle exercise. When we exercise, the higher demand for energy stimulates mitochondrial replication. The healthiest mitochondria now replicate fastest, and this regenerates the stock of viable mitochondria. As usual, there is a catch: vigorous exercise often causes more oxidative damage than it cures, and it is hard to know at what point we begin to do harm; gentle aerobic exercise, like walking or swimming, is probably about right. I wonder whether something similar applies to mental exercise. Education and mental activity tend to protect against Alzheimer's disease; *why* is unknown. It is feasible that intellectual exercise might keep the brain's mitochondrial population turning over, rejuvenating stock.

Mitochondrial medicine is a dynamic field set to expand in the coming years, and the tangible excitement is offset only by the humbling experience of antioxidant interventions in the past. We have learned a hard lesson: it is not good enough just to 'throw in' an antioxidant and hope for the best. We need to find a way of targeting the mitochondrial membranes, whether it be through metabolic boosters such as carnitine, antioxidants such as lipoic acid or coenzyme Q, hormones such as melatonin or thyroxine, or some factor I cannot even begin to imagine. It will almost certainly require an integrated physiological approach. We have a lot to learn about the way in which mitochondria work, and must expect setbacks, but I believe that here we are finally getting close to the heart of the problem. If we ever succeed in extending our lifespan to a healthy 130, I would be surprised if the big strides forward had not begun in mitochondrial medicine.

Viewing evolution through the prism of oxygen gives us some surprising perspectives on our own lives and deaths. If water is the foundation of life, then oxygen is its engine. Without oxygen, life on Earth would never have got beyond a slime in the oceans, and the Earth would probably have ended its days in the sterility of Mars or Venus. With oxygen, life has flourished in all its wonderful variety: animals, plants, sex, sexes, consciousness itself. With it, too, came the evolution of ageing and death.

We cannot hope to understand the complex degenerative diseases of old age unless we have an evolutionary grasp of their cause. Evolutionary theory can take us so far, but will fail unless backed by empirical evidence. In the same way, the sixteenth-century scientist Francis Bacon famously argued that philosophy could never answer the great questions of life and death without the guiding light of experiments. We should not forget that science was born from philosophy, in other words from a system of ideas about the world. Experiments allow us to weigh the value of competing ideas that cannot be discriminated on a logical basis; but for science to be meaningful, experiments must be conducted within the framework of an idea — a hypothesis — about how the world works. Science does not work by induction — by trawling piles of miscellaneous data in the hope of finding patterns or facts — but by hypothesis and refutation. Today, medical research is in danger of becoming too empirical, of accumulating tremendous piles of data without giving them due thought. There is an uncomfortable gap between the hundreds of crazy theories about ageing and disease, which are rarely supported by coherent data, and the headlong rush of medical research, which rarely finds time to interpret new findings in a wider context. In this age of excessive healthcare spending and failing healthcare systems we need to ask whether medical research is taking us in the right direction.

Genetic research has transformed our understanding of biology, health and disease. Many of the ideas in this book would have been unthinkable without the great advances in molecular genetics. But we should not mistake the tool for the solution. Insofar as there is any guiding philosophy behind medical research, it is that genes go wrong and cause disease. We celebrate completion of the human genome project because it tells us far more about which genes might go wrong. The time and money spent chasing defective genes for particular diseases dwarfs research into the underlying processes of ageing itself: there are thousands of specialized disciplinary journals, but just a handful devoted to the science of ageing. We get frustrated with the slow pace of research — a 'breakthrough' now may come to fruition in 20 years — but accept that this is so because the effects of genes are complex and intractable: we must just wait. Will the promises ever come to fruition, or are we being sold a line? The only way we can hazard a guess is by thinking in evolutionary terms, and this has the added bonus of giving us a clearer idea of what kind of approach might actually work.

The idea that oxygen might accelerate ageing is not new: it was implicit in Joseph Priestley's suggestion that we might 'burn out' faster, like a candle, if we breathed his pure oxygen. On the basis of experiments alone, we might reasonably claim that oxygen free radicals contribute to ageing and to some diseases, and are perhaps a consequence of others. From an empirical point of view, the failure of antioxidants to extend life or cure diseases suggests that the role of free radicals is limited, just one factor among many. An evolutionary perspective opens up quite a different vista. We see that life has learned to cope with oxygen through a myriad of adaptations, from behaviour to size to sex. The logic of the evolutionary view can be tested in unexpected ways, through predictions about the evolution of two sexes and the development of egg cells in the follicle, to the failure of cloning experiments or the impact of malaria on the diseases of old age. I hope I have convinced you, from this perspective, that oxygen is not just the engine of evolution and life, but also the single most important cause of ageing and age-related disease.

The crispness of this view is satisfying and helps us to see our place in nature. It is hopeful, as it shows us that ageing is neither programmed nor inevitable, even if it cannot easily be put off. It is corrective, for it shows us the fallacy of chasing 'susceptibility' genes for the diseases of old age. It is constructive, in that it points us to the fields of research that might best tackle the problem of ageing — immune modulation and mitochondrial medicine. And it is practical, for it offers us a rational guide to good health in old age: eat widely, but not too much, don't be obsessively clean or get overly stressed, don't smoke, take regular exercise, and keep an active mind. Start now! If all the advances of biology and medicine can do no more than explain the wisdom of our grandparents, may that restore some lost dignity to wise old age.

Further Reading

GENERAL TEXTS

Brown, G. *The Energy of Life*. HarperCollins, London, 1999.

Cairns-Smith, G. *Seven Clues to the Origin of Life*. Cambridge University Press, Cambridge, 1985.

Cowen, R. *History of Life*. Blackwells, New York, 2000.

Davies, P. *The Fifth Miracle. The Search for the Origin of Life*. Penguin Books, London, 1998.

Dawkins, R. *The Selfish Gene*. Oxford University Press, Oxford, 1989.

Djerassi, C. and Hoffman, R. *Oxygen*. Wiley-VCH, Weinheim, 2001.

Dyson, F. *Origins of Life*. Revised Edition. Cambridge University Press, Cambridge, 1999.

Emsley, J. *Molecules at an Exhibition*. Oxford University Press, Oxford, 1998.

Fenchal, T. and Finlay, B. J. *Ecology and Evolution in Anoxic Worlds*. Oxford University Press, Oxford, 1995.

Fortey, R. *Life: An Unauthorised Biography*. HarperCollins, London, 1997.

Fortey, R. *Trilobite!* Flamingo, London, 2001.

Gould, S. J. *Wonderful Life. The Burgess Shale and the Nature of History*. Penguin Books, London, 1989.

Hager, T. *Linus Pauling and the Chemistry of Life*. Oxford University Press, Oxford, 2000.

Halliwell, B. and Gutteridge, J. M. C. *Free Radicals in Biology and Medicine*, Third Edition. Oxford University Press, Oxford, 1999.

Holliday, R. *Understanding Ageing*. Cambridge University Press, Cambridge, 1995.

Hughes, R. E. *Vitamin C. Cambridge World History of Food* (Eds. Kiple, K. F. and Ornelas, K. C). Cambridge University Press, Cambridge, 2000.

Jacob, F. *Of Flies, Mice and Men*. Harvard University Press, Cambridge, 2001.

Jones, S. *The Language of the Genes*. Second edition. Flamingo, London, 2000.

Kirkwood, T. *The End of Age. Reith Lectures 2001*. Profile Books, London, 2001.

Kirkwood, T. *Time of Our Lives. Why Ageing is Neither Inevitable nor Necessary*. Phoenix, London, 2000.

Lovelock, J. *The Ages of Gaia: A Biography of Our Living Earth*. Oxford University Press, Oxford, 1995.

Margulis, L. and Sagan, D. *Microcosmos. Four Billion Years of Microbial Evolution*. University of California Press, Berkeley, 1986.

Maynard Smith, J. and Eörs Szathmáry, E. *The Origins of Life: From the Birth of Life to the Origin of Language*. Oxford University Press, Oxford, 1999.

Medawar, P. *An Unsolved Problem of Biology*. HK Lewis, London, 1952.

Nesse, R. M. and Williams, G. C. *Evolution and Healing*. Phoenix, London, 1995.

Porter, R. *The Greatest Benefit to Mankind*. HarperCollins, London, 1999.

Ridley, M. *Genome*. Fourth Estate, London, 1999.
Stearns, S. C. (Ed.). *Evolution in Health and Disease*. Oxford University Press, Oxford, 1999.
Tudge, C. *The Variety of Life*. Oxford University Press, Oxford, 2000.
Watson, J. *The Double Helix*. Penguin Books, London, 1999.
Weatherall, D. *Science and the Quiet Art*. Oxford University Press, Oxford, 1995.
Willcox, B. J., Willcox, C. and Suzuki, M. *The Okinawa Way*. Mermaid Books, London, 2001.

CHAPTER 1

Discovery of oxygen

Lavoisier, A. *Elements of Chemistry*. Dover Publications, New York, 1965 (first published Paris, 1789).
Priestley, J. *Experiments and Observations on Different Kinds of Air*. Birmingham, 1775.
Szydlo, Z. A new light on alchemy. *History Today* **47:** 17–24; 1997.
Szydlo, Z. *Water Which Does Not Wet Hands. The Alchemy of Michael Sendivogius*. Polish Academy of Sciences, Warsaw, 1994.
Bernard Jaffe. *Crucibles*. Newton Publishing Co, New York, 1932.

Oxygen therapies

Haldane, J. S. *Respiration*. Yale University Press, New Haven, 1922.
Greif, R., Akca, O., Horn, E. P., Kurz, A., and Sessler, D. I. Supplemental perioperative oxygen to reduce the incidence of surgical-wound infection. *New England Journal of Medicine* **342:** 161–167; 2000.

Diving and barometric pressure

Martin, L. *Scuba Diving Explained: Questions and Answers on Physiology and Medical Aspects of Scuba Diving*. Best Publishing Co, Flagstaff, AZ, 1999.
Ashcroft, F. *Life at the Extremes: The Science of Survival*. HarperCollins, London, 2000.
Bert, P. *La Pression Barometrique*. Paris, 1878.
Haldane, J. B. S. *Possible Worlds and Other Essays*. Chatto and Windus, London, 1930.

CHAPTER 2

Factors controlling oxygen in the atmosphere

Berner, R. A. Biogeochemical cycles of carbon and sulfur and their effect on atmospheric oxygen over Phanerozoic time. *Palaeogeography, Palaeoclimatology, Palaeoecology* **75:** 97–122; 1988.

Oxygen and evolution

Cloud, P. Atmospheric and hydrospheric evolution on the primitive earth. *Science* **160:** 729–736; 1968.

Knoll, A. H. and Holland, H. D. Oxygen and Proterozoic evolution: an update. In *Effects of Past Global Change on Life* (Eds.: Panel on Effects of Past Global Change on Life). National Academy of Sciences, Washington, DC, 1995.

CHAPTER 3

Spiegelman's monsters and loss of complexity

Spiegelman, S. An *in vitro* analysis of a replicating molecule. *American Scientist* **55:** 3–68; 1967.

First signs of life and carbon isotopes

Mojzsis, S. J., Arrhenius, G., McKeegan, K. D., Harrison, T. M., Nutman, A. P. and Friend, C. R. L. Evidence for life on earth before 3,800 million years ago. *Nature* **384:** 55–59; 1996.

Molecular fossils of cyanobacteria and eukaryotes

Brocks, J. J., Logan, G. A., Buick, R. and Summons, R. E. Archean molecular fossils and the early rise of eukaryotes. *Science* **285:** 1033–1036; 1999.

Knoll, A. H. A new molecular window on early life. *Science* **285:** 1025–1026; 1999.

Canfield, D. E. A breath of fresh air. *Nature* **400:** 503–504; 1999.

Banded iron formations (see also General texts)

Widdel, F., Schnell, S., Heising, S., Ehrenreich, A., Assmus, B. and Schink, B. Ferrous iron oxidation by anoxygenic phototrophic bacteria. *Nature* **362:** 834–836; 1993.

Noah's Flood and the Black Sea

Ryan, W., Pitman, W. and Haxby, W. (illustrator). *Noah's Flood: the New Scientific Discoveries about the Event that Changed History*. Simon and Schuster, New York, 1999.

Sulphur isotopes, iron pyrite and oxygen

Canfield, D. E. A new model of Proterozoic ocean chemistry. *Nature* **396:** 450–452; 1998.

Canfield, D. E, Habicht, K. S. and Thamdrup, B. The Archean sulfur cycle and the early history of atmospheric oxygen. *Science* **288:** 658–661; 2000.

Natural nuclear reactors in Gabon

Cowan, G. A. A natural fission reactor. *Scientific American* **235:** 36–41; 1976.

Snowball Earth and Kalahari manganese field

Kirschvink, J. L., Gaidos, E. J., Bertani, L. E., Beukes, N. J., Gutzmer, J., Maepa, L. N. and Steinberger, R. E. Paleoproterozoic snowball earth: extreme climatic and geochemical global change and its biological consequences. *Proceedings of the National Academy of Sciences USA* **97:**1400–1405; 2000.

Oxygen and eukaryotic evolution (see also General texts)

Rye, R. and Holland, H. D. Paleosols and the evolution of atmospheric oxygen: a critical review. *American Journal of Science* **298:** 621–672; 1998.

Knoll, A. H. The early evolution of eukaryotes: a geological perspective. *Science* **256:** 622–627; 1992.

Kurland, C. G. and Andersson, S. G. E. Origin and evolution of the mitochondrial proteome. *Microbiology and Molecular Biology Reviews* **64:** 786–820; 2000.

CHAPTER 4

Evolution of early animals (see also General texts)

Nash, M. When life exploded. *Time Magazine* **146:** 66–74; 4 December, 1995.

Briggs D. E. G. and Fortey, R. A. The early radiation and relationships of the major arthropod groups. *Science* **246:** 241–243; 1989.

Knoll, A. H and Carroll, S. B. Early animal evolution: emerging views from comparative biology and geology. *Science* **284:** 2129–2137; 1999.

Valentine, J. W. Late Precambrian bilatarians: grades and clades. *Proceedings of the National Academy of Sciences USA* **91:** 6751–6757; 1994.

Molecular clocks

Conway Morris, S. Molecular clocks: defusing the Cambrian explosion? *Current Biology* **7:** R71–R74; 1997.

Bromham, L., Rambaut, A., Fortey R., Cooper, A. and Penny, D. Testing the Cambrian explosion hypothesis by using a molecular dating technique. *Proceedings of the National Academy of Sciences USA* **95:** 612386–612389; 1998.

Ayala, J., Rzhetsky, A. and Ayala, F. J. Origin of metazoan phyla: molecular clocks confirm paleontological estimates. *Proceedings of the National Academy of Sciences USA* **95:** 606–611; 1998.

Snowball Earth

Hoffman, P. F., Kaufman, A. J., Halverson, G. P. and Schrag, D. P. A Neoproterozoic snowball Earth. *Science* **281:** 1342–1346; 1998.

Hoffman, P. F. and Schrag, D. P. Snowball Earth. *Scientific American* January 2000.

Walker, G. Snowball Earth. *New Scientist* 6th November 1999.

Isotope ratios and oxygen

Canfield, D. E. and Teske, A. Late Proterozoic rise in atmospheric oxygen concentration inferred from phylogenetic and sulphur-isotope studies. *Nature* **382:** 127–132; 1996.

Knoll, A. H. Breathing room for early animals. *Nature* **382:** 111–112; 1996.

Kaufman, A. J., Jacobsen, S. B. and Knoll, A. H. The Vendian record of C- and Sr-isotopic variations: Implications for tectonics and paleoclimate. *Earth and Planetary Science Letters* **120:** 409–430; 1993.

Brasier, M. D., Shields, G. A., Kuleshov, V. N. and Zhegallo, E. A. Integrated chemo- and bio-stratigraphic calibration of early animal evolution: Neoproterozoic — early Cambrian of southwest Mongolia. *Geological Magazine* **133:** 445–485; 1996.

Logan, G. A., Hayes, J. M., Hieshima, G. B. and Summons, R. E. Terminal Proterozoic reorganization of biogeochemical cycles. *Nature* **376:** 53–56; 1995.

CHAPTER 5

Giant dragonflies

Rutten, M. G. Geologic data on atmospheric history. *Palaeogeography, Palaeoclimatology, Palaeoecology* **2:** 47–57; 1966.

Wakeling, J. M. and Ellington, C. P. Dragonfly flight. III. Lift and power requirements. *Journal of Experimental Biology* **200:** 583–600; 1997.

Fires and methane generation

Watson, A., Lovelock, J. E. and Margulis L. Methanogenesis, fires and the regulation of atmospheric oxygen. *Biosystems* **10:** 293–298; 1978.

Photorespiration

Beerling, D. J., Woodward, F. I., Lomas, M. R., Wills, M. A., Quick, W. P. and Valdes P. J. The influence of Carboniferous palaeo-atmospheres on plant function: an experimental and modelling assessment. *Philosophical Transactions of the Royal Society of London. B.* **353:** 131–140; 1998.

Beerling, D. J. and Berner, R. A. Impact of a Permo-Carboniferous high O_2 event on the terrestrial carbon cycle. *Proceedings of the National Academy of Sciences USA* **97:** 12428–12432; 2000.

Carbon burial and calculation of atmospheric oxygen (see also references in Chapter 2)

Berner, R. A. and Canfield D. E. A new model for atmospheric oxygen over Phanerozoic time. *American Journal of Science* **289:** 333–361:1989.

Gas bubbles in amber

Berner, R. A. and Landis, P. Gas bubbles in fossil amber as possible indicators of the major gas composition of ancient air. *Science* **239:** 1406–1409; 1988. Technical comments on Berner and Landis. *Science* **241:** 717–724; 1988.

Carbon isotopes and calculation of atmospheric oxygen

Berner, R. A., Petsch, S. T., Lake, J. A., Beerling, D. J., Popp, B. N., Lane, R. S., Laws, E. A., Westley, M. B., Cassar, N., Woodward, F. I. and Quick, W. P. Isotopic fractionation and atmospheric oxygen: implications for Phanerozoic O_2 evolution. *Science* **287:** 1630–1633; 2000.

Plant adaptations to fire and fossil charcoal

Robinson, J. M. Phanerozoic O_2 variation, fire and terrestrial ecology. *Palaeogeography, Palaeoclimatology, Palaeoecology* **75:** 223–240; 1989.

Jones, T. P. and Chaloner, W. G. Fossil charcoal, its recognition and palaeoatmospheric significance. In: Kump, L. R., Kasting, J. F. and Robinson, J. M.

(Eds.), *Atmospheric Oxygen Variation through Geologic Time. Global and Planetary Change* **5**: 39–50; 1991.

K–T boundary and global firestorm and tsunami

Wolbach, W. S., Lewis, R. S., Anders, E., Orth, C. J. and Brooks, R. R. Global fire at the Cretaceous–Tertiary boundary. *Nature* **334**: 665–669; 1988.

Kruger, M. A., Stankiewicz, B. A., Crelling, J. C., Montanari, A. and Bensley, D. F. Fossil charcoal in Cretaceous–Tertiary boundary strata: evidence for catastrophic firestorm and megawave. *Geochimica et Geophysica Acta* **58**: 1393–1397; 1994.

Flight mechanics of dragonflies in high-oxygen atmospheres

Graham, J. B., Dudley, R., Aguilar, N. M. and Gans, C. Implications of the late Palaeozoic oxygen pulse for physiology and evolution. *Nature* **375**: 117–120; 1995.

Dudley, R. Atmospheric oxygen, giant paleozoic insects and the evolution of aerial locomotor performance. *Journal of Experimental Biology* **201**: 1043–1050; 1998.

Harrison, J. F. and Lighton J. R. B. Oxygen-sensitive flight metabolism in the dragonfly *Erythemis simplicicollis*. *Journal of Experimental Biology* **201**: 1739–1744; 1998.

Polar gigantism and oxygen

Chapelle, G. and Peck, L. S. Polar gigantism dictated by oxygen availability. *Nature* **399**: 114–115; 1999.

CHAPTER 6

Life of Marie Curie

Quinn, S. *Marie Curie: A Life*. Simon & Schuster, New York, 1995.

Radiation poisoning, radium girls and Hiroshima

Clark, C. *Radium Girls: Women and Industrial Health Reform, 1910–1935*. University of North Carolina Press, Chapel Hill, 1997.

Hersey, J. *Hiroshima*. Penguin Books, London, 1990.

Radiation chemistry

Von Sonntag, C. *Chemical Basis of Radiation Biology*. Taylor and Francis, London, 1987.

Oxygen free radicals

Fridovich, I. Oxygen is toxic! *Bioscience* **27**: 462–466; 1977.

Gerschman, R., Gilbert, D. L., Nye, S. W., Dwyer, P. and Fenn W. O. Oxygen poisoning and X-irradiation: A mechanism in common. *Science* **119**: 623–626; 1954.

Gilbert, D. L. Fifty years of radical ideas. *Annals of the New York Academy of Science* **899**: 1–14; 2000.

Liquefaction of oxygen

Wilson, D. *Supercold. An Introduction to Low Temperature Technology.* Faber and Faber, London, 1979.

Free radical damage from breathing

Shigenaga, M. K., Gimeno, C. J. and Ames B. N. Urinary 8-hydroxy-2′-deoxyguanosine as a biological marker of *in vivo* oxidative DNA damage. *Proceedings of the National Academy of Sciences USA* **86:** 9697–9701; 1989.

Radiation tolerance in bacteria

Hoyle, F. *The Intelligent Universe.* Michael Joseph, London, 1983.

White, O., Eisen, J. A. and Heidelberg J. F., *et al.* Genome sequence of the radioresistant bacterium *Deinococcus radiodurans* R1. *Science* **286:** 1571–1577; 1999.

Surface of Mars

Oyama, V. I. and Berdahl B. J. The Viking gas exchange experiment results from Chryse and Utopia surface samples. *Journal of Geophysical Research* **82:** 4669–4676; 1977.

CHAPTER 7

Evolution of photosynthesis

Des Marais, D. When did photosynthesis emerge on Earth? *Science* **289:** 1703–1705; 2000.

Xiong, J., Fischer, W. M., Inoue, K., Nakahara, M. and Bauer, C. E. Molecular evidence for the early evolution of photosynthesis. *Science* **289:** 1724–1730; 2000.

Hartman, H. Photosynthesis and the origin of life. *Origins of Life and Evolution of the Biosphere* **28:** 515–521; 1998.

Schiller, H., Senger, H., Miyashita, H., Miyachi, S. and Dau, H. Light-harvesting in *Acaryochloris marina* — spectroscopic characterization of a chlorophyll *d*-dominated photosynthetic antenna system. *FEBS Letters* **410:** 433–436; 1997.

Hoganson, C. W., Pressler, M. A., Proshlyakov, D. A. and Babcock, G. T. From water to oxygen and back again: mechanistic similarities in the enzymatic redox conversions between water and dioxygen. *Biochimica et Biophysica Acta* **1365:** 170–174; 1998.

Catalase and the oxygen-evolving complex

Blankenship, R. E. and Hartman, H. The origin and evolution of oxygenic photosynthesis. *Trends in Biological Sciences* **23:** 94–97; 1998.

Ioannidis, N., Schansker, G., Barynin, V. V. and Petrouleas, V. Interaction of nitric oxide with the oxygen evolving complex of photosystem II and manganese catalase: a comparative study. *Journal of Biological and Inorganic Chemistry* **5:** 354–563; 2000.

Hydrogen peroxide on the early Earth

Kasting, J., Holland, H. D. and Pinto, J. P. Oxidant abundances in rainwater and the evolution of atmospheric oxygen. *Journal of Geophysical Research* **90:** 10497–10510; 1985.

Kasting, J. F. Earth's early atmosphere. *Science* **259:**920–926; 1993.

McKay, C. P. and Hartman, H. Hydrogen peroxide and the evolution of oxygenic photosynthesis. *Origins of Life and Evolution of the Biosphere* **21:** 157–163; 1991.

CHAPTER 8

Chimpanzee and human genomes

Chen, F. C. and Li, W. H. Genomic divergences between humans and other hominoids and the effective population size of the common ancestor of humans and chimpanzees. *American Journal of Human Genetics* **68:** 444–456; 2001.

Eukaryotes and mitochondria

Gray, M. W., Burger, G. and Lang, B. F. Mitochondrial evolution. *Science* **283:** 1476–1481; 1999.

Kurland, C. G. and Andersson, S. G. E. Origin and evolution of the mitochondrial proteome. *Microbiology and Molecular Biology Reviews* **64:** 86–820; 2000.

Last Universal Common Ancestor

Woese, C. Interpreting the universal phylogenetic tree. *Proceedings of the National Academy of Sciences USA* **97:** 8392–8396; 2000.

Woese, C. The universal ancestor. *Proceedings of the National Academy of Sciences USA* **95:** 6854–6859; 1998.

Doolittle, W. F. and Brown, J. R. Tempo, mode, the progenote, and the universal root. *Proceedings of the National Academy of Sciences USA* **91:** 6721–6728; 1994.

Evolution of cytochrome oxidase and aerobic respiration

Castresana, J. and Saraste, M. Evolution of energetic metabolism: the respiration-early hypothesis. *Trends in Biological Sciences* **20:** 443–448; 1995.

Castresana, J. and Moreira, D. Respiratory chains in the last common ancestor of living organisms. *Journal of Molecular Evolution* **49:** 453–460; 1999.

Castresana, J., Lübben, M. and Saraste, M. New Archaebacterial genes coding for redox proteins: implications for the evolution of aerobic metabolism. *Journal of Molecular Biology* **250:** 202–210; 1995.

Castresana, J., Lübben, M., Saraste, M. and Higgins, D. G. Evolution of cytochrome oxidase, an enzyme older than atmospheric oxygen. *EMBO Journal* **13:** 2516–2525; 1994.

Hoganson, C. W., Pressler, M. A., Proshlyakov, D. A. and Babcock, G. T. From water to oxygen and back again: mechanistic similarities in the enzymatic redox conversions between water and dioxygen. *Biochimica et Biophysica Acta* **1365:** 170–174; 1998.

Haemoglobins and cytochrome oxidase

Preisig, O., Anthamatten, D. and Hennecke H. Genes for a microaerobically induced oxidase complex in *Bradyrhizobium japonicum* are essential for a nitrogen-fixing endosymbiosis. *Proceedings of the National Academy of Sciences USA* **90:** 3309–3313; 1993.

Shaobin, H., Larsen, R. W., Boudko, D., *et al.* Myoglobin-like aerotaxis transducers in Archaea and Bacteria. *Nature* **403:** 540–544; 2000.

Trotman, C. Life: All the time in the world? *The Biologist* **45:** 76–80; 1998.

CHAPTER 9

Fruit, vegetables and vitamin C

Key, T. J., Thorogood, M., Appleby, P. N. and Burr, M. L. Dietary habits and mortality in 11,000 vegetarians and health-conscious people: results of a 17-year follow up. *British Medical Journal* **313:** 775–779; 1996.

Gutteridge, J. M. C. and Halliwell, B. Free radicals and antioxidants in the year 2000: a historical look to the future. *Annals of the New York Academy of Sciences* **899:** 136–147; 2000.

Khaw, K. T., Bingham, S., Welch, A., Luben, R., Wareham, N., Oakes, S. and Day, N. Relation between plasma ascorbic acid and mortality in men and women in EPIC-Norfolk prospective study: a prospective population study. *Lancet* **357:** 657–663; 2001.

Recommended daily allowances

Levine, M., Conry-Cantilena, C. and Wang, Y., *et al.* Vitamin C pharmacokinetics in healthy volunteers: Evidence for a recommended dietary allowance. *Proceedings of the National Academy of Sciences USA* **93:** 3704–3709; 1996.

Panel on Dietary Antioxidants and Related Compounds. *Dietary Reference Intakes for Vitamin C, Vitamin E, Selenium and Carotenoids.* National Academy Press, Washington, DC, 2000.

Mechanisms and functions of vitamin C (and see General texts)

Levine, M., Dhariwal, K. R., Washko, P. W., Welch, R. W. and Yang, Y. Cellular functions of ascorbic acid: a means to determine vitamin C requirement. *Asia Pacific Journal of Clinical Nutrition* **2** (suppl. 1):5–13; 1993.

Padayatty, S. J. and Levine, M. New insights into the physiology and pharmacology of vitamin C. *Canadian Medical Association Journal* **164:** 353–355; 2001.

Wang, W., Russo, T., Kwon, O., Chanock, S., Rumsey, S. and Levine, M. Ascorbate recycling in human neutrophils: induction by bacteria. *Proceedings of the National Academy of Sciences USA* **94:** 13816–13819; 1997.

McLaran, C. J., Bett, J. H., Nye, J. A. and Halliday J. W. Congestive cardiomyopathy and haemochromatosis — rapid progression possibly accelerated by excessive ingestion of ascorbic acid. *Australia New Zealand Journal of Medicine* **12:** 187–188; 1982.

CHAPTER 10

Avoidance of oxygen (see also General texts)

Bilinski, T. Oxygen toxicity and microbial evolution. *Biosystems* **24:** 305–312; 1991.

Superoxide dismutase

McCord, J. M. and Fridovich, I. Superoxide dismutase. An enzymic function for erythrocuprein (hemocuprein). *Journal of Biological Chemistry* **244:** 6049–6055; 1969.

Fridovich, I. Oxygen toxicity: a radical explanation. *Journal of Experimental Biology* **201:** 1203–1209; 1998.

Lebovitz, R. M., Zhang, H., Vogel, H., Cartwright, J., Dionne, L., Lu, N., Huang, S. and Matzuk M. M. Neurodegeneration, myocardial injury and perinatal death in mitochondrial superoxide dismutase-deficient mice. *Proceedings of the National Academy of Sciences USA* **93:** 9782–9787; 1996.

Peroxiredoxins

Chae, H. Z., Robison, K., Poole, L. B., Church, G., Storz, G., Rhee, S. G. Cloning and sequencing of thiol-specific antioxidant from mammalian brain: alkyl hydroperoxide reductase and thiol-specific antioxidant define a large family of antioxidant enzymes. *Proceedings of the National Academy of Sciences USA* **91:** 7017–7021; 1994.

McGonigle, S., Dalton, J. P. and James, E. R. Peroxidoxins: a new antioxidant family. *Parasitology Today* **14:** 139–145; 1998.

Thiol oxidation, signalling and stress proteins

Arrigo, A. P. Gene expression and the thiol redox state. *Free Radical Biology and Medicine* **27:** 936–944; 1999.

Marshall, H. E., Merchant, K. and Stamler, J. S. Nitrosation and oxidation in the regulation of gene expression. *FASEB Journal* **14:** 1889–1900; 2000.

Groves, J. T. Peroxynitrite: reactive, invasive and enigmatic. *Current Opinion in Chemical Biology* **3:** 226–235; 1999.

Yachie, A., Niida, Y., Wada, T., Igarashi, N., Kaneda, H., Toma, T., Ohta, K., Kasahara, Y. and Koizumi, S. Oxidative stress causes enhanced endothelial cell injury in human heme oxygenase-1 deficiency. *Journal of Clinical Investigation* **103:** 129–135; 1999.

Cai, L., Satoh, M., Tohyama, C. and Cherian, M. G. Metallothionein in radiation exposure: its induction and protective role. *Toxicology* **132:** 85–98; 1999.

Foresti, R., Clark, J. E., Green, C. J. and Motterlini, R. Thiol compounds interact with nitric oxide in regulating heme-oxygenase-1 induction in endothelial cells. Involvement of superoxide and peroxynitrite anions. *Journal of Biological Chemistry*. **272:** 18411–18417; 1997.

Motterlini, R., Foresti, R., Bassi, R., Calabrese, V., Clark, J. E. and Green, C. J. Endothelial heme oxygenase-1 induction by hypoxia: modulation by inducible nitric oxide synthase and S-nitrosothiols. *Journal of Biological Chemistry* **275:** 13613–13620; 2000.

CHAPTER 11

Replication (see also General texts)

Orgel, L. E. The origin of life on the earth. *Scientific American* **271:** 76–83; 1994.

Sexual reproduction (see also General texts)

Atmar, W. On the role of males. *Animal Behaviour* **41:** 195–205; 1991.
Clark, W. *Sex and the Origins of Death*. Oxford University Press, Oxford, 1998.

Disposable soma theory

Kirkwood, T. B. L. Evolution of ageing. *Nature* **270**:301–304; 1977.
Kirkwood, T. B. L. and Holliday, R. The evolution of ageing and longevity. *Proceedings of the Royal Society of London B* **205:** 531–546; 1979.

Modulating lifespan

Austad, S. N. Retarded senescence in an insular population of Virginia opossums (*Didelphis virginiana*). *Journal of Zoology* **229:** 695–708; 1993.
Rose, M. R. Can human aging be postponed? *Scientific American* **281:** 106–111; 1999.
Westendorp, R. G. and Kirkwood, T. B. L. Human longevity at the cost of reproductive success. *Nature* **396:** 743–746; 1998.

CHAPTER 12

Pacific salmon and senescence (see also General texts)

Partridge L. and Barton N. H. Optimality, mutation and the evolution of ageing. *Nature* **362:** 305–311; 1993.

Late-acting genes and antagonistic pleiotropy (see also General texts)

Haldane, J. B. S. *New Paths in Genetics*. Harper, London, 1942.
Williams, G. C. Pleiotropy, natural selection and the evolution of senescence. *Evolution* **11:** 398–411; 1957.
Shokeir, M. H. Investigation on Huntington's disease in the Canadian Prairies. II. Fecundity and fitness. *Clinical Genetics* **7:** 349–353; 1975.
Walker, D. A., Harper, P. S., Newcombe, R. G. and Davies, K. Huntington's chorea in South Wales: mutation, fertility, and genetic fitness. *Journal of Medical Genetics* **20:** 12–17; 1983.

Genes in nematode worms

Friedman, D. B. and Johnson, T. E. A mutation in the *age-1* gene in *Caenorhabditis elegans* lengthens life and reduces hermaphrodite fertility. *Genetics* **118:** 75–86; 1988.
Kenyon, C., Chang, J., Gensch, E., Rudner, A. and Tabtiang, R. A *C. elegans* mutant that lives twice as long as wild type. *Nature* **366:** 404–405; 1993.

Morris, J. Z., Tissenbaum, H. A. and Ruvkun, G. A phosphotidylinositol-3-OH kinase family member regulating longevity and diapause in *Caenorhabditis elegans*. *Nature* **382:** 536–539; 1996.

Kimura, K. D., Tissenbaum, H. A. and Ruvken G. *daf-2*, an insulin-receptor-like gene that regulates longevity and diapause in *Caenorhabditis elegans*. *Science* **277:** 942–946; 1997.

Ogg, S., Paradis, S., Gottlieb, S., Patterson, G. I., Lee, L., Tissenbaum, H. A. and Ruvkun, G. The fork head transcription factor DAF-16 transduces insulin-like metabolic and longevity signals in *C. elegans*. *Nature* **389:** 994–999; 1997.

Insulin and insulin-like growth factors

Tissenbaum, H. A. and Ruvkun, G. An insulin-like signalling pathway affects both longevity and reproduction in *Caenorhabditis elegans*. *Genetics* **148:** 703–717; 1998.

Clancy, D., Gems, D., Harshman, L. G., Oldham, S., Stocker, H., Hafen, E., Leevers, S. J. and Partridge, L. Extension of lifespan by loss of CHICO, a *Drosophila* insulin receptor substrate protein. *Science* **292:** 104–106; 2001.

Thrifty genes and insulin-resistance

Chukwuma, C. Sr. and Tuomilehto, J. The 'thrifty' hypotheses: clinical and epidemiological significance for non-insulin-dependent diabetes mellitus and cardiovascular disease risk factors. *Journal of Cardiovascular Risk* **5:** 11–23; 1998.

Groop, L. C. Insulin resistance: the fundamental trigger of type 2 diabetes. *Diabetes, Obesity and Metabolism* **1** (suppl. 1): S1–S7; 1999.

CHAPTER 13

Rate of living, metabolism and free radicals

Pearl, R. *The Rate of Living*. Knopf, New York, 1928.

Harman, D. Aging: a theory based on free radical and radiation chemistry. *Journal of Gerontology* **11:** 298–300; 1956.

Birds, metabolic rate and free radical production

Austad, S. N. Birds as models of aging in biomedical research. *ILAR Journal* 38: 137–141; 1998.

Barja, G. Mitochondrial free radical production and aging in mammals and birds. *Annals of the New York Academy of Sciences* 854: 224–238; 1998.

Free radicals, stress resistance and ageing

Honda, Y. and Honda, S. The *daf-2* gene network for longevity regulates oxidative stress resistance and Mn-superoxide dismutase gene expression in *Caenorhabditis elegans*. *FASEB Journal* **13:** 1385–1393; 1999.

Barsyte, D., Lovejoy, D. A. and Lithgow, G. J. Longevity and heavy-metal resistance in daf-2 and age-1 long-lived mutants of *Caenorhabditis elegans*. *FASEB Journal* **15:** 627–634; 2001.

Orr, W. C. and Sohal, R. S. Extension of life-span by overexpression of superoxide dismutase and catalase in *Drosophila melanogaster*. *Science* **263:** 1128–1130; 1994.

Gray, M. D., Shen, J. C., Kamath-Loeb, A. S., Blank, A., Sopher, B. L., Martin, G. M., Oshima, J. and Loeb, L. A. The Werner syndrome protein is a DNA helicase. *Nature Genetics* **17:** 100–103; 1997.

Kapahi, P., Boulton, M. E. and Kirkwood, T. B. Positive correlation between mammalian lifespan and cellular resistance to stress. *Free Radical Biology and Medicine*. **26:** 495–500; 1999.

Calorie restriction

Sohal, R. S. and Weindruch, R. Oxidative stress, caloric restriction and aging. *Science* **273:** 59–63; 1996.

Kayo, T., Allison, D., Weindruch, R. and Prolla, T. A. Influences of aging and caloric restriction on the transcriptional profile of skeletal muscle from rhesus monkeys. *Proceedings of the National Academy of Sciences USA* **98:** 5093–5098; 2001.

Mitochondrial theory of ageing

Harman, D. The biological clock: the mitochondria? *Journal of the American Geriatric Society* **20:** 145–147; 1972.

Miquel, J. An update on the oxygen stress–mitochondrial mutation theory of aging: genetic and evolutionary implications. *Experimental Gerontology* **33:** 113–126; 1998.

Richter, C., Park, J. W. and Ames, B. N. Normal oxidative damage to mitochondrial and nuclear DNA is extensive. *Proceedings of the National Academy of Sciences USA* **85:** 6465–6467; 1988.

Beckman, K. B. and Ames, B. N. Endogenous oxidative damage of mitochondrial DNA. *Mutation Research* **424:** 51–58; 1999.

Kirkwood, T. B. and Kowald, A. A network theory of ageing: the interactions of defective mitochondria, aberrant proteins, free radicals and scavengers in the ageing process. *Mutation Research* **316:** 209–236; 1996.

Hayflick limit and telomerase

Hayflick, L. The limited *in vitro* lifetime of human diploid cell strains. *Experimental Cell Research* **37:** 614–636; 1965.

Harley, C. B., Futcher, A. B. and Greider, C. W. Telomeres shorten during ageing of human fibroblasts. *Nature* **345:** 458–460; 1990.

Bodnar, A. G., Ouellette, M., Frolkis, M., Holt, S. H., Chiu, C. P., Morin, G. B., Harley, C. B., Shay, J. W., Lichtsteiner, S. and Wright, W. E. Extension of life-span by introduction of telomerase into normal human cells. *Science* **279:** 349–352; 1998.

Goyns, M. H. and Lavery, W. L. Telomerase and mammalian ageing: a critical appraisal. *Mechanisms of Ageing and Development* **114:** 69–77; 2000.

Mitochondria and cellular differentiation

von Wangenheim, K. H. and Peterson, H. P. Control of cell proliferation by progress in differentiation: clues to mechanisms of aging, cancer causation and therapy. *Journal of Theoretical Biology* **193:**663–678; 1998.

Kowald, A. and Kirkwood, T. B. L. Accumulation of defective mitochondria through delayed degradation of damaged organelles and its possible role in the ageing of post-mitotic and dividing cells. *Journal of Theoretical Biology* **202:** 145–160; 2000.

Mitochondria and gender (see also General texts)

Allen, J. F. Separate sexes and the mitochondrial theory of ageing. *Journal of Theoretical Biology* **180:** 135–140; 1996.

Birky, C. W. Jr. Uniparental inheritance of mitochondrial and chloroplast genes: mechanisms and evolution. *Proceedings of the National Academy of Sciences USA* **92:** 11331–11338; 1995.

Cummins, J. Mitochondrial DNA in mammalian reproduction. *Reviews of Reproduction* **3:** 172–182; 1998.

Sutovsky, P., Moreno, R. D., Ramalho-Santos, J., Dominko, T., Simerly, C. and Schatten, G. Ubiquitin tag for sperm mitochondria. *Nature* **402:** 371–372; 1999.

CHAPTER 14

Infections and oxidative stress

Pahl, H. and Baeuerle, P. Expression of influenza virus hemagglutinin activates transcription factor NF-kappa B. *Journal of Virology* **69:** 1480–1484; 1995.

Pahl, H. and Baeuerle, P. Activation of NF-kappa B by endoplasmic reticulum stress requires both Ca^{2+} and reactive oxygen intermediates as messengers. *FEBS Letters* **392:** 129–136; 1996.

Inflammation in ageing rhesus monkeys

Kayo, T., Allison, D., Weindruch, R. and Prolla, T. A. Influences of aging and caloric restriction on the transcriptional profile of skeletal muscle from rhesus monkeys. *Proceedings of the National Academy of Sciences USA* **98:** 5093–5098; 2001.

Alzheimer's disease

Selkoe, D. J. The origins of Alzheimer disease: A is for amyloid. *Journal of the American Medical Association* **283:** 1615–1617; 2000.

Geula, C., Wu, C. K., Saroff, D., Lorenzo, A., Yuan, M. and Yankner, B. A. Aging renders the brain vulnerable to amyloid beta-protein neurotoxicity. *Nature Medicine* **4:** 827–831; 1998.

Schweers, O., Mandelkow, E. M., Biernat, J. and Mandelkow, E. Oxidation of cysteine-322 in the repeat domain of microtubule-associated protein tau controls the in vitro assembly of paired helical filaments. *Proceedings of the National Academy of Sciences USA* **92:** 8463–8467; 1995.

Sano, M., Ernesto, C., Thomas, R. G., *et al.* A controlled trial of selegiline, alpha-tocopherol, or both as treatment for Alzheimer's disease. The Alzheimer's Disease Cooperative Study. *New England Journal of Medicine* **336:** 1216–1222; 1997.

Down syndrome and Alzheimer's disease

Nunomura, A., Perry, G., Pappolla, M. A., Friedland, R. P., Hirai, K., Chiba, S. and Smith, M. A. Neuronal oxidative stress precedes amyloid-beta deposition in Down syndrome. *Journal Neuropathology and Experimental Neurology* **59:** 1011–1017; 2000.

Herpes simplex infection and Alzheimer's disease

Itzhaki, R. F., Lin, W. R., Shang, D., Wilcock, G. K., Faragher, B. and Jamieson, G. A. Herpes simplex virus type 1 in brain and risk of Alzheimer's disease. *Lancet* **349:** 241–244; 1997.

Inflammation and Alzheimer's disease

McGeer, E. G. and McGeer, P. L. The importance of inflammatory mechanisms in Alzheimer's disease. *Experimental Gerontology* **33:** 371–378; 1998.

Smith, M. A., Rottkamp, C. A., Nunomura, A., Raina, A. K. and Perry, G. Oxidative stress in Alzheimer's disease. *Biochimica et Biophysica Acta* **1502:** 139–144; 2000.

Mattson, M. P. and Camandola, S. NFκB in neuronal plasticity and neurodegenerative disorders. *Journal of Clinical Investigation* **107:** 247–254; 2001.

Cigarette smoke

Kodama, M., Kaneko, M., Aida, M., Inoue, F., Nakayama, T. and Akimoto, H. Free radical chemistry of cigarette smoke and its implication in human cancer. *Anticancer Research* **17:** 433–437; 1997.

Lane, J. D., Opara, E. C., Rose, J. E. and Behm F. Quitting smoking raises whole blood glutathione. *Physiology and Behaviour* **60:** 1379–1381; 1996.

Diabetes, glycoxidation and AGEs

Brownlee, M. Negative consequences of glycation. *Metabolism* **49** (suppl): 9–13; 2000.

Inflammation and atherosclerosis

Becker, A. E., de Boer, O. J. and van Der Wal A. C. The role of inflammation and infection in coronary artery disease. *Annual Review of Medicine* **52:** 289–297; 2001.

Oxidative stress and inflammation in cancer

Kovacic, P. and Jacintho, J. D. Mechanisms of carcinogenesis: focus on oxidative stress and electron transfer. *Current Medical Chemistry* **8:** 773–796; 2001.

Mercurio, F. and Manning, A. M. NFκB as a primary regulator of the stress response. *Oncogene* **18:** 6163–6171; 1999.

CHAPTER 15

Malarial tolerance

Greenwood, B. M. Autoimmune disease and parasitic infections in Nigerians. *Lancet* **ii:** 380–382; 1968.

Clark, I. A., Al-Yaman, F. M., Cowden, W. B. and Rockett K. A. Does malarial tolerance, through nitric oxide, explain the low incidence of autoimmune disease in tropical Africa? *Lancet* **348:** 1492–1494; 1996.

Enwere, G. C., Ota M. O. and Obaro S. K. The host response in malaria and depression of defence against tuberculosis. *Annals of Tropical Medicine and Parasitology* **93:** 669–678; 1999.

Alzheimer's disease in Nigeria

Hendrie, H. C., Ogunniyi, A., Hall, K. S., *et al.* Incidence of dementia and Alzheimer disease in two communities: Yoruba residing in Ibadan, Nigeria, and African Americans residing in Indianapolis, Indiana. *Journal of the American Medical Association* **285:** 739–747; 2001.

Farrer, L. A. Intercontinental epidemiology of Alzheimer disease. A global approach to bad gene hunting. *Journal of the American Medical Association* **285:** 796–798; 2001.

Haem oxygenase and immunosuppression

Taramelli, D., Recalcati, S., Basilico, N., Olliaro, P. and Cairo, G. Macrophage preconditioning with synthetic malaria pigment reduces cytokine production via heme iron-dependent oxidative stress. *Laboratory Investigation* **80:** 1781–1788; 2000.

Soares, M. P., Lin, Y., Anrather, J., *et al.* Expression of heme oxygenase-1 can determine cardiac xenograft survival. *Nature Medicine* **4:** 1073–1077; 1998.

Motterlini, R., Foresti, R., Bassi, R. and Green, C. J. Curcumin, an antioxidant and anti-inflammatory agent, induces heme oxygenase-1 and protects endothelial cells against oxidative stress. *Free Radical Biology and Medicine* **28:** 1303–1312; 2000.

Hygiene hypothesis

Rook, G. A. and Stanford, J. L. Give us this day our daily germs. *Immunology Today* **19:** 113–116; 1998.

Frailty genes in centenarians

Yashin, A. I., De Benedictis, G., Vaupel, J. W., *et al.* Genes and longevity: lessons from studies of centenarians. *Journal of Gerontology* **55A:** B319–B328; 2000.

Mitochondrial variants and ageing

Tanaka, M., Gong, J. S., Zhang, J., Yoneda, M. and Yagi, K. Mitochondrial genotype associated with longevity. *Lancet* **351:** 185–186; 1998.

Vandenbroucke, J. P. Maternal inheritance of longevity. *Lancet* **351:** 1064; 1999.

Ooplasmic transfer and cloning

Barritt, J. A., Brenner, C. A., Malter, H. E. and Cohen, J. Mitochondria in human offspring derived from ooplasmic transplantation. *Human Reproduction* **16:** 513–516; 2001.

Allen J. F. and Allen C. A. A mitochondrial model for premature ageing of somatically cloned mammals. Hypothesis paper. *IUBMB Life* **48:** 369–372; 1999.

Lipid composition of mitochondria

Pamplona, R., Portero-Otín, M., Riba, D., Ruiz, C., Prat, J., Bellmunt, M. J. and Barja, G. Mitochondrial membrane peroxidizability index is inversely related to maximum life span in mammals. *Journal of Lipid Research* **39:** 1989–1994; 1998.

Laganiere, S. and Yu, B. P. Modulation of membrane phospholipid fatty acid composition by age and food restriction. *Gerontology* **39:** 7–18; 1993.

Mitochondrial medicine

Hagen, T., Ingersoll, R. T., Wehr, C. M., Lykkesfeldt, J., Vinarsky, V., Bartholomew, J. C., Song M. H. and Ames B. N. Acetyl-L-carnitine fed to old rats partially restores mitochondrial function and ambulatory activity. *Proceedings of the National Academy of Sciences USA* **95:** 9562–9566; 1998.

Hagen, T. M., Liv, J., Lykkesfeldt, J., Wehr, C. M., Ingersoll, R. T., Vinarsky, V., Bartholomew, J. C. and Ames, B. N. Feeding acetyl-L-carnitine and lipoic acid to old rats significantly improves metabolic function while decreasing oxidative stress. *Proceedings of the National Academy of Sciences USA* **99:** 1870–1875; 2002.

Brierley, E. J., Johnson, M. A., James, O. F. and Turnbull, D. M. Effects of physical activity and age on mitochondrial function. *Quarterly Journal of Medicine* **89:** 251–258; 1996.

Fosslien, E. Mitochondrial medicine — molecular pathology of defective oxidative phosphorylation. *Annals of Clinical Laboratory Science* **31:** 25–67; 2001.

Glossary

AGE (advanced glycation end-product) caramel-like material formed by the reaction of a protein with glucose and oxygen.

age-1 gene that when mutated extends the lifespan of nematode worms.

allele one of (usually) several variants of the same gene.

alpha tocopherol chemical name for the most common form of vitamin E.

amino acids the building blocks that are linked in chains to form proteins in all living things. Twenty different types are found in proteins; the order in which they are linked together is specified by the DNA code.

amyloid protein fragment found in senile plaques in Alzheimer's disease.

anaerobic pertaining to organisms that do not use oxygen for respiration.

anoxygenic photosynthesis ancient form of photosynthesis, which uses sunlight to split hydrogen sulphide or iron salts (instead of water) without generating oxygen.

antagonistic pleiotropy trade-off between opposing (antagonistic) effects of a gene that has more than one effect (pleiotropy).

antigen bacterial protein or other 'foreign' particle recognized by antibodies or cells of the immune system.

antioxidant chemical that hinders the oxidation of other molecules, such as fats or proteins.

ApoE4 a gene variant associated with a greater risk of Alzheimer's disease.

apoptosis programmed cell death, as opposed to necrosis (unplanned or 'violent' cell death).

Archaea one of the three great domains of life. In many respects, archaea are intermediary between eukaryotes (cells with nuclei) and bacteria.

ascorbate chemical name for vitamin C.

ascorbyl radical poorly reactive free radical, formed when vitamin C is partially oxidized.

ATP (adenosine triphosphate) energy 'currency' of cells, generated by all forms of aerobic and anaerobic respiration, as well as photosynthesis.

autoimmune disease disease in which the immune system mistakenly attacks components of the body rather than bacteria or other 'foreign' particles.

bacteriochlorophyll form of chlorophyll found in the most ancient photosynthetic bacteria, which do not generate oxygen.

banded-iron formation rock formation comprising bands of ironstone (such as magnetite or haematite) alternating with quartz or flint.

binary fission bacterial method of cell division by doubling cellular material, then splitting in two.

biomarker biochemical 'fingerprint' that could only have been produced by a particular form of life.

calorie restriction balanced diet with beneficial effects on health and longevity in animals, in which free (unrestricted) calorie intake is restricted by 30 to 40 per cent.

Cambrian geological period from about 543 to 500 million years ago.

Cambrian explosion sudden proliferation of many different types of complex animals around the beginning of the Cambrian period (543 million years ago).

cap carbonates thick belts of limestone capping glacial deposits laid down in the immediate aftermath of global glaciations.

carbon burial burial of organic matter as coal, oil or natural gas, as well as barely noticeable carbon deposits in rocks such as sandstone.

carbon signature imbalance in carbon-isotope ratios in rocks, betraying the activity of life.

carbonate rocks limestone rocks composed mostly of calcium and magnesium carbonates.

Carboniferous geological period from about 360 to 286 million years ago.

cardiolipin lubricating lipid found at high levels in mitochondrial membranes, especially in physiologically active tissues such as heart muscle.

carnitine a molecular 'shuttle' responsible for transferring fatty acids into mitochondria for respiration, and left-over organic acids out again for disposal.

catalase enzyme responsible for breaking down hydrogen peroxide to oxygen and water.

catalyst molecule that speeds up a chemical reaction without being altered permanently itself.

chain-breaking antioxidant chemical that blocks free-radical chain reactions.

chlorophyll plant pigment that captures the energy of the Sun in photosynthesis, converting it into chemical energy.

chloroplast specialized subcellular 'organelle' containing chlorophyll, which is the site of photosynthesis in algae and plants. Chloroplasts originally derived from cyanobacteria.

chromosome strand of DNA encoding a number of genes and wrapped in proteins.

chronic inflammation continuous, unresolved inflammation.

cofactor molecule required for the proper function of an enzyme.

conjugation bacterial equivalent of sex, in which spare genes (usually on small circular chromosomes called plasmids) are passed from one bacterium to another.

cyanobacteria blue-green photosynthetic bacteria (once called blue-green algae). They have been the most important producers of oxygen in the air over evolutionary time.

cytochrome oxidase critical enzyme in oxygen-requiring respiration, which receives electrons and protons (hydrogen atoms) derived from sugars or fats, and combines them with oxygen to form water.

cytoplasm part of the cell outside the nucleus, encompassing both the watery cytosol and the membrane-enclosed structures such as mitochondria.

cytosol watery base solution of the cytoplasm.

cytosolic SOD iron-zinc superoxide dismutase, found in the cytosol of eukaryotic cells.

daf-16 gene that when mutated prolongs the lifespan of nematode worms.

daf-2 gene that when mutated prolongs the lifespan of nematode worms.

dehydroascorbate oxidized form of vitamin C.

differentiation specialization of a cell for a particular task, such as contraction in muscle cells or electrical transmission in neurons.

diploid containing two equivalent sets of chromosomes.

disposable soma theory literally the 'throwaway body' theory. It argues that ageing is the outcome of a trade-off between resources committed to sex and those committed to bodily maintenance.

DNA (deoxyribonucleic acid) genetic material of all cells, twisted into the famous double helix. The sequence of four 'letters', A (adenine), T (thymine), C (cytosine), and G (guanine) encodes the order of amino-acid building blocks in proteins.

dominance power of one gene in a pair of equivalent genes (one inherited from each parent) to make its own effects felt at the expense of the 'weaker' (recessive) gene.

double-agent theory theory arguing that oxygen plays a duplicitous role in health. In youth, oxygen free radicals elicit the inflammatory response to infection, so resolving the infection; in ageing, free radical leakage from mitochondria activates the same inflammatory response, but because mitochondrial leakage is irresolvable, inflammation persists, leading to chronic diseases of old age.

Ediacaran fauna ancient fossils of primitive animals from the Vendian period (about 570 million years ago), first discovered in the Ediacara Hills in Australia.

electromagnetic radiation spectrum of wave-particles of defined energy (depending on the wavelength), including visible light, infrared rays, and ultraviolet rays.

electron subatomic wave-particle with negative charge.

electron donor molecule with chemical tendency to give up one or more electrons to other molecules (also called a reductant).

epithelial cell cell from layer covering internal or external surfaces in the body.

eukaryote organism with cells having a 'true nucleus'. Eukaryotes comprise one of the three domains of life. Animals, plants, fungi, algae and protozoa are eukaryotes.

euxinic stagnant water saturated with hydrogen sulphide, as in the deep waters of the Black Sea (ancient name, the Euxine).

fatty acid molecule with a hydrophilic head and a long hydrophobic hydrocarbon tail. Fatty acids are components of fats, oils and membrane lipids.

Fenton reaction reaction of iron with hydrogen peroxide to form hydroxyl radicals.

fermentation form of anaerobic (oxygen-free) respiration used by yeasts, which produces ethanol as an end-product.

ferritin cage-like protein that locks away iron within cells.

fibroblast connective-tissue cell, found in skin and bodily organs, and important in healing wounds.

free radical atom or molecule with an unpaired electron. In this book, the term mostly refers to reactive forms of oxygen, such as superoxide radicals and hydroxyl radicals.

free-radical scavenger molecule that 'scavenges' (reacts with) free radicals to neutralize them.

free-radical theory of ageing theory arguing that continuous production of oxygen free radicals during oxygen respiration is the root cause of ageing.

gamete sex cell, with half the number of chromosomes of a somatic (body) cell.

gene unit of DNA comprising the coding sequence for a single protein (or RNA molecule).

gene expression active production of the protein or RNA encoded by a gene.

genome the complete set of genes of an organism.

genotype the particular variant of a gene or genes carried by an individual, which can be used to distinguish one individual from another at the genetic level.

germ line the cells responsible for passing genes on to the next generation.

glutathione small sulphur-containing antioxidant that 'polices' the oxidation state of cells.

glycolysis form of anaerobic (oxygen-free) respiration in which glucose is converted to pyruvate with the generation of a small amount of energy. In aerobic cells it is coupled to oxygen respiration to produce more energy.

group selection evolutionary selection of traits that benefit populations rather than individuals. It is a weak selective force in most circumstances.

haem pigment molecule containing iron embedded in a porphyrin ring. It is incorporated into many proteins, including haemoglobin, cytochrome oxidase and catalase.

haem oxygenase important stress protein that breaks down haem to release biologically active products: iron, carbon monoxide (an signalling molecule at low concentrations) and bilirubin (an antioxidant).

haemoglobin haem-containing oxygen-transport molecule packed tightly in red blood cells.

haploid containing half the normal two sets of chromosomes, that is, containing a single set.

Hayflick limit maximum number of divisions that any given type of somatic (body) cell will undergo.

Hox genes 'master-switch' genes that regulate embryonic development in animals as diverse as nematode worms, flies, mice and men.

hydrogen peroxide (H_2O_2) unstable chemical intermediate between oxygen and water. It is especially reactive with iron in the Fenton reaction.

8-hydroxydeoxyguanosine (8-OHdG) oxidized DNA 'letter' derived from attack of hydroxyl radicals on DNA, used as a surrogate measure of free radical damage.

hydroxyl radical (•OH) violently reactive oxygen free radical, which reacts in billiseconds with almost all biological molecules.

hyperbaric oxygen oxygen under high pressure (effectively increasing its concentration).

IGF (insulin-like growth factors) group of closely related hormones, with important effects on sexual maturation, among many other effects.

immunosuppression lowering of responsiveness of the immune system to antigens.

individual selection evolutionary selection of traits that benefit individuals rather than populations. It is by far the most important form of natural selection.

inflammation general defensive reaction to infection or injury, characterized by heat, redness, swelling, and pain. Low-grade, chronic inflammation is almost universal in diseases of old age.

inflammatory cell a cell involved in propagating inflammation, such as a macrophage or neutrophil.

inflammatory messenger chemical signal produced by inflammatory cells to recruit and activate other inflammatory cells from elsewhere in the body.

infrared radiation electromagnetic rays with wavelength longer than about 800 nanometres.

insulin hormone that promotes uptake of glucose from blood, stimulating protein synthesis, fat deposition, weight gain and sexual maturation.

insulin-resistance genetic or acquired resistance to the effects of insulin.

ionizing radiation radiation that dislodges electrons from compounds to produce an electric charge.

iron pyrites (FeS), or fool's gold, formed by the reaction of hydrogen sulphide (from bacteria or volcanoes) with dissolved iron.

isotopes different atomic forms of the same element, with equal numbers of protons (making them chemically equivalent) but different numbers of neutrons (giving them different molecular weights).

junk DNA non-coding DNA which apparently serves no purpose; thought to comprise 'selfish' genes hitching a ride, inserted viral DNA sequences, and defunct genes.

knock-out mice genetically manipulated mice, in which specific genes are mutated so that their protein products are not expressed.

lateral gene transfer 'horizontal' movement of genes between individuals in a population, as opposed to 'vertical' inheritance from parents to offspring.

lignin structural polymer, providing strong, flexible support for woody plants.

LUCA (Last Universal Common Ancestor) the last ancestor common to all known life on Earth, including bacteria, archaea and eukaryotes.

malarial tolerance lack of response to malarial parasite despite infection, leading to absence or suppression of malarial symptoms.

meiosis type of cell division that produces sex cells (gametes) with a single set of chromosomes, rather than the two sets of their diploid progenitors.

messenger RNA a ribonucleic acid that encodes genetic instructions to make a protein. Its sequence of 'letters' corresponds to the DNA template on which it is made. It is used to convey genetic instructions from the DNA to the protein-building apparatus (ribosomes).

metabolic rate the oxygen consumption of an organism at rest (basal metabolic rate) or when active.

metallothionein sulphur-rich stress protein, which protects against physical stresses such as radiation and oxygen poisoning.

microglia inflammatory cells resident in the brain.

mitochondria (singular **mitochondrion**) the energy 'power-houses' of eukaryotic cells, in which oxygen respiration takes place. They were originally symbiotic purple bacteria, and still retain bacterial traits.

mitochondrial DNA genetic material present inside mitochondria. It is comparable to bacterial DNA in structure and genetic sequence.

mitochondrial leakage escape of oxygen free radicals from mitochondria during respiration.

mitochondrial SOD manganese superoxide dismutase, found in mitochondria of eukaryotic cells (and in many bacteria).

mitochondrial theory of ageing theory arguing that damage to mitochondrial DNA by free radical leakage from adjacent respiratory proteins is the root cause of ageing.

mitosis type of cell division in eukaryotic cells in which the chromosomes are doubled and then separated to produce two daughter cells genetically identical to the parent cell.

molecular clock estimates of the timing of evolutionary events based on rates of sequence divergence between equivalent genes in different species.

mutation an alteration in coding sequence of a gene that is passed on to the next generation.

myoglobin oxygen-binding protein containing haem pigment, similar to haemoglobin and found in muscle cells of mammals.

neutrophil inflammatory cell that engulfs and digests bacteria and other 'foreign' particles. Often called the 'foot-soldiers' of the immune system because of their large numbers, unspecialized attributes and dispensability.

NFκB (nuclear factor kappa B) important transcription factor that stimulates expression of inflammatory and antioxidant genes.

nitric oxide (NO) gaseous signalling molecule with profound physiological effects on blood vessels, immune system, nervous system and sexual arousal.

nitric oxide synthase umbrella term for several enzymes that generate nitric oxide gas.

Nrf-2 (nuclear factor erythroid-2-related factor 2) important transcription factor that stimulates expression of antioxidant genes but suppresses expression of inflammatory genes.

nucleic acid generic term for DNA (deoxyribonucleic acid) and RNA (ribonucleic acid).

nucleus central 'control centre' of eukaryotic cells, containing genetic material (DNA) combined with proteins, separated from the rest of the cell by a double membrane.

operon genetic unit in bacteria, comprising genes with related function that are all transcribed and expressed together.

organelle tiny specialized organ within a cell, such as a mitochondrion or chloroplast.

organic carbon carbon in biological molecules, such as carbohydrates, fats, and nucleic acids; also carbon in material of biological origin, such as coal, oil and natural gas.

oxidation loss of electrons to an oxidant such as oxygen; the opposite of reduction.

oxidative damage damage to biological molecules as a result of oxidation.

oxidative stress distortion of chemical balance in cells towards the oxidized state, as a result of an imbalance between the rate of formation of oxygen free radicals, and their elimination by antioxidants.

oxygen poisoning toxic effects of breathing high levels of oxygen, caused by the formation of oxygen free radicals.

oxygen-evolving complex enzyme used in photosynthesis to extract electrons and protons from water, releasing oxygen gas as a waste product.

oxygenic photosynthesis form of photosynthesis that uses the energy of light to split water, releasing oxygen as a waste product.

Permian geological period from about 286 to 245 million years ago.

peroxiredoxins group of sulphur-containing antioxidant enzymes that break down hydrogen peroxide to water using thioredoxin as an electron donor.

Phanerozoic the 'modern' age of plants, animals and fungi, stretching from the Cambrian explosion, 543 million years ago, to the present day.

photon electromagnetic wave-particle with a defined amount of energy.

photorespiration complex series of biochemical reactions that stunt plant growth in sun-light. The reactions parallel oxygen respiration by consuming oxygen and releasing carbon dioxide, but do not generate energy, and are thought to protect plants against oxygen toxicity.

photosynthesis synthesis of carbohydrates and other organic molecules from carbon dioxide and water, using the energy of sunlight.

plaque pathological conglomerate of proteins and inflammatory cells. Senile plaques in Alzheimer's disease are composed largely of amyloid, microglial cells, and damaged nerve endings.

pleiotropy multiple effects or outcomes.

polymorphic alleles different versions of the same gene found in a population.

Precambrian geological period accounting for nine tenths of Earth's history, from its formation about 4.6 billion years ago until the Cambrian period 543 million years ago.

prokaryote cell without nucleus, such as a bacterium.

pro-oxidant opposite of an antioxidant, a molecule that promotes the oxidation of other molecules.

protein large molecule made of a long chain of amino acids folded into a three-dimensional shape that dictate's its function. Cells make many different kinds of proteins, which make up most of a cell's structure and carry out all its functions. The amino acid sequences of proteins are encoded in the genes.

proton positively charged particle within the nucleus of an atom. The hydrogen nucleus is a single proton.

radiation poisoning toxic effects of radiation, many of which are caused by formation of oxygen free radicals.

rate-of-living theory theory arguing that lifespan depends on metabolic rate.

recombination random swapping of different alleles of the same gene between chromosomes, which generates new combinations of alleles on a chromosome.

reduction addition of electrons to a molecule; opposite of oxidation.

respiration generation of energy from biochemical reactions.

respiratory chain chain of electron-transporting proteins responsible for generating energy in mitochondria and bacteria.

ribosomal RNA ribonucleic acid constituent of ribosomes. Comparisons of ribosomal RNA from different species have enabled the construction of an evolutionary tree of life.

ribosome protein-building machinery present in all cells.

RNA (ribonucleic acid) single-stranded thread of nucleic acid, which resembles DNA in its sequence of letters. There are various types of RNA (messenger RNA, ribosomal RNA and transfer RNA), which are all essential to cells.

Rubisco (ribulose-1,5-bisphosphate carboxylase/oxygenase) enzyme that binds carbon dioxide (CO_2) in photosynthesis, incorporating it into carbohydrate molecules. It can also bind oxygen, leading to photorespiration.

saturated fat fat composed of fatty acids without any double bonds between carbon atoms.

selection pressure likelihood that a disadvantageous trait will be eliminated from a population by natural selection. Traits that jeopardize reproduction are not passed on, and so disappear from the population; mildly disadvantageous traits may be offset by hidden benefits.

singlet oxygen reactive form of molecular oxygen, in which the spin of an electron is flipped so that it enters a higher-energy orbital.

snowball Earth global glaciation.

SNP (single-nucleotide polymorphism) single letters in the genetic code that vary between individuals, giving many slightly different versions of the same genes.

SOD (superoxide dismutase) antioxidant enzyme that converts superoxide radicals into oxygen and hydrogen peroxide.

soma the body, as opposed to the germ line (the sex cells).

somatic mutation theory theory arguing that accumulation of mutations in DNA of somatic (body) cells during a lifetime is the root cause of ageing.

stem cell unspecialized progenitor cell, which divides by mitosis to replenish populations of specialized (differentiated) cells.

stomata pores in the surface of a leaf that allow exchange of gases between the air and the plant's tissues.

stress protein protein produced in response to physical stress (such as radiation, heat or infection), which counters the stress.

stress response synchronized production of stress proteins, which coordinate resistance to physical stress or its recurrence. The effect persists for days, possibly much longer.

sulphate-reducing bacteria anaerobic bacteria that generate energy from organic matter using sulphate instead of oxygen as the electron acceptor. Their waste product of respiration is hydrogen sulphide gas (H_2S) instead of water (H_2O).

superoxide radical ($O_2^{\bullet-}$) mildly reactive oxygen free radical, which tends to behave as an electron donor, giving up a single electron to revert to oxygen.

symbiosis intimate relationship between two organisms in which each partner gains a benefit from the other.

tangle pathological feature of Alzheimer's disease, in which neurons die off, leaving behind tangled fibrils of oxidized tau protein.

tau protein that maintains structure and function of microtubules in neurons. It pathologically oxidized and phosphorylated in Alzheimer's disease.

telomerase enzyme that renews telomere caps on chromosomes, preventing 'fraying' of chromosome ends and loss of coding genes during DNA replication. When the telomerase gene is permanently active, it conveys 'immortality' on cells in culture.

telomere non-coding DNA that caps ends of the chromosomes in eukaryotes, preventing loss of gene-coding sequences during DNA replication.

thiol (–SH), the sulphur-containing group of the amino acid cysteine. It is an important molecular switch in many transcription factors, 'reporting' on the oxidation state of the cell.

thioredoxin small sulphur-rich protein, which gives up electrons to regenerate antioxidant enzymes such as peroxiredoxins.

thrifty genotype particular genetic configuration that encourages the 'hoarding' of energy during times of plenty.

tocopheryl radical poorly reactive free radical, formed when vitamin E is partially oxidized.

transcription copying of the genetic code for a gene from DNA into messenger RNA, in readiness for protein synthesis.

transcription factor protein that binds to DNA, either stimulating or inhibiting the transcription of one or more genes.

transgenic describes organisms that have had one or more genes replaced or added through genetic engineering.

translation conversion of RNA code into amino-acid sequence as a protein is synthesized.

tumour necrosis factor important protein in inflammation, which attracts and activates inflammatory cells. It was originally identified and named for its toxic effect on tumours.

ultraviolet radiation electromagnetic rays with wavelength shorter than about 400 nanometres.

unsaturated fat fat composed of fatty acids with one or more double bonds between carbon atoms.

Vendobionts first large animals (up to 1 metre in diameter), which evolved during the Vendian period, about 570 million years ago. They were mostly radially symmetrical, quilted bags of protoplasm, similar to jellyfish.

Index

D

N

O